NEW INSIGHTS INTO FIBROMYALGIA

Edited by **William S. Wilke**

New Insights into Fibromyalgia
Edited by William S. Wilke

Published by InTech

Janeza Trdine 9, 51000 Rijeka, Croatia

Publishing Process Manager Adriana Pecar

Technical Editor Teodora Smiljanic

Cover Designer InTech Design Team

Additional hard copies can be obtained from orders@intechopen.com

New Insights into Fibromyalgia, Edited by William S. Wilke
p. cm.
ISBN 978-953-307-407-8

For EU product safety concerns:
IN TECH d.o.o., Prolaz Marije Krucifikse Kozulić 3, 51000 Rijeka, Croatia,
info@intechopen.com or visit our website at intechopen.com.

Meet the editor

Dr. William S Wilke received a Bachelor of Science degree from the University of Notre Dame and a Doctor of Medicine degree from the Medical College of Wisconsin. His medical internship and residency were completed at Akron City Hospital. Afterward, he finished rheumatology training and remained at the Cleveland Clinic foundation for 37 years, retiring in April, 2011. Throughout the span of his career, Dr. Wilke was Chairman of the Medical Grand Rounds Committee and the Pharmacy and Therapeutics Committee. He was also head of Subspecialty Clinics from 1981-1995. Since 1990, he has been Associate Editor of the Cleveland Clinic Journal of Medicine. Outside activities included annual "Meet the Professor" meetings at the American College of Rheumatology from 1991-1993, and participation as a member of the United States Pharmacopeia General Committee from 1995-2000. Dr. Wilke is the author of 121 various publications, editor of Methotrexate Therapy in Rheumatic Disease: Marcel Dekker Inc, 1989, and authored The Cleveland Clinic Guide to Fibromyalgia: Kaplan Publishing, 2010.

Contents

Preface

I try, not always successfully, to read most of each week's edition of *The Economist*. In the October 15th-21st science section a report Transporter of Delight, discusses new information about the genetic underpinning of happiness, and confirmed some of my biases about the biopsychosocial nature of fibromyalgia. The story reported that Jan-Emanuel De Neve and colleagues performed a rigorous case control study involving 1,939 adolescents (De Neve J-E et al, 2011). Genetic information was collected and correlated with a validated satisfaction questionnaire. They controlled for other important potentially causative factors including education, economic status, religiosity, and many others, and discovered a gene that regulates happiness, the same gene that is found in people with fibromyalgia. The magazine story is the genesis for this polemic.

A little background

Human DNA is composed of approximately 21,000 genes, distinct regions arranged in 23 pairs - one each from the mother and father. The two components of each separate gene have subtle differences of amino acid distribution, which affect their function. These slightly different functional variants of the same gene are called alleles. The gene of interest is called 5-HTTLPR and modulates serotonin concentrations in the nervous system. This gene consists of a short and long allele, based on the number of amino acid residues in their structures - the long one produces more serotonin transporter proteins, which produce more serotonin than the shorter allele.

The researchers found that individuals with long alleles and those with higher serotonin concentrations in the nervous system were statistically more likely to be very satisfied with their life, compared to individuals with short alleles ($p=0.012$).

Serotonin is a chemical, which is released by neurons and governs the magnitude of neural activity. Low serotonin concentrations in the central nervous system is causally linked to affective disorders (Thieme K et al, 2004; Abeles AM et al, 2007). This same alteration of the serotonergic system is a crucial factor underlying the severity and pathogenesis of FMS (Russell IJ et al 1992; Stahl SM et al, 2009).

Familial aggregation occurs in fibromyalgia. For example, among 533 first degree relatives of 78 fibromyalgia patients compared to 272 first degree relatives of 40 rheumatoid arthritis patients, the odds ratio for having fibromyalgia was 8.5 in fibromyalgia relatives versus the rheumatoid arthritis relatives (Arnold LM et al, 2004). Others have demonstrated an increased frequency of the short allele of the serotonin transporter gene 5-HTTLPR in fibromyalgia (Offenbacher M et al, 1999). In the general population, the short allele confers vulnerability to a spectrum of illnesses including anxiety disorder, depression, and bipolar disorder (Lucki I, 1998). Furthermore, increased amygdala activation to environmental stresses such as facial expression, proven by functional magnetic resonance imaging (Hariri AR et al, 2006), and other methods (Munafo MR et al, 2008) has been linked to this same allele. The amygdala is responsible for negative interpretation of environmental stimuli; increased activation equals higher "fear factor".

Next piece, temperament

In 1993, C. Robert Cloninger postulated that temperament, which can be best understood as automatic emotional responses to situations in society by an individual, might be primarily due to inheritance (Cloninger CR et al, 1993). He constructed a model of four descriptive personality dimensions, which included novelty seeking, harm avoidance, reward dependency and persistence. People with harm avoidance were described as having "... pessimistic worry in anticipation of future problems, ...fear of uncertainty, shyness of strangers and rapid fatigability." Those with novelty seeking were curious, disorganized, quick-tempered and impulsive. Reward dependence traits include sensitivity and a need for societal contact. Not surprisingly, people with the persistence trait were described as industrious and hard working. He suggested that all of our personalities are a mixture of these traits, and in some individuals, one or another trait dominates.

He further predicted that each of these traits was due to different genes which were intimately related to brain chemistry metabolism. Harm avoidance was related to serotonin pathways, novelty seeking to dopamine pathways, and the last two to norepinephrine pathways. Research since 1992 has shown that these relationships are a bit more complicated, but the general concept is largely confirmed. Of significance, a recent publication has shown a higher frequency of both the short 5-HTTLPR allele and harm avoidance temperament trait in patients with fibromyalgia compared to controls (Cohan H et al, 2002).

The genetic plus side

At this point, I have associated unhappiness, 5-HTTPR short allele, harm avoidance due to amygdala activation, low central nervous system serotonin with fibromyalgia. This relationship makes sense. In the medical and in psychology and economic literature, people with harm avoidance are prone to increased situational fear responses, depression, and low self direction, as well as fibromyalgia (Celikel FC et al, 2009; Glazer Y et al, 2010). However, they are also more likely to be creative, more spiritual, more likely to participate in the arts (Bachner-Melman R et al, 2005), and perform better academically (Calapoglu M et al, 2011). In my own clinic, patients with FMS tell me that they always lock their doors, have insurance, pay their bills on time, wear their seat belts and vote regularly. They are good, careful people, people with high harm avoidance. The long allele is not associated with harm avoidance-they are happy, but not necessarily very careful people. They take chances.

Our first attempts at literature were epic poems that celebrated heroes. These were people like Beowulf, who went off into the dark woods to slay the monster Grendel, and Odysseus, who sailed off beyond the sight of land to great adventure. Why were these people celebrated? Because they came back. Most people who entered the great woods alone or sailed beyond the sight of land never returned. Who were these people? Certainly not the ones with high harm avoidance traits. The harm avoidance people stayed home and procreated. This was a good trait. They were a bit afraid, but they weren't stupid. They planned ahead, put up stocks and were ready for winter. They brought order and law to society. What's more, in times of societal stress, they slept lightly and gave the first alarms. With their enhanced central sensitivity, they were the first to hear the enemy, to perceive signs of the attacker in the woods, to smell the hunting animal's approach. It mattered little to the genes how these harm avoiders felt during times of stress. All that was important was getting through the danger safely. Anyway, most of those ancient stressful events resolved quickly and were short-lived. For most of our history, having these genes was predominantly a good thing.

Societal prospective

If we consider the evolution of our species during the last 100,000 years, we live in very special, very different times. We no longer respect diurnal variation, the dictates of the sun and moon. We have efficient artificial lighting. We're shift workers with little respect for sleep. Unprecedented sensory input from radio, television, PCs, iPods, Blackberries, cell phones, and beepers allow no down time and facilitates "multi-tasking" as never before. We're told by advertisers that we need more and more of these things so that we can watch Reality Shows which demean contestants as a substitute for entertainment, tune in to news shows which compete under the mantra, "If it bleeds, it leads", and experience the negative radio and television sound bites for politicians who will protect us from the terrorists who, incidentally, are everywhere (Hassett AL et al, 2002). We, especially in the United States, are stressed and in trouble.

Expenditures for health care per capita in the United Kingdom are less than half the monies spent in the United States ($2164 versus $5274), but people in the United States are relatively less healthy according to Banks and colleagues (Banks J et al, 2006). They analyzed self-reported illness profiles completed by approximately 10,000 US residents and approximately 5,500 UK residents all middle-aged or older, and discovered that the usual risk factors for diabetes, atherosclerotic heart and vascular disease, which include *traditional* definitions of lower socioeconomic status, obesity, tobacco, and alcohol use, did not fully explain these differences. From the same study, we also learned that C- reactive protein (CRP), which is also associated with the risk of heart and vascular disease (Ridker PM et al, 2002), was 20 percent higher in US residents, and also not explained by those previously mentioned traditional disease risk factors. They concluded that there must be some other important risk factor(s).

In Banks' discussion, we are reminded that psycho-social factors (also known as stress) play a major role in conferring poorer health for people in lower socio-economic stratas of society. Lower social status has been shown to increase the risk of the Metabolic Syndrome (Wannamethee SG et al, 2007), which is at the root of diabetes and cardiovascular disease. In addition, the serum level of CRP rises as socio-economic status falls (Steptoe A et al, 2003). Banks points out that we are progressively losing our middle class in the US (Banks J et al, 2006). This places more and more of us, if not in the *traditional* lower class, but still at accelerating "socio-economic risk". Therefore, increasing stress, an important cause of FMS, and which increases as social status declines, also mediates elevation of CRP, which is at the very least a surrogate measure of declining overall health.

One valid interpretation of Banks' important report is that people who live in the US experience greater stress than do people who live in the UK, hence the higher CRP, and that stress makes people in the US less healthy than their UK counterparts - US residents, therefore, live in a stressful and not very healthy society. The concept that origins of most diseases are at least in part due to social inequities and mood/affective disorders is still novel for most physicians, but hardly a new or unique interpretation (Syme SL et al, 1976; Moussavi S et al 2007). Moreover, certain illness profiles might preferentially manifest in the genetically susceptible.

Genes determine magnitude of response to stress, susceptibility to depression, and are a major determinant of who manifests fibromyalgia and who doesn't (Martinez-Lavin M et al, 2007; Caspi A et al, 2003).

As we already know, genetic background dictates susceptibility to depression, and the short allele contributes (Munafo MR et al, 2006)). When distress begets depression, it gradually reduces health scores (Moussavi S et al, 2007), accelerates atherosclerosis (Faramawi MF et al, 2007), and by shortening telomeres, may contribute to premature cellular senescence (Simon NM et al, 2006). Stress also raises markers of inflammation and is an important player in the pathogenesis of FMS (Chandola T et al, 2006; Abales AM et al, 2007).

An analysis by Wolfe and co-workers might help us to understand this better (Wolfe F et al, 2006). In Wolfe's "Survey Criteria", used to diagnose fibromyalgia, if numerical values are given to two symptoms, pain and fatigue, and the values added, a numerical Symptom Intensity Scale is produced. When this scale was applied to 25,417 patients in his data base, regardless of the presence or absence of fibromyalgia, higher scores on the scale were associated with more severe medical illnesses of all kinds, greater mortality, and more socioeconomic disadvantage. So, having cardinal symptoms of stress, pain, and fatigue, often a surrogate for depression, is clearly associated with poor health.

Furthermore, this study allows us to generalize the concept to the entire population, a population in which as we are told by economists, the middle class is shrinking. We're all getting increasingly stressed, our moods worsened, and some of us manifest this situation as the symptoms of pain and fatigue. Add a new study, which shows that negative mood mediates sleep abnormalities in people with chronic pain, and the conceptual illness fibromyalgia is complete (O'Brien EM et al, 2010).

Canaries

Who are the first people, the early warning system, who notice this toxic environmental state of affairs? The harm avoiders of course, and with their sensitive serotonin genes, they experience it viscerally, with symptoms. People with fibromyalgia, genetically determined neurologic central sensitivity, have real physiologic symptoms, which constitute illness by definition when they seek medical care. The pathogenesis of their symptoms is multifactorial and includes chronic distress/stress from living in a society that would have been described as a dystopia in 1950's science fiction literature.

Genes interacting with society; no wonder it's so hard to treat fibromyalgia. These sensitive, relatively unhappy people whose fear nerves fire at lower thresholds, are the canaries in our society's coal mine. Canaries first. Eventually, we are all at risk.

Back to the book

This discussion is a polemic, a persuasion and a hypothesis about how the various pieces fit and interact in the puzzle that is fibromyalgia, and reduced to the most fundamental, happiness and society. Since we're ultimately dealing with science, it is one approximation of "the truth". This book provides further pieces to the puzzle. We'll keep trying to get it right.

Dr. William S. Wilke, M.D.
Department of Rheumatic and Immunologic Diseases,
A50 Cleveland Clinic, Orthopedic and Rheumatologic Institute,
Cleveland, Ohio,
USA

References

Abales AM, Pillinger MH, Soltar BM, Abeles M: Narrative review: The pathophysiology of fibromyalgia. Ann Intern Med 2007; 146; 726-734

Arnold LM, Hudson JI, Hess EV et al: Family study of fibromyalgia. Arthritis Rheum 2004; 50: 944-952

Bachner-Melman R, Dina C, Zohar AH: AVPR1a and SLC6A4 gene polymorphisms are associated with creative dance performance. PLOS Genetics 2005; 1: 0394-0403

Banks J, Marmot M, Oldfield Z, Smith JP: Disease and disadvantage in the United States and in England. JAMA 2006; 295: 2037-2045

Calapoglu M, Sahin-Calapoglu N, Karacop A et al: Serotonin transporter bi- and triallelic genotypes and their relationship with anxiety and academic performance: a preliminary study. Neuropsychobiology 2011; 63: 103-111

Caspi A, Sugden K, Moffit TE et al: Influence of life stresses on depression: Moderation of a polymorphism of the 5-HTT gene. Science 2003; 301: 386-389

Celikel FC, Kose S, Cumurcu BE, et al: Cloninger's temperament and character dimensions of personality I patients with major depressive disorder. Comp Psyciatry 2009; 50: 556-561

Chandola T, Brunner E, Marmot M: Chronic stress at work and the metabolic syndrome: prospective study. BMJ 2006; 332: 521-525

Cloninger CR, Svakic DM, Przybeck TR: A psychobiological model of temperament and character. Arch Gen Psychiatry 1993; 50: 975-990

Cohan H, Buskila D, NeumannL, et al: Confirmation of an association between fibromyalgia and serotonin transporter promoter region (5-HTTLPR) polymorphism and relationship to anxiety-related personality traits. Arthritis Rheum 2002; 46: 845-847

De Neve J-E, Christakis N A, Fowler JH, Frey B S: Genes, economics, and happiness: GenesEconomicsHappiness_RESTAT_20sept11.tex date: September 20, 2011

Faramawi MF, Gustat J, Wildman RP, et al: Relationship between depressive symptoms and common carotid artery atherosclerosis in american persons =/>65 years of age. Am J Cardiology 2007; 99: 1610-1613

Glazer Y, Buskila D, Cohen H, et al:Differences in personality profile of fibromyalgia patients and their relatives with and without fibromyalgia. Clin Exp Rheumatol 2010; 28 (Suppl): S27-S32

Hariri AR, Drabant EM, Weinberger DR: Imaging genetics: Perspectives from studies from genetically driven variation in serotonin function and corticolimbic affective processing. Biol Psychiatry 2006; 59: 888-897

Hassett AL, Sigal LH: Unforeseen consequences of terrorism. Medically unexplained symptoms in a time of fear. Arch Intern Med. 2002; 162:1809-1813

Lucki I: The spectrum of behaviors influenced by serotonin. Biol Psychiatry 1998; 44: 151-162

Martinez-Lavin M: Stress, the stress response and fibromyalgia. Arthritis Res Therapy 2007; 9: 216-223

Moussavi S, Chatterji S, Verdes E, et al: Depression, chronic diseases, and decrements in health: Results from the World Health Surveys. Lancet 2007; 370: 851-858

Munafo MR, Brown SM, Hariri AR: Serotonin transporter (5-HTTLPR) genotype and amygdala activation: A meta-analysis. Biologic Psych 2008; 63: 852-857

Munafo MR, Clark TG, Roberts KH, Johnstone EC: Neuroticism mediates the association of serotonin transporter gene with lifetime major depression. Neuropsychobiology 2006; 53: 1-8

O'Brien EM, Waxenberg LB, Atchison JW et al: Negative mood mediates the effect of poor sleep on pain among chronic pain patients. Clin J Pain 2010; 26: 310-319

Offenbacher M, Bondy B, de Jong S, et al: Possible association of fibromyalgia with a polymorphism in the serotonin transporter promoter regulatory region. Arthritis Rheum 1999; 42: 2482-2488

Ridker PM, Rifai N, Rose L, et al: Comparison of c-reactive protein and low-density lipoprotein cholesterol level in the prediction of first cardiovascular events. N Engl J Med 2002; 347: 1557-1565

Russell IJ, Vaeroy H, Javors M, et al: Cerebrospinal fluid biogenic amine metabolites in fibromyalgia/fibrositis syndrome and rheumatoid arthritis. Arthritis Rheum 1992; 35: 550-556

Simon NM, Smoller JW, McNamara KL, et al: Telomere shortening and mood disorders: Preliminary support for a chronic stress model of accelerated aging. Biol Psychiatry 2006; 60: 432-435

Stahl SM: Fibromyalgia-pathways and transmitters. Human Psychopharmacol 2009; 24: S11-S17

Steptoe A, Kunz-Ebrecht S, Owen N, et al: Influence of socioeconomic status and job control on plasma fibrinogen responses to acute mental stress. Psychosom Med 2003; 65: 137-144

Syme SL, Berkman LF: Social class, susceptibility, and sickness. Am J Epidemiol 1976; 104: 1-8

Thieme K, Turk DC, Flor H: Comorbid depression and anxiety in fibromyalgia syndrome: Relationship to somatic and psychosocial variables. Psychosomatic Medicine 2004; 66: 837-844

Wannamethee SG, Whincup PH, Rumley A, Lowe GDO: Inter-relationships of interleukin-6, cardiovascular risk factors and the metabolic syndrome among older men. J Thromb Haemost 2007; 5: 1637-1643

Watanabi MAE, Nunes SOV, Amarante MK et al: Genetic polymorphism of serotonin transporter 5-HTTLPR: involvement in smoking behavior. J Genetics 2011; 90:179-185

Wolfe F, Rasker JJ: The Symptom Intensity Scale, fibromyalgia, and the meaning of fibromyalgia-like symptoms. J Rheumatol 2006; 33: 2291-2299

Part 1

Pathogenesis of Fibromyalgia

1

Sleep and Fibromyalgia

Fumiharu Togo[1], Akifumi Kishi[2] and Benjamin H. Natelson[3]
[1]*Educational Physiology Laboratory, Graduate School of Education, The University of Tokyo, Tokyo,*
[2]*Division of Pulmonary, Critical Care and Sleep Medicine, Department of Medicine, NYU School of Medicine, New York, NY,*
[3]*Pain and Fatigue Study Center, Beth Israel Medical Center, Albert Einstein Medical Center, New York, NY,*
[1]*Japan*
[2,3]*USA*

1. Introduction

Fibromyalgia (FM) is a medically unexplained illness characterized by four quadrant pain lasting at least 3 months and accompanied by multiple areas of tenderness on palpation of the body using 4 kg force. FM occurs more often in women than men but is quite common in both sexes, occurring in approximately 3% of the population. Although sleep difficulties are not part of standard diagnostic criteria, insomnia complaints of poor and nonrestorative sleep are common and have been associated with intense of pain, fatigue, sleepiness, and cognitive difficulties in FM.

FM frequently occurs in conjunction with chronic fatigue syndrome (CFS). CFS is a medically unexplained condition characterized by persistent or relapsing fatigue lasting at least 6 months, which substantially reduces normal activity. In addition to severe fatigue, one of the eight symptoms used for diagnosing CFS is "unrefreshing sleep", and this sleep-related problem is the most common complaint among CFS patients.

Although, FM and CFS often have similar symptoms, including sleep-related complaints, differences between FM and CFS exist. In this chapter, we will review studies on sleep in FM and CFS patients in order to better understand differences between them. Polysomnographic studies have shown sleep problems in FM by using simple descriptive statistics, for instance, increased non-rapid eye movement (non-REM) Stage 1 sleep, reduced slow-wave (Stages 3 and 4) sleep, more arousals, prolonged sleep onset, reduced sleep efficiency, etc. Sleep problems in CFS shown by polysomnographic studies are quite similar to those in FM. However, we have shown that dynamic aspects of sleep, a new way of assessing sleep, are different between patients with CFS alone compared to those with CFS+FM. The probability of transition from rapid eye movement (REM) sleep to waking in CFS is greater than in healthy controls. Probabilities of transitions from waking, Stage 1 sleep, and REM sleep to Stage 2 and those from slow-wave sleep to waking and Stage 1 sleep are greater in FM+CFS than in healthy controls.

Over the course of the many decades, sleep researchers have used simple descriptive statistics to characterize and summarize sleep architecture. While this methodology has been extremely useful in defining the abnormalities that currently constitute sleep pathology, this approach does not explain specific patient complaints of disturbed and unrefreshing sleep. However, a dynamic analysis complements the classical approach by allowing an analysis of transition dynamics between sleep stages and shows that FM and CFS may be different illnesses associated with different problems in sleep regulation.

2. Sleep studies and FM

2.1 Sleep and symptoms of FM

Although sleep difficulties are not part of standard diagnostic criteria (Wolfe et al., 1990), insomnia complaints of poor and nonrestorative sleep are common in patients with FM. An early study shows that 65.7% of patients with FM reported nonrestorative sleep (White et al., 1999). Recently, two epidemiologic studies (Bigatti et al., 2008; Theadom et al., 2007) reported that more than 90% of patients with FM complain about sleep problems such as difficulty falling asleep, difficulty falling back to sleep after waking up during nocturnal sleep, and unrefreshing sleep. Sleep is also one of the domains which associate most strongly with the patients' overall impression of improvement (Arnold et al., 2011).

Data strongly suggest that FM-like symptoms develop following sleep disruption in healthy volunteers. Four studies in normal healthy controls (Lentz et al., 1999; Moldofsky et al., 1975; Moldofsky & Scarisbrick, 1976; Onen et al., 2001) have reported increases in musculoskeletal pain and/or decreases in pain threshold after a period of sleep disruption or deprivation, while one study did not find this result (Older et al., 1998). Moldofsky's group found that Stage 4 sleep deprivation was associated with increasing in tenderness, musculoskeletal symptoms, and mood disturbances (Moldofsky et al., 1975; Moldofsky & Scarisbrick, 1976). In addition, healthy volunteers with disrupted sleep produced experimentally by sound pulses every 2 minutes but with normal total sleep time had a decrease in day-time energy levels. Moreover, their ability to do complex auditory monitoring tasks was also impaired (Martin et al., 1996). These data indicate that partial sleep deprivation can produce the hallmark symptoms of FM – namely, musculoskeletal achiness, marked daytime fatigue, and cognitive problems.

One study reported that sleep disturbances led to exacerbation of pain in patients with FM (Affleck et al., 1996). One recent study reported that negative mood (i.e., depression and anxiety), which are common among chronic pain patients or poor sleepers, almost fully mediated the relationship between sleep and pain in chronic pain patients (O'Brien et al., 2010). Moderating impact of depressive symptoms on the relationship between sleep and pain was also reported in another study (O'Brien et al., 2011).

2.2 Sleep disorders in patients with FM

One group suggested that as many as 33% of individuals with FM had the restless leg syndrome (Viola-Saltzman et al., 2010). Another recent study (Gold et al., 2004) reported a high rate of sleep disturbed breathing in patients with FM (i.e., 96%). The prevalence of overweight women (Moldofsky, 2002) may contribute to sleep disturbed breathing, such as sleep apnea and inspiratory airflow limitation with arousals (Gold et al., 2004). However, one study found that patients with FM had the same frequency of sleep apnea as normal controls (Molony et al., 1986). Unpublished data from our laboratory finds rates of FM in patients with polysomnography-documented obstructive sleep apnea to be similar to those found in the community. A genetic study found common genetic characteristics between FM and narcolepsy (Spitzer & Broadman, 2010).

2.3 Sleep abnormality in patients with FM

In contrast to studies on sleep pathology, a host of studies strongly suggest that the pattern of sleep is abnormal in many FM patients. The most consistent abnormality is significantly increased Stage 1 sleep compared to healthy controls (Anch et al., 1991; Cote & Moldofsky, 1997; Drewes et al., 1994; Landis et al., 2004; Leventhal et al., 1995; Molony et al., 1986; Shaver et al., 1997). Sleep disturbance in patients with FM is obvious because polysomnographic studies have shown longer sleep latencies (Drewes et al., 1994; Horne & Shackell, 1991; Landis et al., 2004), more wakefulness (Drewes et al., 1994), reduced sleep efficiency (i.e., the proportion of time spent sleeping relative to the time available for sleeping) (Drewes et al., 1994; Landis et al., 2004), reduced Stage 2 sleep (Landis et al., 2004), and reduced Stage 4 sleep (Anch et al., 1991; Lashley, 2003) in FM patients compared to healthy control subjects of similar age. Patients with FM awaken more easily (Perlis et al., 1997) and compared to healthy controls have higher levels of physical activity during the night (Affleck et al., 1996; Korszun et al., 2002).

Although sleep efficiency was comparable to that of controls, FM patients showed more arousals (Jennum et al., 1993; Molony et al., 1986) and Stage 1 sleep (Molony et al., 1986). Molony et al. reported that patients with FM had three times more microarousals (brief sleep interruptions lasting 5-19 seconds) per hour than did healthy controls (Molony et al., 1986). These results indicate that patients with FM have poor sleep quality with fragmented sleep.

An alpha-EEG anomaly during non-REM sleep has been considered a biologic correlate of chronic pain and a possible basis of nonrestorative sleep complains in patients with FM (Branco et al., 1994; Moldofsky et al., 1975; Moldofsky & Scarisbrick, 1976; Moldofsky, 1989; Roizenblatt et al., 2001). The alpha-EEG anomaly is excessive alpha wave intrusion which has been interpreted as a heightened arousal state during non-REM sleep (Moldofsky, 1989; Scheuler et al., 1983). However, this has not been found consistently across studies (Horne & Shackell, 1991). Alpha-delta sleep is an abnormal sleep EEG rhythm characterized by alpha activity that is superimposed on delta waves of Stages 3 and 4 sleep (McNamara, 1993). Horne and Shackell found that the mean alpha activity in Stages 2, 3, and 4 sleep were greater for the patients with FM than in healthy controls (Horne & Shackell, 1991). Branco et al. studied alpha and delta activity and the alpha-delta ratio across sleep cycles in patients with FM and healthy controls (Branco et al., 1994). The alpha-delta sleep anomaly occurred in almost all patients who had fragmented sleep; this anomaly was not observed in any of the healthy controls. Perlis et al. found that the alpha-EEG sleep associated with perception of shallow sleep and an increased tendency to display arousal in response to external auditory stimuli (Perlis et al., 1997).

Most studies on alpha-EEG anomaly in patients with FM have been based on visual and hence relatively subjective analysis of the EEG. Using spectral analysis, a quantitative measurement is provided not only for alpha component of EEG, but also for other existing frequency components. Drewes et al. examined spectral EEG patterns and found that patients with FM showed more power in the alpha (higher frequency) band and a decrease in the lower frequency bands in Stages 2, 3, and 4 sleep and all sleep cycles (Drewes et al., 1995). However, the alpha-EEG anomaly is not specific for patients with FM in that it also occurs in healthy individuals (Horne & Shackell, 1991; Scheuler et al., 1983; Shaver et al., 1997) and in patients with disorders such as rheumatoid arthritis and CFS (Moldofsky et al., 1983; 1988).

A task force of the American Sleep Disorders Association has defined a cortical arousal (American Sleep Disorders Association, 1992) as a return to alpha or fast frequency EEG activity, well differentiated from the background, lasting at least 3 seconds. Cortical microarousals are briefer arousals lasting at least 1.5 seconds (Martin et al., 1997). While the major focus of sleep researchers studying arousals has been on EEG measures, one group (Pitson & Stradling, 1998) suggested that non-EEG markers might be an important and even more reliable sign of arousals than cortical arousal as reflected by the EEG. For example, it is known that somatosensory and auditory stimulation during sleep can produce alterations in cardiac, respiratory, and somatic measures without overt EEG desynchronization (Carley et al., 1997; Halasz, 1993; Winkelman, 1999). These changes are thought to reflect activation of the brainstem or subcortical arousal system without affecting the cortex. Hence current thinking is that there are different levels of arousal responses generated from subcortical and cortical areas of the brain (Sforza et al., 2000, 2002).

One study (Sforza et al., 2000) showed that bursts of K-complexes and delta waves, expressions of an activation of subcortical arousal system, represent a real arousal response inducing cardiac activation similar to that found during cortical arousals (microarousal and phases of transitory activity). We have investigated sleep microstructure in young healthy men with no sleep complaints (Togo et al., 2006). We found increases in delta wave power in both cortical and subcortical arousals relative to just before the onset of the arousals; increases in delta power might be an even better measure of arousals than alpha wave changes.

Symptoms of unrefreshing sleep are reported to be greater when the cyclic alternating pattern (CAP, periodic appearance of delta waves and K-complexes) of EEG occupies a greater percent of sleep (Terzano & Parrino, 2000). Sforza et al. suggested that bursts of delta waves and K-complexes were expressions of subcortical arousals representing a real arousal response with tachycardia similar to that seen during cortical arousals (Sforza et al., 2000). Patients with FM have increased amounts of CAP - more so in the more severely symptomatic patients (Rizzi et al., 2004).

3. Sleep studies and CFS

3.1 Sleep disorders in patients with CFS

One of the symptoms used for diagnosing CFS is unrefreshing sleep, and, in fact, this sleep-related problem is the most common complaint among patients with severe medically unexplained fatigue (Unger et al., 2004). Partial sleep deprivation in healthy people can produce marked daytime fatigue (Martin et al., 1996), cognitive problems (Martin et al., 1996), and musculoskeletal achiness (Lentz et al., 1999; Moldofsky & Scarisbrick, 1976; Onen et al., 2001), which are the hallmark symptoms of CFS.

Several early studies suggested that as many as one-half of individuals with CFS have mild sleep apnea syndrome (five or more episodes per hour of apnea/hypopnea), periodic leg movements, or the restless leg syndrome (Buchwald et al., 1994; Krupp et al., 1993). Other studies with more stringent criteria for these disorders either did not find this result (Krupp et al., 1993; Le Bon et al., 2000; Sharpley et al., 1997; Togo et al., 2008).

3.2 Sleep abnormality in patients with CFS

Polysomnographic studies suggest that the sleep architecture is abnormal in CFS patients. The most consistent abnormality is significantly reduced sleep efficiency when compared to controls (Fischler et al., 1997; Krupp et al., 1993; Morriss et al., 1993; Sharpley et al., 1997);

the reported average values range from clearly abnormal (i.e., 76.5%) (Fischler et al., 1997) to those within the normal range (i.e., 90%) (Morriss et al., 1993). From one study providing data on individual patients' sleep efficiencies, one can estimate that 75% of CFS patients have reduced sleep efficiences (Krupp et al., 1993). Sleep disturbance in these patients is obvious because they often show increases in time needed to fall asleep (Morriss et al., 1993; Sharpley et al., 1997) and multiple periods of awakenings or arousals (Fischler et al., 1997; Morriss et al., 1993; Sharpley et al., 1997). Decrease in total duration of Stage 4 sleep has also been reported (Fischler et al., 1997).

We have recently reported the sleep architecture of a sample of female CFS patients during a fixed period of their menstrual cycle and after excluding patients with diagnosable sleep disorders and co-existing major depressive disorder to reduce patient pool heterogeneity (Togo et al., 2008). These patients differed significantly from matched controls in showing evidence of sleep disruption in the form of significantly reduced total sleep time, reduced sleep efficiency, and shorter bouts of sleep than healthy controls. In comparison with controls, sleep in CFS had little effect on either self-reported sleepiness or fatigue. And, interestingly, for patients only, ratings of sleepiness and fatigue correlated well with total sleep duration and efficiency. Dichotomizing the patients into a group that felt sleepier after a night's sleep than before sleep [a.m. sleepier] and a group that felt less sleepy after a night's sleep [a.m. less sleepy] reduced the variability of the sleep records considerably (Togo et al., 2008).

Those patients reporting less sleepiness after a night's sleep had sleep structures similar to those for healthy controls except for a shorter total sleep time and a commensurate reduction in Stage 2 sleep; moreover, they reported their fatigue and pain to diminish following sleep. In contrast, patients in the a.m. sleepier group had the greatest abnormalities of sleep architecture, including poor sleep efficiency, longer sleep latency, and more disrupted sleep as manifested by a higher percentage of short-duration sleep runs, than either controls or patients in the a.m. less sleepy group.

As the time since awakening from sleep increases, sleep latency decreases (Devoto et al., 1999), and one early study of young adults reported an average sleep latency of 30 seconds after a night of sleep deprivation (Carskadon & Dement, 1979). We have determined latency to fall asleep for patients with CFS and healthy controls, previously habituated to sleeping in a sleep lab, after such a night of sleep deprivation in our laboratory (Nakamura et al., 2010). Nine healthy subjects fell asleep within 5 minutes, however 3 subjects took longer – falling asleep within 9 minutes. The CFS patients as a group showed a significantly longer latency to fall asleep after sleep deprivation, but the study population fell out into two groups with the largest group of 10 patients falling asleep within 5 minutes. However, the remaining 5 patients remained awake for a longer period than any control, suggesting that they may have a disorder of arousal. Sleep latency following sleep deprivation correlated inversely with sleep efficiency on the normal sleep night for the patients with CFS. Our results indicate that some CFS patients may have a disorder of arousal which interferes with normal sleep and may, at least in part, be responsible for their disabling fatigue.

3.3 Exercise and sleep in patients with CFS

Exercise elevates core body temperature and increases total duration of slow-wave sleep in the night following exercise in healthy people (Horne & Staff, 1983). To our knowledge, only our study (Togo et al., 2010) has compared sleep in CFS patients before and after exercise.

Exertion is a particularly interesting thing to study in CFS because a disabling and characteristic feature of CFS patients is that even minimal exertion produces a dramatic worsening of symptoms (Komaroff & Buchwald, 1991). No such effect occurs in healthy controls and, in fact, some reports, although anecdotal, suggest that acute exercise can actually improve sleep (Youngstedt et al., 1997).

We have used a standard cardiac-type stress test to probe effects of exertion on symptoms in CFS patients in other studies too. First we found that CFS patients reported more fatigue as much as four days after the exercise stress test (Sisto et al., 1996). Next, we used actigraphy to monitor activity before and after exercise and found that activity levels also fell significantly four days after the exercise stress test (Sisto et al., 1998). We recently replicated and extended this finding using real-time assessment techniques and demonstrated that CFS symptoms do worsen several days after maximal exercise but that neither mood nor cognitive function was affected (Yoshiuchi et al., 2007). We interpreted these changes in activity to support the patient complaint of worsening of symptoms induced by exercise or effort.

We recently investigated the influence of an acute bout of exercise on polysomnography and self-reported measures of sleep (Togo et al., 2010). CFS patients as a group have disrupted sleep characterized by significantly poorer quality sleep than controls. However, the patients as a group showed evidence of improved sleep after exercise. The results were clearer after we used the same stratification strategy that we had used in our earlier work (Togo et al., 2008), that is, splitting subjects into those who were either sleepier or less sleepy after a night's sleep. As expected, exercise improved the sleep quality of healthy controls who had reported decreased morning sleepiness after the baseline sleep night. Contrary to expectation, it had the same result in CFS patients with decreased morning sleepiness. However, patients who reported increased morning sleepiness showed no improvement in sleep disruption, but exercise did not exacerbate their sleep pathology. These patients also had the lowest average sleep efficiency of any of the groups studied. Because exercise did not produce a significant worsening of sleep morphology in CFS, the complaints of symptom worsening, which are reported to occur the next day after exertion, cannot be explained by disruption in sleep. After exercise, approximately half the patients actually sleep better than on their baseline study night, whereas the rest simply did not improve.

4. Sleep dynamics

4.1 Sleep dynamics in healthy humans

Most sleep studies have been performed based upon sleep stage scoring according to the traditional standardized criteria established by Rechtschaffen and Kales (Rechtstchaffen & Kales, 1968). While this methodology has been extremely useful to describe sleep architecture, sleep stage analysis has been limited to simple descriptive statistics, such as total sleep time, sleep efficiency, the number of awakenings, latencies to sleep onset and REM sleep, and the total duration of each sleep stage.

Recently, sleep dynamics, such as transition probabilities among sleep stages and duration distributions of each sleep stage, has been reported by some studies in which the importance of dynamical aspects of sleep has been pointed out (Comte et al., 2006; Kishi et al., 2008, in press; Lo et al., 2002). Yassouridis et al. (Yassouridis et al., 1999) studied survival time statistics of a particular sleep stage ended by other different sleep stages with their event history analysis, a modification of the Cox regression analysis of life-tables (Cox, 1972), by assuming an exponential decay of sleep stage durations,

$$P(t) \sim e^{-t/\tau} \quad (1)$$

where the $P(t)$ is a probability distribution of durations t of a sleep stage and the τ is a constant. Lo et al. studied the dynamics of two-state asleep-awake transitions during sleep in humans, focusing on the duration distributions, and found entirely different behavior in the periods awake and asleep (Lo et al., 2002). Subsequently, they expanded their investigations in humans to other mammalian species, i.e., mice, rats, and cats (Lo et al., 2004). Durations of awake during sleep exhibited a power-law distribution for all species, while durations of sleep episode followed exponential distributions. Comte et al. investigated the transition probabilities and duration distributions of three sleep stages (awake, non-REM, and REM sleep) in rats (Comte et al., 2006). Duration statistics of REM sleep in rats took a power-law probability distribution,

$$P(t) \sim t^{-\alpha} \quad (2)$$

where the a is a constant, partially devaluing the exponential survival time analysis. Finding a power-law relation in REM sleep and waking durations rather than an exponential decay characteristic of random survival times (Lo et al., 2002, 2004) points to the presence of an underlying complex mechanism governing sleep stage transitions because power-law or heavy-tailed distributions of survival times are often observed in a variety of complex systems (Sethna et al., 2001; Sornette, 2004).

We have recently investigated transition dynamics in humans for six sleep stages (awake, Stages 1, 2, 3, and 4 sleep, and REM sleep), the entire set of sleep states in humans (Kishi et al., 2008). Duration of slow-wave sleep follows a power-law probability distribution function, while the durations of Stage 1 sleep take an exponential function, those of Stage 2 sleep obey a stretched exponential form characteristic of a multifactorial decay (Sornette, 2004), and REM sleep durations follow an exponential function. We have also found a substantial number of REM to non-REM sleep transitions in humans, while this transition is reported to be virtually nonexistent in rats (Comte et al., 2006). These features likely reflect stage-specific neural activities (De Gennaro & Ferrara, 2003; Hobson et al., 1986; Koyama & Hayaishi, 1994; McCarley, 2007), and theories explaining different duration or survival time distributions (Sornette, 2004) might give deeper insights into the underlying mechanisms governing sleep stage regulations.

4.2 Sleep dynamics in patients with FM and CFS

FM and CFS share considerable overlapping symptoms, including sleep-related complaints. However, differences between FM and CFS have been reported, and research focusing on uncovering differences between these medically unexplained illnesses is helpful to understand them, rather than focusing on their similarities (Lange & Natelson, 2009). Polysomnographic studies have shown that sleep problems in FM and CFS are quite similar, for instance, increased Stage 1 sleep, reduced slow-wave sleep, more arousals, prolonged sleep onset, reduced sleep efficiency, and the alpha-EEG anomaly, as shown in the previous sections (Fischler et al., 1997; Moldofsky, 2008; Sharpley et al., 1997; Van Hoof et al., 2007). However, these observations are not consistent between studies for both FM and CFS, and there are even cases not showing any statistical differences in normal sleep parameters between healthy controls and FM or CFS patients (Afari & Buchwald, 2003; Chervin et al., 2009; Fischler, 1999;

Reeves et al., 2006). In our study (Togo et al., 2008), after excluding patients with diagnosable sleep disorders, such as sleep-disturbed breathing and leg movement disorders, patients with CFS plus FM had sleep architecture similar to those of patients with CFS alone.

We have found robust differences in sleep dynamics between healthy controls and patients with CFS (Kishi et al., 2008). Although the duration distributions of each sleep stage are not different between healthy controls and patients with CFS, probabilities of transition from both Stage 1 sleep and REM sleep to awake are significantly greater in patients with CFS than healthy controls, indicating that the influence of factors interfering with the continuation of Stage 1 sleep and REM sleep may be different between healthy controls and CFS patients, while the fundamental mechanisms determining durations of each sleep stage are similar. CFS patients might not have a dysfunction in systems maintaining each sleep stage, but they may have a disturbed switching mechanism governing sleep stage transitions. Our data suggest that the major complaint of CFS patients of "unrefreshing sleep" may be derived from this sudden arousal from both Stage 1 sleep and REM sleep.

One study (Burns et al., 2008) showed that sleep stage dynamics was different between patients with FM and healthy controls. Patients with FM showed a parameter that reflects shortened durations of Stage 2 sleep periods. Although shorter Stage 2 sleep durations did not predict daytime sleepiness, they did predict pain which is the main symptom in FM. Short Stage 2 sleep durations may associate with sleep fragmentation or pressure for recovery sleep.

We have recently compered dynamical aspects of sleep, such as transitions probabilities between sleep stages between CFS alone and CFS+FM patients and found differences between them, although sleep architecture did not differ between the groups (Kishi et al., in press). CFS alone has greater probabilities of transitions from REM sleep to awake than healthy controls. This result could be interpreted as a lower sleep pressure in CFS alone. In contrast, CFS+FM has greater probabilities of transitions from waking, Stage 1 sleep, and REM sleep to Stage 2 sleep and from Stage 2 sleep to slow-wave sleep than healthy controls, suggesting the increased sleep pressure in CFS+FM. Transitions from waking and REM sleep to Stage 2 sleep are unusual transitions in healthy humans (Kishi et al., 2008). CFS+FM also has greater probabilities of transitions from slow-wave sleep to waking and Stage 1 sleep, suggesting that this may be the specific sleep problem of CFS+FM.

There are reports of decreased level of central serotonin in FM patients (Juhl, 1998; Neeck & Riedel, 1994). On the other hand, it has been observed that central serotonin responses are upregulated in CFS patients (Afari & Buchwald, 2003; Weaver et al., 2010). We have recently reported that the administration of central monoaminergic (serotonergic and dopaminergic) antagonist alters dynamical sleep stage transitions from Stage 2 sleep to slow-wave sleep; probability of transition from Stage 2 sleep to slow-wave sleep was significantly increased when central serotonergic and dopaminergic antagonist was administered (Kishi et al., 2010). Such monoaminergic systems are closely related with pain modulation (Bannister et al., 2009). Thus, the imbalance of central monoaminergic (serotonergic) systems in FM patients would lead to abnormalities of pain modulations and sleep regulations.

5. Conclusion

Patients with FM and CFS often have sleep-related complaints. Polysomnographic studies have shown sleep problems in FM, i.e., increased Stage 1 sleep, reduced slow-wave sleep, more arousals, prolonged sleep onset, reduced sleep efficiency, the alpha-EEG anomaly during sleep. Although these problems are also shown in patients with CFS, dynamic aspects of sleep show different patterns between FM and CFS patients. Patients with CFS+FM had greater probabilities of transitions from waking, Stage 1 sleep, and REM sleep to Stage 2 sleep and from Stage 2 sleep to slow-wave sleep than healthy controls, suggesting the increased sleep pressure in CFS+FM. In contrast, CFS alone has greater probabilities of transitions from REM sleep to awake than healthy controls, suggesting the lower sleep pressure in CFS alone. Finding such differences is support for the thesis that FM is different illness from CFS, associated with different problems in sleep regulation.

6. Acknowledgment

This work is supported, in part, by National Institute of Health Grant AI-54478 and by Grant-in-Aid for Scientific Research (C) (22500690).

7. References

Afari N. & Buchwald D. (2003). Chronic Fatigue Syndrome: a Review. *The American Journal of Psychiatry*, Vol.160, No.2, (February 2003), ISSN 0002-953X

Affleck G.; Urrows S.; Tennen H.; Higgins P. & Abeles M. (1996). Sequential Daily Relations of Sleep, Pain Intensity, and Attention to Pain Among Women With Fibromyalgia. *Pain*, Vol.68, No.2-3, (December 1996), ISSN 0304-3959

American Sleep Disorders Association. (1992). EEG Arousals: Scoring Rules and Examples: a Preliminary Report From the Sleep Disorders Atlas Task Force of the American Sleep Disorders Association. *Sleep*, Vol.15, No.2, (April 1992), ISSN 0161-8105

Anch A.M.; Lue F.A.; MacLean A.W. & Moldofsky H. (1991). Sleep Physiology and Psychological Aspects of the Fibrositis (Fibromyalgia) Syndrome. *Canadian Journal of Psychology*, Vol.45, No.2, (June 1991), ISSN 0008-4255

Arnold L.M.; Zlateva G.; Sadosky A.; Emir B. & Whalen E. (2011). Correlations Between Fibromyalgia Symptom and Function Domains and Patient Global Impression of Change: a Pooled Analysis of Three Randomized, Placebo-Controlled Trials of Pregabalin. *Pain Medicine*, Vol.12, No.2, (February 2011), ISSN 1526-2375

Bannister K.; Bee L.A. & Dickenson A.H. (2009) Preclinical and Early Clinical Investigations Related to Monoaminergic Pain Modulation. *Neurotherapeutics : The Journal of the American Society for Experimental NeuroTherapeutics*, Vol.6, No.4, (October 2009), ISSN 1933-7213

Bigatti S.M.; Hernandez A.M.; Cronan T.A. & Rand K.L. (2008). Sleep Disturbances in Fibromyalgia Syndrome: Relationship to Pain and Depression. *Arthritis and Rheumatism*, Vol.59, No.7, (July 2008), ISSN 0004-3591

Branco J.; Atalaia A. & Paiva T. (1994). Sleep Cycles and Alpha-Delta Sleep in Fibromyalgia Syndrome. *The Journal of Rheumatology*, Vol.21, No.6, (June 1994), ISSN 0315-162X

Buchwald D.; Pascualy R.; Bombardier C. & Kith P. (1994). Sleep Disorders in Patients With Chronic Fatigue. *Clinical Infectious Diseases*, Vol.18 Suppl 1, (January 1994), ISSN 1058-4838

Burns J.W.; Crofford L.J. & Chervin R.D. (2008). Sleep Stage Dynamics in Fibromyalgia Patients and Controls. *Sleep Medicine*, Vol.9, No.6, (August 2008), ISSN 1389-9457

Carley D.W.; Applebaum R.; Basner R.C.; Onal E. & Lopata M. (1997). Respiratory and Arousal Responses to Acoustic Stimulation. *Chest*, Vol.112, No.6, (December 1997), ISSN 0012-3692

Carskadon M.A. & Dement W.C. (1979). Effects of Total Sleep Loss on Sleep Tendency. *Perceptual and Motor Skills*, Vol.48, No.2, (April 1979), ISSN 0031-5125

Chervin R.D.; Teodorescu M.; Kushwaha R.; Deline A.M.; Brucksch C.B.; Ribbens-Grimm C.; Ruzicka D.L.; Stein P.K.; Clauw D.J. & Crofford L.J. (2009). Objective Measures of Disordered Sleep in Fibromyalgia. *The Journal of Rheumatology*, Vol.36, No.9, (September 2009), ISSN 0315-162X

Comte J.C.; Ravassard P. & Salin P.A. (2006). Sleep Dynamics: a Self-Organized Critical System. *Physical Review E, Statistical, Nonlinear, and Soft Matter Physics*, Vol.73, No.5 Pt 2, (May 2006), ISSN 1539-3755

Cote K.A. & Moldofsky H. (1997). Sleep, Daytime Symptoms, and Cognitive Performance in Patients With Fibromyalgia. *The Journal of Rheumatology*, Vol.24, No.10, (October 1997), ISSN 0315-162X

Cox D.R. (1972). Regression Models and Life-Tables. *Journal of the Royal Statistical Society Series B (Methodological)*, Vol.34, No.2, (March 1972), ISSN 00359246

De Gennaro L. & Ferrara M. (2003). Sleep Spindles: an Overview. *Sleep Medicine Reviews*, Vol.7, No.5, (October 2003), ISSN 1087-0792

Devoto A.; Lucidi F.; Violani C. & Bertini M. (1999). Effects of Different Sleep Reductions on Daytime Sleepiness. *Sleep*, Vol.22, No.3, (May 1999), ISSN 0161-8105

Drewes A.M.; Nielsen K.D.; Taagholt S.J.; Bjerregard K.; Svendsen L. & Gade J. (1995). Sleep Intensity in Fibromyalgia: Focus on the Microstructure of the Sleep Process. *British Journal of Rheumatology*, Vol.34, No.7, (July 1995), ISSN 0263-7103

Drewes A.M.; Svendsen L.; Nielsen K.D.; Taagholt S.J. & Bjerregard K. (1994). Quantification of Alpha-EEG Activity During Sleep in Fibromyalgia: A Study Based on Ambulatory Sleep Monitoring. *Journal of Musculoskeletal Pain*, Vol.2, No.4, (1994), ISSN 1058-2452

Fischler B. (1999). Review of Clinical and Psychobiological Dimensions of the Chronic Fatigue Syndrome: Differentiation From Depression and Contribution of Sleep Dysfunctions. *Sleep Medicine Reviews*, Vol.3, No.2, (June 1999), ISSN 1087-0792

Fischler B.; Le Bon O.; Hoffmann G.; Cluydts R.; Kaufman L. & De Meirleir K. (1997). Sleep Anomalies in the Chronic Fatigue Syndrome. A Comorbidity Study. *Neuropsychobiology*, Vol.35, No.3, (1997), ISSN 0302-282X

Gold A.R.; Dipalo F.; Gold M.S. & Broderick J. (2004). Inspiratory Airflow Dynamics During Sleep in Women With Fibromyalgia. *Sleep*, Vol.27, No.3, (May 2004), ISSN 0161-8105

Halasz P. (1993). Arousals Without Awakening--Dynamic Aspect of Sleep. *Physiology & Behavior*, Vol.54, No.4, (October 1993), ISSN 0031-9384

Hobson J.A.; Lydic R. & Baghdoyan H.A. (1986). Evolving Concepts of Sleep Cycle Generation: From Brain Centers to Neuronal Populations. *Behavioral and Brain Sciences*, Vol.9, No.3, (1986), ISSN 0140-525X

Horne J.A. & Shackell B.S. (1991). Alpha-Like EEG Activity in Non-REM Sleep and the Fibromyalgia (Fibrositis) Syndrome. *Electroencephalography and Clinical Neurophysiology*, Vol.79, No.4, (October 1991), ISSN 0013-4694

Horne J.A. & Staff L.H. (1983). Exercise and Sleep: Body-Heating Effects. *Sleep*, Vol.6, No.1, (1983), ISSN 0161-8105

Jennum P.; Drewes A.M.; Andreasen A. & Nielsen K.D. (1993). Sleep and Other Symptoms in Primary Fibromyalgia and in Healthy Controls. *The Journal of Rheumatology*, Vol.20, No.10, (October 1993), ISSN 0315-162X

Juhl J.H. (1998). Fibromyalgia and the Serotonin Pathway. *Alternative Medicine Review : a Journal of Clinical Therapeutic*, Vol.3, No.5, (October 1998), ISSN 1089-5159

Kishi A.; Natelson B.H.; Togo F.; Struzik Z.R.; Rapoport D.M.& Yamamoto Y. (In press). Sleep-Stage Dynamics in Patients With Chronic Fatigue Syndrome With or Without Fibromyalgia. *Sleep*, (In press), 0161-8105

Kishi A.; Struzik Z.R.; Natelson B.H.; Togo F. & Yamamoto Y. (2008). Dynamics of Sleep Stage Transitions in Healthy Humans and Patients With Chronic Fatigue Syndrome. *American Journal of Physiology Regulatory, Integrative and Comparative Physiology*, Vol.294, No.6, (June 2008), ISSN 0363-6119

Kishi A.; Yasuda H.; Matsumoto T.; Inami Y.; Horiguchi J.; Struzik Z.R. & Yamamoto Y. (2010). Sleep Stage Transitions in Healthy Humans Altered by Central Monoaminergic Antagonist. *Methods of Information in Medicine*, Vol.49, No.5, (2010), ISSN 0026-1270

Komaroff A.L. & Buchwald D. (1991). Symptoms and Signs of Chronic Fatigue Syndrome. *Reviews of Infectious Diseases*, Vol.13 Suppl 1, (January 1991), ISSN 0162-0886

Korszun A.; Young E.A.; Engleberg N.C.; Brucksch C.B.; Greden J.F. & Crofford L.A. (2002). Use of Actigraphy for Monitoring Sleep and Activity Levels in Patients With Fibromyalgia and Depression. *Journal of Psychosomatic Research*, Vol.52, No.6, (June 2002), ISSN 0022-3999

Koyama Y. & Hayaishi O. (1994). Firing of Neurons in the Preoptic/Anterior Hypothalamic Areas in Rat: Its Possible Involvement in Slow Wave Sleep and Paradoxical Sleep. *Neuroscience Research*, Vol.19, No.1, (February 1994), ISSN 0168-0102

Krupp L.B.; Jandorf L.; Coyle P.K. & Mendelson W.B. (1993). Sleep Disturbance in Chronic Fatigue Syndrome. *Journal of Psychosomatic Research*, Vol.37, No.4, (May 1993), ISSN 0022-3999

Landis C.A.; Lentz M.J.; Tsuji J.; Buchwald D. & Shaver J.L. (2004). Pain, Psychological Variables, Sleep Quality, and Natural Killer Cell Activity in Midlife Women With and Without Fibromyalgia. *Brain, Behavior, and Immunity*, Vol.18, No.4, (July 2004), ISSN 0889-1591

Lange, G. & Natelson, B.H. (2009). Chronic Fatigue Syndrome, In: *Functional Pain Syndromes: Presentation and Pathophysiology*, 1st ed., E.A.Mayer, M.C.Bushnell, (Ed.), 245-261, IASP Press, ISBN 978-0931092756, Seattle, USA

Lashley F.R. (2003). A Review of Sleep in Selected Immune and Autoimmune Disorders. *Holistic Nursing Practice*, Vol.17, No.2, (March 2003), ISSN 0887-9311

Le Bon O.; Fischler B.; Hoffmann G.; Murphy J.R.; De Meirleir K.; Cluydts R. & Pelc I. (2000). How Significant Are Primary Sleep Disorders and Sleepiness in the Chronic Fatigue Syndrome? *Sleep Research Online : SRO*, Vol.3, No.2, (2000), ISSN 1096-214X

Lentz M.J.; Landis C.A.; Rothermel J. & Shaver J.L. (1999). Effects of Selective Slow Wave Sleep Disruption on Musculoskeletal Pain and Fatigue in Middle Aged Women. *The Journal of Rheumatology*, Vol.26, No.7, (July 1999), ISSN 0315-162X

Leventhal L.; Freundlich B.; Lewis J.; Gillen K.; Henry J. & Dinges D. (1995). Controlled Study of Sleep Parameters in Patients With Fibromyalgia. *Journal of Clinical Rheumatology : Practical Reports on Rheumatic & Musculoskeletal Diseases*, Vol.1, No.2, (April 1995), ISSN 1076-1608

Lo C.C.; Amaral L.A.N.; Havlin S.; Ivanov P.C.; Penzel T.; Peter J.H. & Stanley H.E. (2002). Dynamics of Sleep-Wake Transitions During Sleep. *Europhysics Letters*, Vol.57, No.5, (March 2002), ISSN 0295-5075

Lo C.C.; Chou T.; Penzel T.; Scammell T.E.; Strecker R.E.; Stanley H.E. & Ivanov P.C. (2004). Common Scale-Invariant Patterns of Sleep-Wake Transitions Across Mammalian Species. *Proceedings of the National Academy of Sciences of the United States of America*, Vol.101, No.50, (December 2004), ISSN 0027-8424

Martin S.E.; Engleman H.M.; Deary I.J. & Douglas N.J. (1996). The Effect of Sleep Fragmentation on Daytime Function. *American Journal of Respiratory and Critical Care Medicine*, Vol.153, No.4 Pt 1, (April 1996), ISSN 1073-449X

Martin S.E.; Engleman H.M.; Kingshott R.N. & Douglas N.J. (1997). Microarousals in Patients With Sleep Apnoea/Hypopnoea Syndrome. *Journal of Sleep Research*, Vol.6, No.4, (December 1997), ISSN 0962-1105

McCarley R.W. (2007). Neurobiology of REM and NREM Sleep. *Sleep Medicine*, Vol.8, No.4, (June 2007), ISSN 1389-9457

McNamara M.E. (1993). Alpha Sleep: a Mini Review and Update. *Clinical EEG (Electroencephalography)*, Vol.24, No.4, (October 1993), ISSN 0009-9155

Moldofsky H. (1989). Sleep and Fibrositis Syndrome. *Rheumatic Diseases Clinics of North America*, Vol.15, No.1, (February 1989), ISSN 0889-857X

Moldofsky H. (2002). Management of Sleep Disorders in Fibromyalgia. *Rheumatic Diseases Clinics of North America*, Vol.28, No.2, (May 2002), ISSN 0889-857X

Moldofsky H. (2008). The Significance of the Sleeping-Waking Brain for the Understanding of Widespread Musculoskeletal Pain and Fatigue in Fibromyalgia Syndrome and Allied Syndromes. *Joint, Bone, Spine : Revue du Rhumatisme*, Vol.75, No.4, (July 2008), ISSN 1297-319X

Moldofsky H.; Lue F.A. & Smythe H.A. (1983). Alpha EEG Sleep and Morning Symptoms in Rheumatoid Arthritis. *The Journal of Rheumatology*, Vol.10, No.3, (June 1983), ISSN 0315-162X

Moldofsky H.; Saskin P. & Lue F.A. (1988). Sleep and Symptoms in Fibrositis Syndrome After a Febrile Illness. *The Journal of Rheumatology*, Vol.15, No.11, (November 1988), ISSN 0315-162X

Moldofsky H. & Scarisbrick P. (1976). Induction of Neurasthenic Musculoskeletal Pain Syndrome by Selective Sleep Stage Deprivation. *Psychosomatic Medicine*, Vol.38, No.1, (January 1976), ISSN 0033-3174

Moldofsky H.; Scarisbrick P.; England R. & Smythe H. (1975). Musculosketal Symptoms and Non-REM Sleep Disturbance in Patients With "Fibrositis Syndrome" and Healthy Subjects. *Psychosomatic Medicine*, Vol.37, No.4, (July 1975), ISSN 0033-3174

Molony R.R.; MacPeek D.M.; Schiffman P.L.; Frank M.; Neubauer J.A.; Schwartzberg M. & Seibold J.R. (1986). Sleep, Sleep Apnea and the Fibromyalgia Syndrome. *The Journal of Rheumatology*, Vol.13, No.4, (August 1986), ISSN 0315-162X

Morriss R.; Sharpe M.; Sharpley A.L.; Cowen P.J.; Hawton K. & Morris J. (1993). Abnormalities of Sleep in Patients With the Chronic Fatigue Syndrome. *British Medical Journal*, Vol.306, No.6886, (May 1993), ISSN 0959-8138

Nakamura T.; Togo F.; Cherniack N.S.; Rapoport D.M. & Natelson B.H. (2010). A Subgroup of Patietns With Chronic Fatigue Syndrome May Have a Disorder of Arousal. *The Open Sleep Journal* No.3, (2010), ISSN 1874-6209

Neeck G. & Riedel W. (1994). Neuromediator and Hormonal Perturbations in Fibromyalgia Syndrome: Results of Chronic Stress? *Bailliere's Clinical Rheumatology*, Vol.8, No.4, (November 1994), ISSN 0950-3579

O'Brien E.M.; Waxenberg L.B.; Atchison J.W.; Gremillion H.A.; Staud R.M.; McCrae C.S. & Robinson M.E. (2011). Intraindividual Variability in Daily Sleep and Pain Ratings Among Chronic Pain Patients: Bidirectional Association and the Role of Negative Mood. *The Clinical Journal of Pain*, Vol.27, No.5, (June 2011), ISSN 0749-8047

O'Brien E.M.; Waxenberg L.B.; Atchison J.W.; Gremillion H.A.; Staud R.M.; McCrae C.S. & Robinson M.E. (2010). Negative Mood Mediates the Effect of Poor Sleep on Pain Among Chronic Pain Patients. *The Clinical Journal of Pain*, Vol.26, No.4, (May 2010), ISSN 0749-8047

Older S.A.; Battafarano D.F.; Danning C.L.; Ward J.A.; Grady E.P.; Derman S. & Russell I.J. (1998). The Effects of Delta Wave Sleep Interruption on Pain Thresholds and Fibromyalgia-Like Symptoms in Healthy Subjects; Correlations With Insulin-Like Growth Factor I. *The Journal of Rheumatology*, Vol.25, No.6, (June 1998), ISSN 0315-162X

Onen S.H.; Alloui A.; Gross A.; Eschallier A. & Dubray C. (2001). The Effects of Total Sleep Deprivation, Selective Sleep Interruption and Sleep Recovery on Pain Tolerance Thresholds in Healthy Subjects. *Journal of Sleep Research*, Vol.10, No.1, (March 2001), ISSN 0962-1105

Perlis M.L.; Giles D.E.; Bootzin R.R.; Dikman Z.V.; Fleming G.M.; Drummond S.P. & Rose M.W. (1997). Alpha Sleep and Information Processing, Perception of Sleep, Pain, and Arousability in Fibromyalgia. *The International Journal of Neuroscience*, Vol.89, No.3-4, (February 1997), ISSN 0020-7454

Pitson D.J. & Stradling J.R. (1998). Autonomic Markers of Arousal During Sleep in Patients Undergoing Investigation for Obstructive Sleep Apnoea, Their Relationship to EEG Arousals, Respiratory Events and Subjective Sleepiness. *Journal of Sleep Research*, Vol.7, No.1, (March 1998), ISSN 0962-1105

Rechtstchaffen A. & Kales A. (1968). *A Manual of Standardized Terminology, Techniques and Scoring System for Sleep States of Human Subjects*. US Government Printing Office, National Institute of Health Publication, Washington (DC), USA

Reeves W.C.; Heim C.; Maloney E.M.; Youngblood L.S.; Unger E.R.; Decker M.J.; Jones J.F. & Rye D.B. (2006). Sleep Characteristics of Persons With Chronic Fatigue Syndrome and Non-Fatigued Controls: Results From a Population-Based Study. *BMC Neurology*, Vol.6, (2006), ISSN 1471-2377

Rizzi M.; Sarzi-Puttini P.; Atzeni F.; Capsoni F.; Andreoli A.; Pecis M.; Colombo S.; Carrabba M. & Sergi M. (2004). Cyclic Alternating Pattern: a New Marker of Sleep Alteration in Patients With Fibromyalgia? *The Journal of Rheumatology*, Vol.31, No.6, (June 2004), ISSN 0315-162X

Roizenblatt S.; Moldofsky H.; edito-Silva A.A. & Tufik S. (2001). Alpha Sleep Characteristics in Fibromyalgia. *Arthritis and Rheumatism*, Vol.44, No.1, (January 2001), ISSN 0004-3591

Scheuler W.; Stinshoff D. & Kubicki S. (1983). The Alpha-Sleep Pattern. Differentiation From Other Sleep Patterns and Effect of Hypnotics. *Neuropsychobiology*, Vol.10, No.2-3, (1983), ISSN 0302-282X

Sethna J.P.; Dahmen K.A. & Myers C.R. (2001). Crackling Noise. *Nature*, Vol.410, No.6825, (March 2001), ISSN 0028-0836

Sforza E.; Jouny C. & Ibanez V. (2000). Cardiac Activation During Arousal in Humans: Further Evidence for Hierarchy in the Arousal Response. *Clinical Neurophysiology*, Vol.111, No.9, (September 2000), ISSN 1388-2457

Sforza E.; Juony C. & Ibanez V. (2002). Time-Dependent Variation in Cerebral and Autonomic Activity During Periodic Leg Movements in Sleep: Implications for Arousal Mechanisms. *Clinical Neurophysiology*, Vol.113, No.6, (June 2002), ISSN 1388-2457

Sharpley A.; Clements A.; Hawton K. & Sharpe M. (1997). Do Patients With "Pure" Chronic Fatigue Syndrome (Neurasthenia) Have Abnormal Sleep? *Psychosomatic Medicine*, Vol.59, No.6, (November 1997), ISSN 0033-3174

Shaver J.L.; Lentz M.; Landis C.A.; Heitkemper M.M.; Buchwald D.S. & Woods N.F. (1997). Sleep, Psychological Distress, and Stress Arousal in Women With Fibromyalgia. *Research in Nursing & Health*, Vol.20, No.3, (June 1997), ISSN 0160-6891

Sisto S.A.; LaManca J.; Cordero D.L.; Bergen M.T.; Ellis S.P.; Drastal S.; Boda W.L.; Tapp W.N. & Natelson B.H. (1996). Metabolic and Cardiovascular Effects of a Progressive Exercise Test in Patients With Chronic Fatigue Syndrome. *The American Journal of Medicine*, Vol.100, No.6, (June 1996), ISSN 0002-9343

Sisto S.A.; Tapp W.N.; LaManca J.J.; Ling W.; Korn L.R.; Nelson A.J. & Natelson B.H. (1998). Physical Activity Before and After Exercise in Women With Chronic Fatigue Syndrome. *QJM : Monthly Journal of the Association of Physicians*, Vol.91, No.7, (July 1998), ISSN 1460-2725

Sornette D. (2004). *Critical Phenomena in Natural Sciences: Chaos, Fractals, Self-Organization and Disorder: Concepts and Tools.* 2nd ed. Springer, ISBN 978-3540308829, Berlin, Germany

Spitzer A.R. & Broadman M. (2010). A Retrospective Review of the Sleep Characteristics in Patients With Chronic Fatigue Syndrome and Fibromyalgia. *Pain Practice*, Vol.10, No.4, (July 2010), ISSN 1530-7085

Terzano M.G. & Parrino L. (2000). Origin and Significance of the Cyclic Alternating Pattern (CAP). REVIEW ARTICLE. *Sleep Medicine Reviews*, Vol.4, No.1, (February 2000), ISSN 1087-0792

Theadom A.; Cropley M. & Humphrey K.L. (2007). Exploring the Role of Sleep and Coping in Quality of Life in Fibromyalgia. *Journal of Psychosomatic Research*, Vol.62, No.2, (February 2007), ISSN 0022-3999

Togo F.; Cherniack N.S. & Natelson B.H. (2006). Electroencephalogram Characteristics of Autonomic Arousals During Sleep in Healthy Men. *Clinical Neurophysiology*, Vol.117, No.12, (December 2006), ISSN 1388-2457

Togo F.; Natelson B.H.; Cherniack N.S.; FitzGibbons J.; Garcon C. & Rapoport D.M. (2008). Sleep Structure and Sleepiness in Chronic Fatigue Syndrome With or Without Coexisting Fibromyalgia. *Arthritis Research & Therapy*, Vol.10, No.3, (2008), ISSN 1478-6362

Togo F.; Natelson B.H.; Cherniack N.S.; Klapholz M.; Rapoport D.M. & Cook D.B. (2010). Sleep Is Not Disrupted by Exercise in Patients With Chronic Fatigue Syndromes. *Medicine and Science in Sports and Exercise*, Vol.42, No.1, (January 2010), ISSN 0195-9131

Unger E.R.; Nisenbaum R.; Moldofsky H.; Cesta A.; Sammut C.; Reyes M. & Reeves W.C. (2004). Sleep Assessment in a Population-Based Study of Chronic Fatigue Syndrome. *BMC Neurology*, Vol.4, (April 2004), ISSN 1471-2377

Van Hoof E.; De Becker P.; Lapp C.; Cluydts R. & De Meirleir K. (2007). Defining the Occurrence and Influence of Alpha-Delta Sleep in Chronic Fatigue Syndrome. *The American Journal of the Medical Sciences*, Vol.333, No.2, (February 2007), ISSN 0002-9629

Viola-Saltzman M.; Watson N.F.; Bogart A., Goldberg J. & Buchwald D. (2010). High Prevalence of Restless Legs Syndrome Among Patients With Fibromyalgia: a Controlled Cross-Sectional Study. *Journal of Clinical Sleep Medicine*, Vol.6, No.5, (October 2010), ISSN 1550-9389

Weaver S.A.; Janal M.N.; Aktan N.; Ottenweller J.E. & Natelson B.H. (2010). Sex Differences in Plasma Prolactin Response to Tryptophan in Chronic Fatigue Syndrome Patients With and Without Comorbid Fibromyalgia. *Journal of Women's Health*, Vol.19, No.5, (May 2010), ISSN 1059-7115

White K.P.; Speechley M.; Harth M. & Ostbye T. (1999). The London Fibromyalgia Epidemiology Study: Comparing the Demographic and Clinical Characteristics in 100 Random Community Cases of Fibromyalgia Versus Controls. *The Journal of Rheumatology*, Vol.26, No.7, (July 1999), ISSN 0315-162X

Winkelman J.W. (1999). The Evoked Heart Rate Response to Periodic Leg Movements of Sleep. *Sleep*, Vol.22, No.5, (August 1999), ISSN 0161-8105

Wolfe F.; Smythe H.A.; Yunus M.B.; Bennett R.M.; Bombardier C.; Goldenberg D.L.; Tugwell P.; Campbell S.M.; Abeles M.; Clark P. et al. (1990). The American College of Rheumatology 1990 Criteria for the Classification of Fibromyalgia. Report of the Multicenter Criteria Committee. *Arthritis and Rheumatism*, Vol.33, No.2, (February 1990), ISSN 0004-3591

Yassouridis A.; Steiger A.; Klinger A. & Fahrmeir L. (1999). Modelling and Exploring Human Sleep With Event History Analysis. *Journal of Sleep Research*, Vol.8, No.1, (March 1999), ISSN 0962-1105

Yoshiuchi K.; Cook D.B.; Ohashi K.; Kumano H.; Kuboki T.; Yamamoto Y. & Natelson B.H. (2007). A Real-Time Assessment of the Effect of Exercise in Chronic Fatigue Syndrome. *Physiology & Behavior*, Vol.92, No.5, (December 2007), ISSN 0031-9384

Youngstedt S.D.; O'Connor P.J. & Dishman R.K. (1997). The Effects of Acute Exercise on Sleep: a Quantitative Synthesis. *Sleep*, Vol.20, No.3, (March 1997), ISSN 0161-8105

2

Central Sensitization and Descending Facilitation in Chronic Pain State

Emiko Senba[1], Keiichiro Okamoto[2] and Hiroki Imbe[1]
[1]*Wakayama Medical University,*
[2]*University of Minnesota, School of Dentistry,*
[1]*Japan*
[2]*USA*

1. Introduction

Physiological pain is inevitable for us to avoid harmful stimuli as one of the self-defense mechanisms. On the other hand, pathological pain, especially deep tissue pain often persists and develops into chronic pain states. Patients suffering chronic pain often complain of pain and tenderness in various parts of the body, and chronic stress often exacerbates their pain. Fibromyalgia represents the extreme end of the spectrum of chronic widespread pain syndromes. In this chapter, we address questions as follows; why deep tissue pain tends to pass into a chronic state (is deep tissue pain more unpleasant than superficial pain?), how or why the whole body aches in these patients, and how chronic stress exacerbate their pain. We speculate that persistent pain-induced brain sensitization underlies these chronic pain states. Areas involved in the emotional aspects of pain, such as the anterior cingulate cortex (ACC), insular cortex (IC), amygdala, may be sensitized. Memory of pain based on sensitization of brain areas involved in emotional response is useful for us to avoid future damage. However, in some pathological conditions, descending pain control system may switch from inhibition to facilitation, with unknown mechanism, and facilitate pain sensation and behavior. Serotonin (5-HT) may be involved in the descending facilitation in pathological pain conditions. Understanding of the etiology, pathophysiology and treatment of chronic pain states is an urgent issue from the view points of patients' quality of life and socio-economy.

2. Clinical features of chronic pain states

2.1 Clinical features of fibromyalgia syndrome

Fibromyalgia (FM) or fibromyalgia syndrome (FMS) is an intractable widespread pain disorder of unknown etiology that affects about 2-3% of population and is most frequently diagnosed in women. One unique feature of FM is its wide spread nature of pain and tenderness. Despite extensive investigations, no distinct tissue damage, structural abnormalities, or evidence for a source of chronic stimulation of pain afferents have been detected in FM patients (Meeus & Nijs, 2007). Furthermore, FM pain is diffuse and multifocal, lacks a distinct spatial localization, often waxes and wanes and is frequently migratory in nature. These features have led to the hypothesis that hyperexcitability of the central nervous system or dysfunction of the central inhibitory system may exist in these patients. Central hyperexcitability could explain exaggerated pain and withdrawal reflex of FM patients with minimal and undetectable tissue damage, in that the nociceptive signals are amplified by the hyperexcitable neurons (Banic et al., 2004). Indeed, more numerous regions are activated in the brain of fibromyalgia patients by the same intensity of noxious heat stimuli compared to pain-free controls (Cook et al., 2004). Cognitive-behavioral therapy was shown to attenuate nociceptive flexion reflex threshold in FM patients (Ang et al., 2010). On the other hand, subtle peripheral tissue abnormalities, such as increased levels of substance P in muscle tissue, increased IL-1 levels in cutaneous tissue have also been demonstrated in FM patients (for review, see Staud & Rodriguez, 2006). Sympathetic hyperactivity, abnormal heart rate variability, has been postulated in FM patients, which may explain the multisystem features of FM and symptoms such as sleep disorders, anxiety and constant fatigue. Some people assume that FM is a generalized form of reflex sympathetic dystrophy (CRPS type I) (Martinez-Lavin, 2001, 2007). Both conditions affect mostly females and have frequent post-traumatic onset. Moreover, many features of FMS resemble those of posttraumatic stress disorder (PTSD). PTSD is highly associated with FMS in male FMS patients (Amital et al., 2006). Thus, peripheral impulse input may also play an important role in maintaining central sensitization (Staud et al, 2009). But it does not necessarily extensive, because central sensitization seems to require little sustained input for the maintenance of chronic pain state.

Several animal models for FMS or chronic widespread pain have been produced to elucidate the underlying mechanisms. Biogenic amine depletion by repeated administration of reserpine (once daily for 3 consecutive days) caused muscle and cutaneous hyperalgesia, and increased immobility time in forced swim test (Nagakura et al., 2009). Vagal dysfunction induced by subdiaphragmatic vagotomy, caused muscle and visceral, but not cutaneous, hyperalgesia (Furuta et al., 2009). Mice subjected to intermittent cold stress (4°C) exhibited prolonged bilateral allodynia (Nishiyori & Ueda, 2008).

FM is often accompanied by a variety of other symptoms (Bennett et al., 2007; Clauw, 2009; Moldofsky, 2008). Common disorders associated with fibromyalgia include chronic fatigue syndrome (21-80%), irritable bowel syndrome (32-80%), temporomandibular disorder (TMD) (75%), headache (tension/migraine) (10-80%), major depressive disorder (62%), insomnia (60-90%) and urinary disturbance (interstitial cystitis) (20-60%).

Clinical features of chronic pain states, such as FM, low back pain, TMD etc., can be summarized as follows.

1. Patients usually suffer deep tissue (musculo-skeletal) pain and tenderness.
2. Their pain and hyperalgesia is often bilateral and widespread in nature.
3. Their symptoms are aggravated by psychological/emotional stress.
4. They often accompany major depression and sleep disturbance.
5. Their symptoms are often attenuated by anti-depressant (amitriptyline) or anti-convulsant (gabapentin).

We have attempted to discuss possible mechanisms that underlie these symptoms using some animal models of chronic pain states

2.2 Why deep tissues are more frequently affected in chronic pain state?

Brain imaging studies indicate the network of somatosensory (S1, S2, IC), limbic (IC, ACC) and associative structures, such as prefrontal cortex (PFC), receiving parallel inputs from multiple nociceptive pathways (Apkarian et al., 2005). The clinical pain processing may be different from experimental pain processing, as well as acute pain perception in normal subjects is distinct from that seen in chronic clinical pain conditions. Chronic pain engages brain regions critical for cognitive/emotional assessments (Apkarian et al., 2005; Hsieh et al., 1995). Schweinhardt et al. (2006) revealed that activation patterns of anterior insular cortex (AIC) were different in experimental and clinical pains. They divided the AIC into two parts; rostral AIC (rAIC) and caudal AIC (cAIC). Clinical pain is preferentially processed in rAIC, while experimental pain in healthy volunteers predominantly evoked cAIC. Henderson et al. (2007) compared muscular pain and cutaneous pain, by injecting 5% saline solution into muscle and subcutaneous, respectively. They found that muscular pain evoked more rostral part of the AIC than cutaneous pain did, suggesting that muscular deep pain is more uncomfortable and intractable in nature. These findings may provide us with some important cues to explain why muscular or deep tissue pain easily develops into chronic pain state.

2.3 Mirror-image pain (pain or hyperalgesia in unaffected side or area)

2.3.1 Mirror-image pain in the literature

In the literature, we can find many examples of bilateral hyperalgesia and allodynia due to unilateral neuropathy (Erichsen & Blackburn-Munro, 2002; Li et al., 2006 Yasuda et al., 2005; Yu et al., 1996;), cancer pain (Mao-Ying et al., 2006) or inflammation (Milligan et al., 2003, 2005; Shenker, 2003; Schreiber et al., 2008) in animal models and neuropathic pain (Becerra L et al., 2006; Hatashita et al., 2008; Wasner et al., 2008) , capsaicin-induced experimental pain (Shenker et al., 2008) and toothache (Khan et al., 2007) in human cases (for review see Huang and Yu, 2010). An experimental pain induced by unilateral intramuscular injection of low pH saline caused bilateral long-lasting hyperalgesia in rats (Sluka et al., 2001). They speculated that this contralateral spread of hyperalgesia was mediated by central sensitization. Milligan et al. (2003, 2005) produced sciatic inflammatory neuropathy induced by perineural injection of Zymozan to cause localized inflammation of the sciatic nerve of rats (Chacur et al., 2001). It should be noted that these animals showed unilateral (low-dose Zymozan) or bilateral mechanical allodynia (high-dose Zymozan) depending upon the intensity of the sciatic nerve inflammation induced. They also used chronic constriction injury (CCI) model, a classic partial nerve injury, to show a typical bilateral mechanical allodynia. They speculated that mirror-image pain is caused by the spreading of inflammation to the contralateral spinal cord.

2.3.2 A unilateral nerve injury-induced bilateral hyperalgesia

We reported an animal model of unilateral nerve injury induced bilateral hyperalgesia (Yasuda et al., 2005). In the course of experiments using CCI model rats, we noticed that the nociceptive threshold to mechanical stimulation was decreased bilaterally when the unilateral sciatic nerve was accidentally tightly ligated. We also found that unilateral axotomy of the sciatic nerve exhibited bilateral hyperalgesia. Then we focused our investigation on the axotomy model. The left common sciatic nerve was exposed and tightly ligated at two locations and the sciatic nerve was cut between the ligatures (injured paw). Mechanical nociceptive thresholds were assessed by loading pressure stimulation. The lateral dorsal surface of the injured paws was hypoalgesic, while the medial dorsal paw (innervated by intact saphenous nerve) on the injured side was hyperalgesic. Hyperalgesia in the injured paw may be mediated by intact saphenous nerve as previously described as adjacent neuropathic hyperalgesia (Markus et al., 1984). To our surprise, dorsal paws on the opposite side also showed hyperalgesia on the day of nerve injury, and these levels were maintained throughout the 14 days of experimental period (Fig. 1).

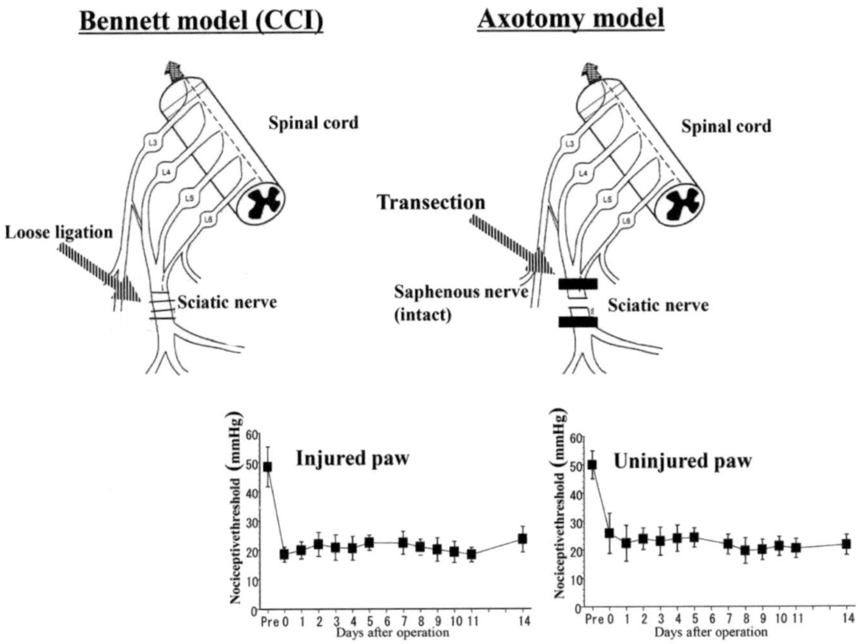

Fig. 1. Bilateral mechanical hyperalgesia observed in unilateral axotomy model rats.

Daily administration of amitriptyline resulted in significant dose-dependent normalization of the nociceptive thresholds in both paws. However, morphine was ineffective. Only the highest dose of morphine was tentatively effective (Fig. 2 A, B). Treatment with gabapentin resulted in significant dose-dependent normalization of the nociceptive thresholds in both paws, while, indomethacin, even at excessive dose was not effective at all (Fig. 2 C, D). Tail-flick latency was reduced at 4 h after axotomy, and it was maintained throughout the experiment, indicating the existence of systemic thermal hyperalgesia. We produced the same model in mice and they also showed bilateral thermal hyperalgesia.

The most prominent feature of this axotomy model is bilateral and systemic hyperalgesia in response to pressure and heat, which appeared immediately after transection of the sciatic nerve. In terms of symptoms and drug efficacy, this axotomy model resembles those seen in human patients with neuropathic pain. First, these animals exhibited mechanical and heat hyperalgesia spreading to unaffected areas, which is also observed in various chronic pain conditions in humans. Second, tricyclic amitriptyline and gabapentin, which have been accepted as first-line agents for treatment of neuropathic pain in humans, also attenuated hyperalgesia in this model. We examined the activation of microglia in the ipsi- and contralateral dorsal horn. They were activated more slowly than the appearance of the symptom and got the peak at 1 week after the transaction (Fig. 3).

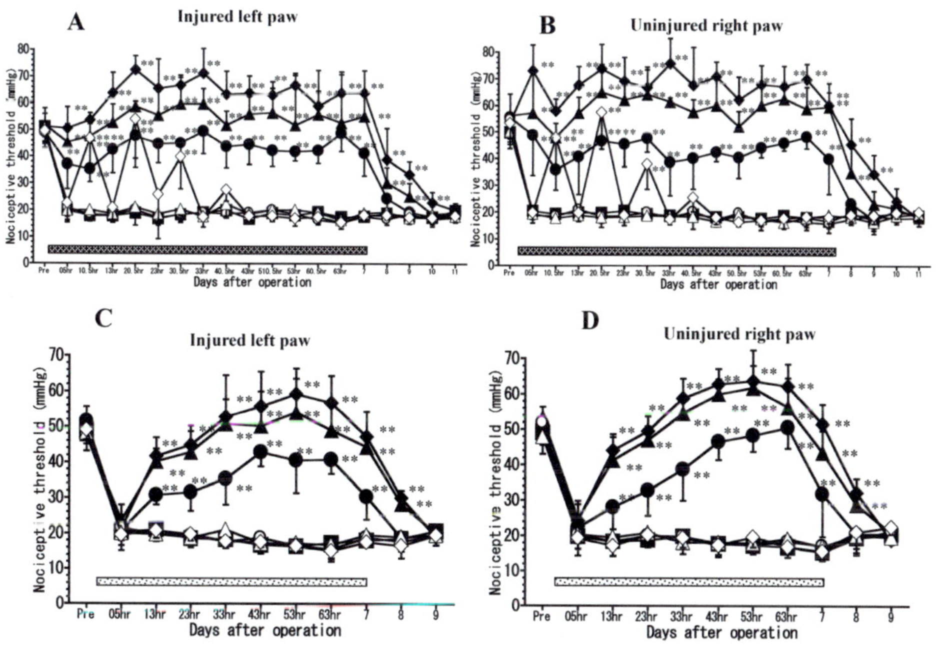

Fig. 2. Inhibitory effects of amitriptyline (A,B) and gabapentin (C,D) on axotomy-induced hyperalgesia. **P<0.01 vs. control, Dunnett's multiple comparison test.

■: control, amitriptyline ●: 25 mg/kg, p.o. ▲: 50 mg/kg, p.o. ◆: 100 mg/kg, p.o.
morphine ○: 3 mg/kg, s.c. △: 10 mg/kg, s.c. ◇: 30 mg/kg, s.c. (A,B)
gabapentin ●: 30 mg/kg, p.o. ▲: 100 mg/kg, p.o. ◆: 300 mg/kg, p.o.
indomethacin ○: 1 mg/kg, p.o. △: 3 mg/kg, p.o. ◇: 10 mg/kg, p.o.(C,D)

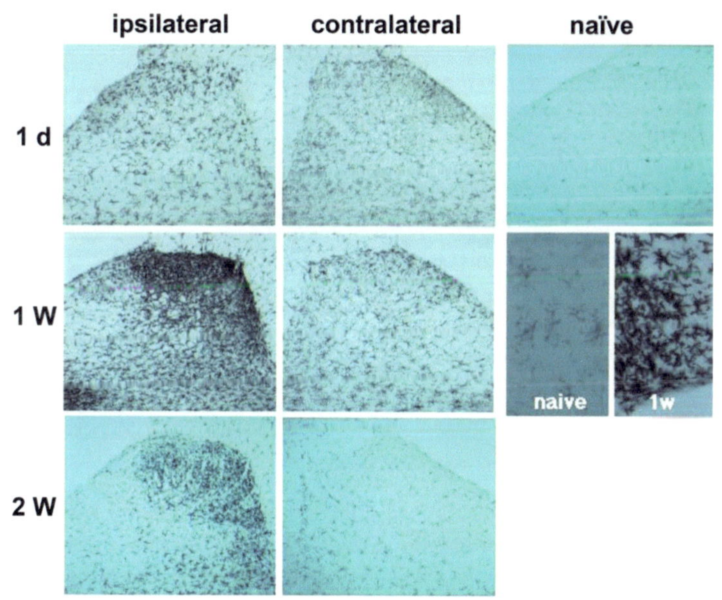

Fig. 3. Activation of microglia (OX-42-positive) in the dorsal horn after sciatic nerve injury.

2.3.3 Possible mechanisms for Mirror-image pain

The underlying mechanisms of Mirror-image pain are still obscure, but sensitization of pain processing in the central nervous system, including the spinal cord and descending pain control system, in addition to peripheral mechanisms, have been postulated.

1. Peripheral mechanisms: Contralateral effects may be mediated by circulating factors such as cytokines, chemokines and other chemical mediators, produced in the injured tissue and/or sympatho-adrenal hormones released in response to traumatic stress.
2. Spinal mechanisms:
   #1 Neural theory: Contralateral effects may be mediated by afferent fibers projecting to the contralateral side or interposed interneurons (Koltzenburg et al., 1999). These systems may be silent in normal condition, and remodeling can be triggered by the injury.
   #2 Immune theory: Contralateral effects may be mediated by immune and glial cells, such as microglia and astrocytes, and chemical mediators derived from these cells may contribute to central sensitization through activating NMDA receptors (DeLeo et al., 2006). However, as far as our axotomy model is concerned, this theory does not seem to fully explain a prominent bilateral hyperalgesia, since only weak microglial activation was observed in the contralateral dorsal horn much later than the appearance of hyperalgesia.

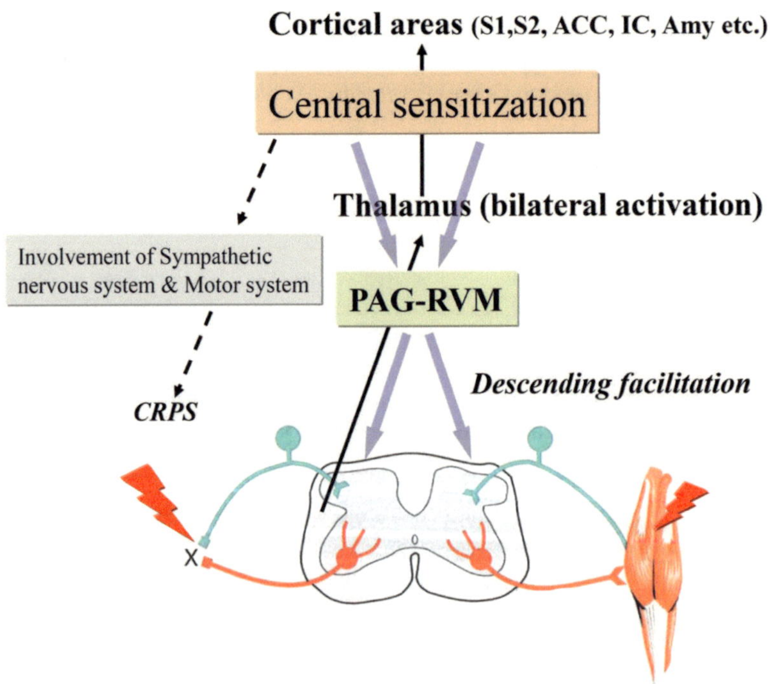

Fig. 4. Central mechanisms of Mirror-image pain.

3. Supraspinal mechanisms: Contralateral effects may be mediated by brain activation and descending facilitation in chronic pain state. Persistent noxious inputs from the periphery may sensitize certain brain areas involved in the central pain circuitry, such as ACC, IC and amygdala (Ikeda et al., 2007) (for review, see Meeus & Nijs, 2007). Activation of these areas may cause bilateral hyperalgesia via descending pain control system (Fig. 4).

Central autonomic system and motor system may also be activated and exert their influences on target tissues via visceral and somatic efferent pathways, respectively. Such pathological features may establish a syndrome called CRPS (complex regional pain syndrome) due to peripheral tissue or nerve injury.

As described above, the descending pain modulatory system may lose balance between inhibition and facilitation, turning the balance in favor of facilitation, in pathological pain conditions. There is a growing body of evidence to show that this descending facilitation plays a role in the establishment and maintenance of chronic pain state in various pathological conditions (Fig. 4).

3. How TMD patients exhibit widespread pain and hyperalgesia?

3.1 Clinical features of TMD

TMD patients, as well as those suffering FM or low back pain, often complain of persistent pain in multiple body areas (Turp et al., 1998), and evidence for generalized hyperalgesia in TMD patients has been reported (Sarlani et al., 2003). The diagnostic criteria for TMD include a pain in the TMJ and associated masticatory muscles such as masseter muscles. Several clinical studies have demonstrated decreased nociceptive threshold in remote areas in TMD patients and suggested a deficit of the endogenous pain inhibitory systems for the pathophysiology of TMD. A deficit of endogenous pain inhibitory systems including diffuse noxious inhibitory control (DNIC), has been suggested also in fibromyalgia patients (Julien et al., 2005). DNIC contributes to enhance the biologically valuable pain signals by reducing other irrelevant noise in the pain transmission system.

3.2 Hyperalgesia in remote areas in a craniofacial pain model

We developed a deep craniofacial pain model that partially mimics symptoms of TMD patients by injecting complete Freund's adjuvant (CFA) into temporomandibular joint (TMJ). Then, the influence of persistent TMJ inflammation on nociceptive responses of remote bodily areas of the rat was investigated (Okamoto et al., 2006a). Von Frey test revealed mechanical hypersensitivity in these rats at 8, 10 and 14 days after CFA injection compared to non-CFA group that had not been treated with CFA. When formalin was injected into the left hindpaw, these rats showed significantly enhanced nocifensive behavior at 10 and 14 days after CFA injection compared to non-CFA control group. The numbers of Fos-positive neurons in the ipsilateral lumbar dorsal horn were also significantly increased compared to those in non-CFA group. These findings clearly indicate that persistent TMJ inflammation may enhance nociceptive perception and nocifensive behavior in remote bodily areas.

3.3 Descending facilitation mediated by 5-HT and 5-HT3 receptors

It is well known that 5-HT plays important roles in modulating spinal nociceptive transmission. Activation of the descending serotonergic bulbospinal system modulates responses of dorsal horn neurons to noxious stimuli. Spinal cord dorsal horn as well as trigeminal subnucleus caudalis (Vc) are the major sites of 5-HT released from descending fibers from the RVM. However, controversy remains as to which types of 5-HT receptor are involved in mediating serotonergic pronociceptive or anti-nociceptive effects in this system.

Nociceptive transmission in the spinal cord is modulated by the nucleus raphe magnus (NRM) and adjacent structures of the rostral ventromedial medulla (RVM). The RVM

receives projections from the periaqueductal gray matter (PAG) and sends projections to the spinal dorsal horn largely along the dorsolateral funiculus. The NRM is a major source of the descending serotonergic pathways that participate in spinal nociceptive modulation. These descending serotonergic pathways exert bi-directional control of nociception.
We further demonstrated that 5-HT3 receptors play a role in descending facilitation from the RVM to spinal dorsal horn.
In the next experiment, we injected formalin into masseter muscle in rats suffering persistent TMJ inflammation. By means of electrophysiology, we showed exaggerated responses of WDR neurons in the Vc/C2 (trigeminal subnucleus caudalis/ upper cervical spinal junction) region, and these responses were attenuated when 5-HT3 receptors were blocked (Okamoto et al., 2005b). By extracellular recordings in the Vc/C2 region, we identified two types of units; Deep-wide dynamic range (WDR) units and Skin-WDR units. Deep-WDR units have mechanoreceptive fields in the deep craniofacial tissues including masseter muscle but do not have cutaneous mechanoreceptive fields. As shown in previous studies, formalin-induced neural activities in the dorsal horn and trigeminal caudalis neurons showed a biphasic time course that is similar to that of formalin-induced behavior. Deep-WDR unit discharges evoked by the formalin injection into masseter muscle were significantly enhanced in the late phase in CFA-injected day 7 group. Discharges of Skin-WDR units evoked by the noxious pinch stimulation to facial skin in CFA-injected day 7 group were also significantly enhanced compared with those in non-CFA-injected group. Topical administration of central 5-HT3 receptor antagonist, tropisetron, onto trigeminal Vc/C2 region significantly reduced both formalin-evoked Deep-WDR unit and pinch-evoked Skin-WDR unit discharges in non-CFA and CFA day 7 groups (Fig. 5).

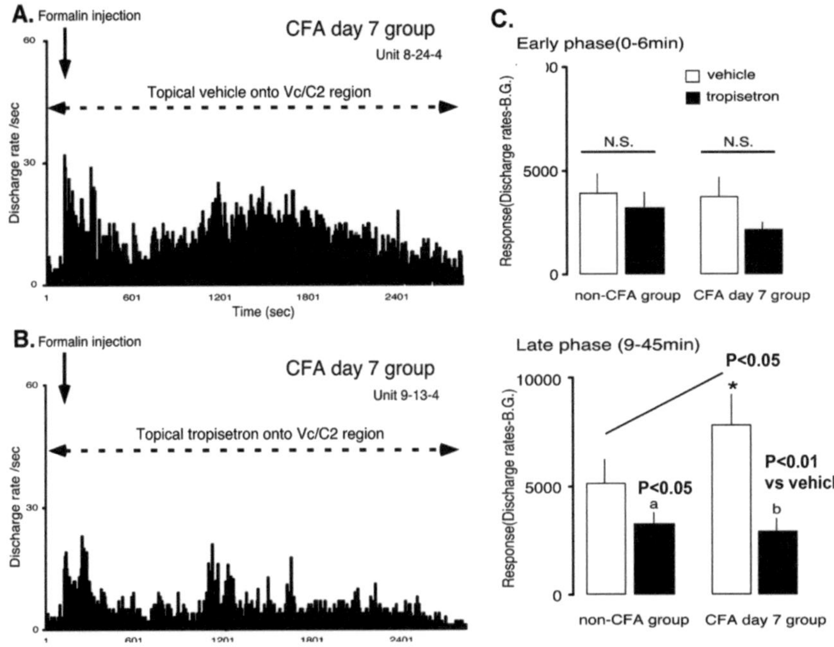

Fig. 5. The effects of topical application of tropisetron onto the Vc/C2 region on the early and late phases of Deep-WDR unit responses evoked by formalin injection.

Therefore, it is concluded that 5-HT derived from descending fibers from the RVM activated spinal/trigeminal dorsal horn neurons via 5-HT3 receptors and facilitated pain behavior (Okamoto et al., 2005b). Recently, Wei et al. (2010) showed the essential role of 5-HT in this system in pathological pain conditions by selectively blocking 5-HT production in the RVM. As for 5-HT2A receptors, we demonstrated that this receptor subtype is involved in the suppression of nociceptive processing and behavior using the deep craniofacial pain model (Okamoto et al., 2007). Most of 5-HT2A receptor-immunoreactive neurons in the superficial Vc/C2 region were glutamic acid decarboxylase (GAD)-positive, i.e. inhibitory interneurons (unpublished observation of K.O.). On the other hand, the majority (about 87%) of 5-HT3 receptor-immunoreactive dorsal horn axons seem to be derived from GAD-negative, presumably excitatory interneurons (Maxwell et al., 2003). Therefore, 5-HT2A and 5-HT3 receptor subtypes may exert opposite influences on pain transmission in the dorsal horn (Fig. 6), while we have shown, in the periphery at the level of sensory nerve terminals, these two receptors potentiate hyperalgesia in pathological conditions (Okamoto et al., 2004, 2005a, 2006b). 5-HT2A receptors are exclusively involved in the potentiation of inflammatory pain (Okamoto et al., 2002).

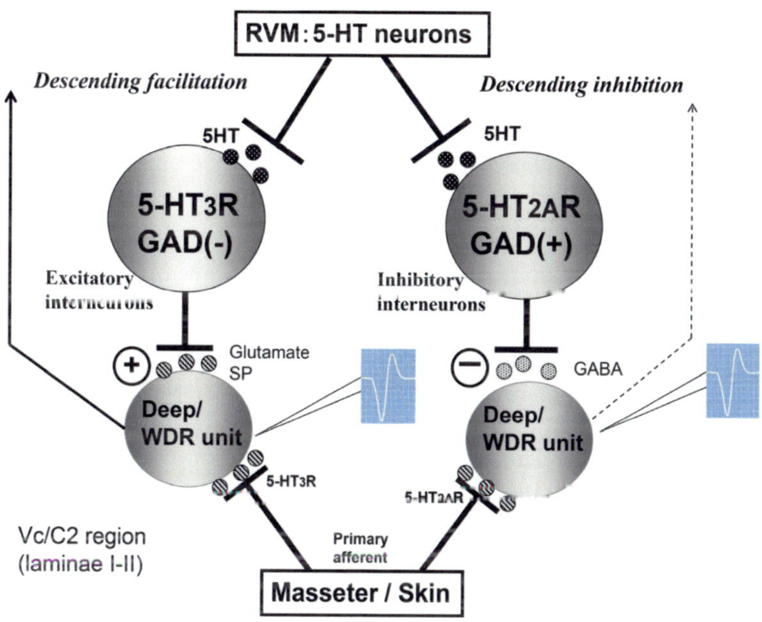

Fig. 6. Descending facilitation and inhibition differentially mediated via 5-HT3 receptor and 5-HT2A receptor expressing interneurons in the dorsal horn.

4. How is chronic pain state aggravated by stress?

4.1 Stress-induced analgesia and hyperalgesia

Acute stress is generally considered to suppress pain, which is called stress-induced analgesia (SIA). We usually feel less pain when we are injured in a battle or in a sport game. On the other hand, repeated exposure to non-noxious situation, such as chronic restraint stress, forced swim stress, cold environment or social defeat, can elicit hyperalgesia and allodynia in experimental animals (for review, see Imbe et al., 2006). It is well known that chronic psychoemotional stress and anxiety enhance pain sensitivity in human (Ashkinazi & Vershinina, 1999; Rhudy & Meagher, 2000). Stress has also been found to exacerbate and could contribute to the etiology of chronic painful disorders, such as, fibromyalgia (Clauw, 2009; Wood, 2004), low back pain (Pincus et al., 2002), irritable bowel syndrome (Delvaux, 1999), rheumatoid arthritis (Herrmann et al., 2000) and headache (Nash & Thebarge, 2006).

A variety of environmental and/or psychological stressful stimuli have been shown to affect pain sensitivity. Some kinds of acute stress and most of chronic stress increase pain sensitivity. These phenomena are termed stress-induced hyperalgesia (SIH) (Imbe et al., 2006). Acute exposure to emotionally arousing non-noxious stress, such as inescapable holding, novel environments or vibration, produces an immediate and transient hyperalgesia (Jorum, 1988; Vidal & Jacob, 1982).

A line of evidence suggests that chronic stress induced by repeated exposure to cold environment (Satoh et al., 1992), restraint (Gameiro et al., 2005, 2006; Imbe et al., 2004) and forced swim (Quintero et al., 2000, 2003; Suarez-Roca et al., 2008) also produces relatively persistent hyperalgesia, which seems to mimic the human chronic pain condition. For example, chronic restraint stress (1h daily for 4 days) produced mechanical allodynia and hyperalgesia in formalin test in rats (Bardin et al., 2009). Repeated swim stress (10-20 min daily for 3 consecutive days) induced hyperalgesic responses to formalin injection into hindpaw and increased c-Fos expression in the spinal dorsal horn compared to those observed in naïve rats or rats subjected to sham stress (Quintero et al., 2003). Repeated cold stress (4^0C or -3^0C for 5 days) induced bilateral deep mechanical hyperalgesia in rats (Nasu et al., 2010). Although in these animal models dysfunctions of several neurotransmitter systems have been shown pharmacologically to be involved in SIH, the responsible neural circuits remain elusive.

4.2 Proposed mechanisms for stress-induced hyperalgesia

There may be two hypotheses to explain the mechanisms of stress-induced hyperalgesia; one is peripheral theory and the other is central theory. Water avoidance stress model, a psychological stress, induced muscle hyperalgesia and increased muscle nociceptor activity, including increased action potentials and conduction velocity (Chen et al., 2011). These authors speculated that these peripheral mechanisms may contribute to the stress-induced chronic widespread pain, like fibromyalgia, and supported peripheral theory. Repeated sound stress also activates hypothalamo-pituitary-adrenal and sympathoadrenal axes, and enhances mechanical hyperalgesia induced by inflammatory mediators. The released glucocorticoids and catecholamines cause long lasting alterations in intracellular signal pathways in primary afferent nociceptor (Khasar et al., 2005, 2008, 2009). Spinal mechanism, i.e. reduced γ-aminobutyric acid (GABA) release in the dorsal horn of rats subjected to repeated forced swim stress is also implicated in the SIH (Suarez-Rosa et al., 2008). As central mechanisms of SIH, descending pain modulatory system from the RVM to the spinal dorsal horn may play a key role. Initially this system was considered to be solely inhibitory (Basbaum & Fields, 1984). However, it gradually became evident that this descending system from the RVM exerts bidirectional (facilitatory and inhibitory) control of nociception (Porreca et al., 2002; Ren & Dubner, 2002). Many studies have reported the participation of descending facilitatory input from the RVM in inflammatory and neuropathic pain conditions. Recent studies have demonstrated that inflammatory (Sugiyo et al., 2005), neuropathic (Burgess et al., 2002), cancer (Donovan-Rodriguez et al., 2006) and visceral pains (Vera-Portocarrero et al., 2006) are linked to the activation of descending facilitatory pathway from the RVM. Burgess et al. (2002) showed that this system maintains, but does not initiate, neuropathic pain. At the cellular level, both on- and off-cells, but not neutral cells, in the RVM are sensitized to mechanical stimuli after nerve injury (Carlson et al., 2007). As to the contribution of the RVM to stress-induced pain modulation, previous studies have shown that the RVM is involved in SIA (Foo & Helmstetter, 2000; Mitchell et al., 1998). Balance between inhibition and facilitation may determine the outcome effects of stress on pain behaviors.

4.3 Descending facilitation in chronic stress and chronic pain state

We have examined possible involvement of descending facilitation in chronic stress-induced hyperalgesia. Briefly, rats were stressed by restraint daily for 6 h. In the acute stress model, they were exposed to a single restraint. In the chronic stress model, they were repeatedly exposed to daily restraint for 1, 2 or 3 weeks. The control group was not subjected to restraint. Tail-flick latency (TFL) and activation of extracellular signal-regulated kinases (ERK) in the RVM were assessed. Acute restraint stress (6 h, 1 day) obviously increased TFLs. Conversely, chronic restraint stress (6 h/day, 2- or 3-week) caused a reduction in TFLs (Fig. 7). Only a few phosphorylated-ERK (p-ERK)-immunoreactive (IR) neurons were observed in the RVM of the control and acutely restraint rats, but 3 week restraint stress produced robust increase in p-ERK-IR in the RVM (Fig. 8). On the other hand, c-Fos expression was increased only in the brain of acutely stressed animals. C-Fos response was dramatically decreased when stressful stimuli were repeated as we have shown previously (Senba & Ueyama, 1997; Umemoto et al., 1996).

Especially 3-week restraint stress induced about 60% increase in the number of p-ERK-IR neurons in the RVM, compared to the control rats. Since about 20% of the total RVM neurons are serotonergic, we examined immunohistochemical colocalization of p-ERK and 5HT. The majority of p-ERK-IR neurons also expressed 5HT-IR. The incidence of p-ERK-IR in RVM 5-HT neurons was doubled after 3-weeks of restraint (30.2%) compared to control rats (14.2%). Western blot analysis showed that the level of TPH in the 3-week restraint rats was significantly increased (130%, n=4) compared to that in the control rats (100%, n=4, $P<0.05$) (Fig. 9) (Imbe et al., 2004).

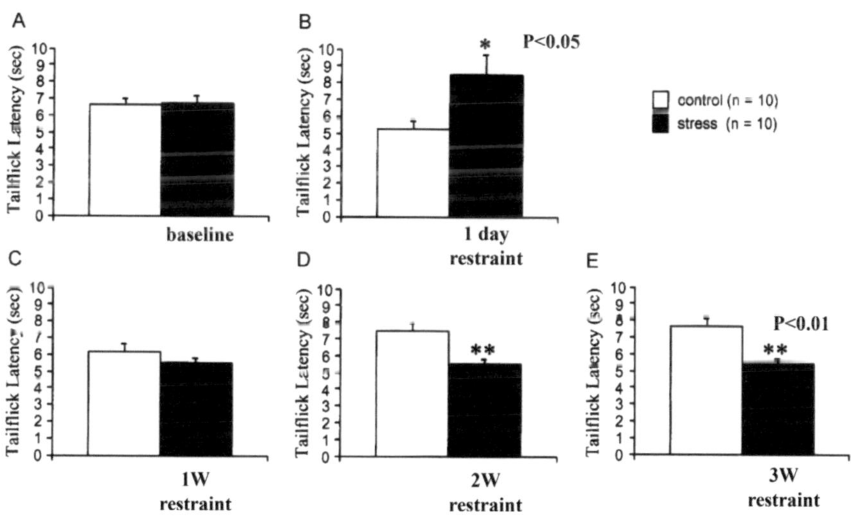

Fig. 7. Effects of acute and chronic restraint stress on tailflick latencies, showing stress-induced analgesia (acute) and stress-induced hyperalgesia (chronic).

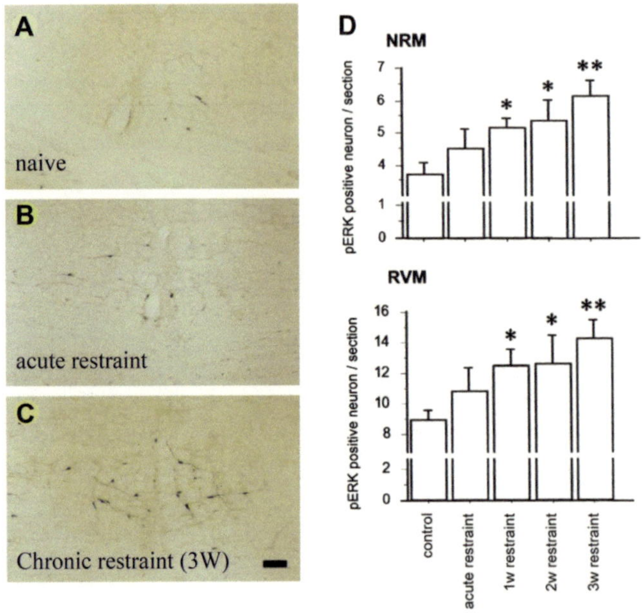

Fig. 8. Numbers of p-ERK-positive neurons in the RVM in control and stressed rats.

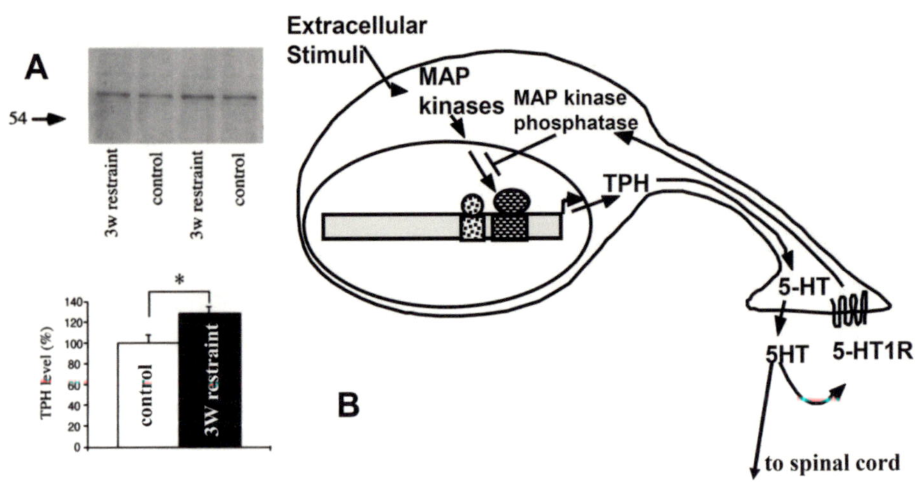

Fig. 9. Effects of 3-week restraint stress on the level of TPH (A). Activation of ERK may increase the transcription of TPH in 5-HT producing RVM neurons (B).

We have previously showed that activation of ERK in the RVM is involved in thermal hyperalgesia during peripheral inflammation (Imbe et al., 2005, 2008, 2011). We also showed that microinjection of MEK inhibitor, U0126, into the RVM attenuated the hyperalgesia due to CFA-induced peripheral inflammation. Recently it has been demonstrated that activation of ERK in the RVM also contributes to persistent neuropathic pain (Geranton et al., 2010). Under the special condition of chronic restraint stress, persistent ERK activation induced by chronic stress may increase the transcription of tryptophan hydroxylase (TPH), a rate-limiting enzyme in serotonin biosynthesis (Wood & Russo, 2001), leading to central sensitization of dorsal horn neurons via descending serotonergic facilitatory projection (Fig. 9). Moreover, in the same study we observed p-ERK-immunoreactivity was dramatically decreased in the locus coeruleus (LC) of chronically stressed animals (Imbe et al., 2004), suggesting that descending noradrenergic inhibition to the spinal dorsal horn is decreased. This may also contribute to the chronic stress-induced hyperalgesia.

We showed another example of stress-induced hyperalgesia and enhancement of pain behavior, in which rats were subjected to forced swim stress for 3 days (Imbe et al., 2010). These animals showed prolonged nocifensive behavior in Formalin test. The destruction of the RVM with ibotenic acid almost completely prevented the enhancement of formalin-evoked nocifensive behavior following the forced swim stress (Fig. 10).

As a summary, chronic restraint stress induced thermal hyperalgesia in rats (Imbe et al., 2004), in which p-ERK and levels of TPH were increased in the RVM. 5HT released from the bulbospinal neurons may exert facilitatory effects on spinal nociceptive processing probably through 5HT3 receptors (Okamoto et al., 2005b; Suzuki et al., 2002, 2004). We also demonstrated that descending facilitation from the RVM is required for the enhancement of formalin-evoked nocifensive behavior following repeated forced swim stress (Imbe et al., 2010). These findings clearly indicate that stress-induced brain sensitization may aggravate pain and hyperalgesia, through PAG-RVM pathway. Molecular and cellular mechanisms of stress-induced central sensitization are still obscure, but it has been demonstrated that chronic stress induces dendritic remodeling of cortical (Radley et al., 2004), hippocampal and amygdaloid neurons (Vyas et al., 2002).

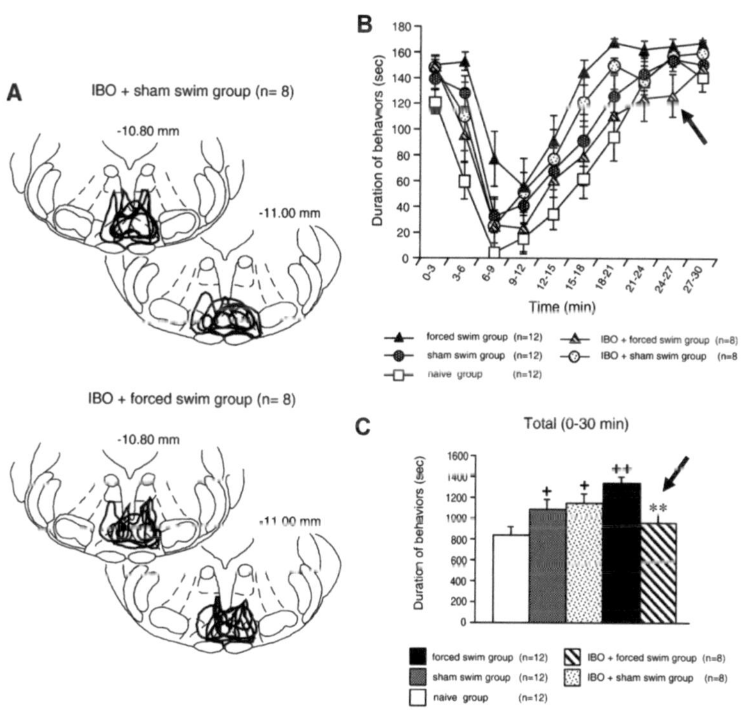

Fig. 10. Destruction of the RVM with ibotenic acid (IBO) (A) blocks the enhancement of formalin-evoked nocifensive behavior in rats subjected to forced swim stress (arrows in B, C).

Now it is obvious that PAG-RVM-spinal (or trigeminal) dorsal horn system is a final common pathway for descending pain modulatory system. A line of experimental evidence suggests that the RVM is a target of descending fibers from upper brain regions to induce SIH. The dorsomedial nucleus of the hypothalamus (DMH) plays important roles in mediating neuroendocrine, cardiovascular and thermogenic responses to emotional stressor. Activation (disinhibition) of the DMH induced a robust activation of ON-cells, suppression of OFF-cells in the RVM and behavioral hyperalgesia in rats (Martenson et al., 2009). Other brain regions that are critical to emotional processes, such as amygdala, lateral hypothalamus, ACC and PFC, also modulate nociceptive behavior by affecting the activity of RVM (Calejesan et al., 2000; Holden and Pizzi, 2008; Hutchison et al., 1996; McGaraughty & Heinricher, 2002, McGaraughty et al., 2004).

Thus, brain activation due to chronic emotional stress in addition to persistent pain in the peripheral tissue may add widespread nature to the chronic pain state. In the situations where threats are repeated, a shift towards descending facilitation of pain may be required to enhance vigilance in order to ensure survival.

5. Plastic changes in the ACC and chronic persistent pain

5.1 Mechanisms of central sensitization

Central sensitization has been defined as "an enhanced responsiveness of nociceptive neurons in the CNS to their normal afferent input" by the International Association for the Study of Pain (IASP). Synaptic plasticity in the cortex as well as in the spinal dorsal horn is believed to be important for the amplification of painful information in chronic pain conditions. The ACC is found to be a key area that links various noxious and painful stimuli to emotional responses. Central sensitization of this area may lead to chronic pain, as well as pain-related cognitive emotional disorders.

There are two major mechanisms that contribute to central plasticity: first, it is well known that LTP, which includes postsynaptic glutamate receptors, such as NMDA and AMPA receptors, and downstream signaling molecules, various kinds of kinases, increased presynaptic glutamate release, provides basic neural mechanism for learning, memory, and chronic pain (Sandkuhler, 2007). The mechanisms of LTP in the ACC have been extensively studied (Kim et al., 2010; Zhuo, 2007, 2008). Second, decreased inhibitory mechanisms by GABA may also contribute to the central sensitization. Recently it has been clearly demonstrated that under neuropathic pain conditions, membrane-bound GABA transporter-3 (GAT-3) on activated astrocytes were significantly increased in the ACC, leading to a decrease in extracellular GABA levels and an increase in neuronal activation in this area (Narita et al., 2011). These long-term synaptic changes help the brain to recognize the changing environment, to gain the ability to respond properly to it and avoid danger in the future. However, in the case of permanent injury, brain fails to distinguish acute physiological pain and chronic pathological pain, and respond similarly, subjecting us to useless, only harmful exaggerated pain state.

5.2 Mechanisms of insomnia in chronic pain patients

One of the common complaints and sufferings of people with chronic pain, including FM, is insomnia (Moldofsky, 2008; O'Brien et al., 2010). Cortical GABAergic neurons play important roles in sleep/wake regulation (Gottesmann, 2004; Kilduff et al., 2011), and disturbance of this system may lead to the pathogenesis of insomnia in these patients. In the experimental model mice for neuropathic pain, an increase in wakefulness and a decrease in non-rapid eye movement sleep during day time, in which mice are supposed to be drowsy or sleeping, have been demonstrated by means of EEG analysis (Takemura et al., 2011). Under these conditions, as mentioned above, GAT-3 on activated astrocytes were increased in the ACC, and extracellular GABA levels in this area after depolarization were rapidly decreased (Narita et al., 2011). Furthermore, sleep disturbance induced by sciatic nerve ligation was improved by the intra-cingulate cortex injection of a GAT-3 inhibitor. These findings provide novel evidence that sciatic nerve ligation decreases extracellular-released GABA in the ACC of mice, which may, at least in part, explain the insomnia in patients with neuropathic pain. On the other hand, it should be noted that sleep disturbance, i.e. REM deprivation, itself may reduce pain threshold (Hakki Onen, 2001).

5.3 Treatment of chronic pain with gabapentin

Recently brain hyperactivity in neuropathic pain model mice in response to painful thermal stimulation has been examined using functional MRI (fMRI) (Takemura et al., 2011). Compared to sham-operated animals, exaggerated BOLD signals in the brain regions, such as the S1, ACC and medial/lateral thalamus, of animals subjected to nerve ligation were observed. Injection of gabapentin (GBP) (60 mg/kg, i.p.) dramatically reduced the nerve injury-induced brain hyperactivity. At the same time, thermal hyperalgesia these animals showed was also normalized by GBP. Changes in EEG patterns identified in these mice during light phase were also attenuated by GBP treatment (Takemura et al., 2011).

The site of action of GBP is still obscure, although GBP was shown to block the function of voltage-gated calcium channels (VGCCs), by binding to the $\alpha2\delta$-1 subunits. VGCCs play an essential role in controlling neurotransmitter release, neuronal excitability, and gene expression in the nervous system. In VGCCs, $\alpha1$ subunit harbors the channel pore and gating machinery, while accessory subunits, $\alpha2\delta$-1, affect channel kinetics, contributing to the trafficking and insertion of main $\alpha1$ subunit into the membrane.

One of the possible sites of action of GBP is spinal dorsal horn, in which $\alpha2\delta$-1 subunits are expressed on the primary afferent terminals (Li C-Y et al., 2004). Supraspinal hypothesis has also been postulated. GBP may inhibit the release of GABA, and disinhibition of LC neurons may potentiate descending inhibition (Tanabe et al., 2008). There may be numerous candidates for possible sites of action of GBP, since $\alpha2\delta$-1 subunit mRNAs are widely distributed in supraspinal pain-related areas, in addition to DRG and spinal dorsal horn (Cole et al., 2005). Indeed, GBP has marked positive and negative effects on BOLD signal intensity in a number of pain-related supraspinal areas (Governo et al., 2008), supporting its well-described analgesic effects both in animal models and patients of chronic pain states.

6. Conclusion

1. Mirror-image pain and stress-induced hyperalgesia may be induced by central sensitization and subsequent descending facilitation.
2. Brain hyperactivation, systemic pain and sleep disturbance were attenuated by gabapentin. Central sensitization could be a treatment target in patients suffering chronic widespread pain, such as fibromyalgia.

7. References

Amital, D., Fostick, L., Polliack, M.L., Segev, S., Zohar, J., Rubinow, A. & Amital, H. Posttraumatic stress disorder, tenderness, and fibromyalgia syndrome: are they different entities? J Psychosom Res. 2006 Nov; 61(5):663-669.

Ang DC, Chakr R, Mazzuca S, France CR, Steiner J, Stump T. Cognitive-behavioral therapy attenuates nociceptive responding in patients with fibromyalgia: a pilot study. Arthritis Care Res (Hoboken). 2010 May;62(5):618-23.

Apkarian, A.V., Bushnell, M.C., Treede, R.D. & Zubieta, J.K. Human brain mechanisms of pain perception and regulation in health and disease. *Eur J Pain*. 2005 Aug; 9(4):463-484.

Ashkinazi, I.Ya. & Vershinina, E.A. Pain sensitivity in chronic psychoemotional stress inhumans. *Neurosci Behav Physiol* 1999; 29:333-337.

Banic, B., Petersen-Felix, S., Andersen, O.K., Radanov, B.P., Villiger, P.M., Arendt-Nielsen, L. & Curatolo, M. Evidence for spinal cord hypersensitivity in chronic pain after whiplash injury and in fibromyalgia. Pain. 2004 Jan; 107(1-2):7-15.

Bardin, L., Malfetes, N., Newman-Tancredi, A. & Depoortère, R. Chronic restraint stress induces mechanical and cold allodynia, and enhances inflammatory pain in rat: Relevance to human stress-associated painful pathologies. *Behav Brain Res.* 2009 Dec ; 205(2):360-366.

Basbaum, A.I. & Fields, H.L. Endogenous pain control systems: brainstem spinal pathways and endorphin circuitry. *Annu Rev Neurosci* 1984; 7:309-338.

Becerra, L., Morris, S., Bazes, S., Gostic, R., Sherman, S., Gostic, J., Pendse, G., Moulton, E., Scrivani, S., Keith, D., Chizh, B. & Borsook, D. Trigeminal neuropathic pain alters responses in CNS circuits to mechanical (brush) and thermal (cold and heat) stimuli. *J Neurosci.* 2006 Oct ;26(42):10646-10657.

Bennett, R.M., Jones, J., Turk, D.C., Russell, I.J. & Matallana, L. An internet survey of 2,596 people with fibromyalgia. *BMC Musculoskelet Disord.* 2007 Mar; 8:27.

Burgess, S.E., Gardell, L.R., Ossipov, M.H., Malan, T.P. Jr, Vanderah, T.W., Lai, J. & Porreca, F. Time-dependent descending facilitation from the rostral ventromedial medulla maintains, but does not initiate, neuropathic pain. *J Neurosci.* 2002 Jun; 22(12):5129-5136.

Calejesan, A.A., Kim, S.J. & Zhuo, M. Descending facilitatory modulation of a behavioral nociceptive response by stimulation in the adult rat anterior cingulate cortex. *Eur J Pain.* 2000; 4(1):83-96.

Carlson, J.D., Maire, J.J., Martenson, M.E. & Heinricher, M.M. Sensitization of pain-modulating neurons in the rostral ventromedial medulla after peripheral nerve injury. *J Neurosci.* 2007 Nov; 27(48):13222-13231.

Chacur, M., Milligan, E.D., Gazda, L.S., Armstrong, C., Wang, H., Tracey, K.J. & Maier, S.F. Watkins LR. A new model of sciatic inflammatory neuritis (SIN): induction of unilateral and bilateral mechanical allodynia following acute unilateral peri-sciatic immune activation in rats. *Pain.* 2001 Dec; 94(3):231-244.

Chen, X., Green, P.G. & Levine, J.D. Stress enhances muscle nociceptor activity in the rat. *Neuroscience.* 2011 Jun; 185:166-173.

Clauw, D.J. Fibromyalgia: an overview. *Am J Med.* 2009 Dec; 122(12 Suppl):S3-S13.

Cole, R.L., Lechner, S.M., Williams, M.E., Prodanovich, P., Bleicher, L., Varney, M.A. & Gu, G. Differential distribution of voltage-gated calcium channel alpha-2 delta ($\alpha 2\delta$) subunit mRNA-containing cells in the rat central nervous system and the dorsal root ganglia. *J Comp Neurol.* 2005 Oct; 491(3):246-269.

Cook, D.B., Lange, G., Ciccone, D.S., Liu, W.C., Steffener, J. & Natelson, B.H. Functional imaging of pain in patients with primary fibromyalgia. J Rheumatol. 2004 Feb; 31(2):364-378.

De Leo, J.A., Tawfik, V.L. & LaCroix-Fralish, M.L. The tetrapartite synapse: path to CNS sensitization and chronic pain. *Pain.* 2006 May; 122(1-2):17-21.

Delvaux, M.M. Stress and visceral perception. *Can. J. Gastroenterol.* 1999; 13 Suppl A: 32A–36A.

Donovan-Rodriguez, T., Urch, C.E. & Dickenson, A.H. Evidence of a role for descending serotonergic facilitation in a rat model of cancer-induced bone pain. *Neurosci Lett.* 2006 Jan; 393(2-3):237-242.

Erichsen, H.K. & Blackburn-Munro, G. Pharmacological characterisation of the spared nerve injury model of neuropathic pain. *Pain.* 2002 Jul; 98(1-2):151-161.

Foo, H. & Helmstetter, F.J.. Activation of kappa opioid receptors in the rostral ventromedial medulla blocks stress-induced antinociception. *Neuroreport.* 2000 Oct; 11(15):3349-3352.

Furuta, S., Shimizu, T., Narita, M., Matsumoto, K., Kuzumaki, N., Horie, S., Suzuki, T. & Narita, M. Subdiaphragmatic vagotomy promotes nociceptive sensitivity of deep tissue in rats. *Neuroscience.* 2009 Dec; 164(3):1252-1262.

Gameiro, G.H., Andrade Ada, S., de Castro, M., Pereira, L.F., Tambeli, C.H. & Veiga, M.C. The effects of restraint stress on nociceptive responses induced by formalin injected in rat's TMJ. *Pharmacol Biochem Behav.* 2005 Oct; 82(2):338-344.

Gameiro, G.H., Gameiro, P.H., Andrade Ada, S, Pereira, L.F., Arthuri, M.T., Marcondes, F.K. & Veiga, M.C. Nociception- and anxiety-like behavior in rats submitted to different periods of restraint stress. *Physiol Behav.* 2006 Apr; 87(4):643-649.

Géranton, S.M., Tochiki, K.K., Chiu, W.W., Stuart, S.A. & Hunt, S.P. Injury induced activation of extracellular signal-regulated kinase (ERK) in the rat rostral ventromedial medulla (RVM) is age dependant and requires the lamina I projection pathway. *Mol Pain.* 2010 Sep; 6:54.

Gottesmann, C. Brain inhibitory mechanisms involved in basic and higher integrated sleep processes. *Brain Res Brain Res Rev.* 2004 Jul; 45(3):230-249.

Governo, R.J., Morris, P.G., Marsden, C.A. & Chapman, V. Gabapentin evoked changes in functional activity in nociceptive regions in the brain of the anaesthetized rat: an fMRI study. Br J Pharmacol. 2008 Apr; 153(7):1558-1567.

Hakki Onen S, Alloui A, Jourdan D, Eschalier A, Dubray C. Effects of rapid eye movement (REM) sleep deprivation on pain sensitivity in the rat. Brain Res. 2001 May; 900(2):261-267.

Hatashita, S., Sekiguchi, M., Kobayashi, H., Konno, S. & Kikuchi, S. Contralateral neuropathic pain and neuropathology in dorsal root ganglion and spinal cord following hemilateral nerve injury in rats. *Spine* (Phila Pa 1976). 2008 May; 33(12):1344-1351.

Henderson, L.A., Gandevia, S.C. & Macefield, V.G. Somatotopic organization of the processing of muscle and cutaneous pain in the left and right insula cortex: a single-trial fMRI study. *Pain.* 2007 Mar; 128(1-2):20-30.

Herrmann, M., Schölmerich, J. & Straub, R.H. Stress and rheumatic diseases. *Rheum Dis Clin North Am.* 2000 Nov; 26(4):737-763, viii.

Holden, J.E. & Pizzi, J.A. Lateral hypothalamic-induced antinociception may be mediated by a substance P connection with the rostral ventromedial medulla. Brain Res. 2008 Jun; 1214:40-49.

Hsie, J.C., Belfrage, M., Stone-Elander, S., Hansson, P. & Ingvar, M. Central representation of chronic ongoing neuropathic pain studied by positron emission tomography. *Pain.* 1995 Nov; 63(2):225-236.

Ikeda, R., Takahashi, Y., Inoue, K. & Kato, F. NMDA receptor-independent synaptic plasticity in the central amygdala in the rat model of neuropathic pain. *Pain.* 2007 Jan; 127(1-2):161-172.

Huang, D. & Yu, B. The mirror-image pain: an unclered phenomenon and its possible mechanism. Neurosci Biobehav Rev. 2010 Mar; 34(4):528-532.

Hutchison, W.D., Harfa, L. & Dostrovsky, J.O. Ventrolateral orbital cortex and periaqu eductal gray stimulation-induced effects on on- and off-cells in the rostral ventromedial medulla in the rat. *Neuroscience.* 1996 Jan;70(2):391-407.

Imbe, H., Murakami, S., Okamoto, K., Iwai-Liao, Y. & Senba, E. The effects of acute and chronic restraint stress on activation of ERK in the rostral ventromedial medulla and locus coeruleus. *Pain.* 2004 Dec; 112(3):361-371.

Imbe, H., Okamoto, K., Okamura, T., Kumabe, S., Nakatsuka, M., Aikawa, F., Iwai-Liao, Y. & Senba, E. Effects of peripheral inflammation on activation of ERK in the rostral ventromedial medulla. *Brain Res.* 2005 Nov; 1063(2):151-158.

Imbe, H., Iwai-Liao, Y. & Senba, E. Stress-induced hyperalgesia: animal models and putative mechanisms. Front Biosci. 2006 Sep; 11:2179-2192.

Imbe, H., Kimura, A., Okamoto, K., Donishi, T., Aikawa, F., Senba, E. & Tamai, Y. Activation of ERK in the rostral ventromedial medulla is involved in hyperalgesia during peripheral inflammation. *Brain Res.* 2008 Jan; 1187:103-110.

Imbe, H., Okamoto, K., Donishi, T., Senba, E. & Kimura, A. Involvement of descending facilitation from the rostral ventromedial medulla in the enhancement of formalin-evoked nocifensive behavior following repeated forced swim stress. *Brain Res.* 2010 May; 1329:103-112.

Imbe, H., Senba, E., Kimura, A., Donishi, T., Yokoi, I. & Kaneoke, Y. Activation of mitogen-activated protein kinase in descending pain modulatory system. *J Signal Transduct.* 2011; 2011:468061.

Jørum, E. Analgesia or hyperalgesia following stress correlates with emotional behavior in rats. *Pain.* 1988 Mar; 32(3):341-348.

Julien, N., Goffaux, P., Arsenault, P. & Marchand, S. Widespread pain in fibromyalgia is related to a deficit of endogenous pain inhibition. *Pain.* 2005 Mar; 114(1-2):295-302.

Khan, A.A., Owatz, C.B., Schindler, W.G., Schwartz, S.A., Keiser, K. & Hargreaves, K.M. Measurement of mechanical allodynia and local anesthetic efficacy in patients with irreversible pulpitis and acute periradicular periodontitis. *J Endod.* 2007 Jul; 33(7):796-799.

Khasar, S.G., Green, P.G. & Levine, J.D. Repeated sound stress enhances inflammatory pain in the rat. *Pain.* 2005 Jul; 116(1-2):79-86.

Khasar, S.G., Burkham, J., Dina, O.A., Brown, A.S., Bogen, O., Alessandri-Haber, N., Green, P.G., Reichling, D.B. & Levine, J.D. Stress induces a switch of intracellular signaling in sensory neurons in a model of generalized pain. *J Neurosci.* 2008 May; 28(22):5721-5730.

Khasar, S.G., Dina, O.A., Green, P.G. & Levine, J.D. Sound stress-induced long-term enhancement of mechanical hyperalgesia in rats is maintained by sympathoadrenal catecholamines. *J Pain.* 2009 Oct; 10(10):1073-1077.

Kilduff, T.S., Cauli, B. & Gerashchenko, D. Activation of cortical interneurons during sleep: an anatomical link to homeostatic sleep regulation? *Trends Neurosci.* 2011 Jan; 34(1):10-19.

Kim, S.S., Descalzi, G. & Zhuo, M. Investigation of molecular mechanism of chronic pain in the anterior cingulate cortex using genetically engineered mice. *Curr Genomics.* 2010 Mar; 11(1):70-76.

Koltzenburg, M., Wall, P.D. & McMahon, S.B. Does the right side know what the left is doing? *Trends Neurosci.* 1999 Mar; 22(3):122-127.

Li, C.Y., Song, Y.H., Higuera, E.S. & Luo, Z.D. Spinal dorsal horn calcium channel alpha2delta-1 subunit upregulation contributes to peripheral nerve injury-induced tactile allodynia. *J Neurosci.* 2004 Sep; 24(39):8494-8499.

Li, D., Yang, H., Meyerson, B.A. & Linderoth, B. Response to spinal cord stimulation in variants of the spared nerve injury pain model. *Neurosci Lett.* 2006 May; 400(1-2):115-120.

Mao-Ying, Q.L., Zhao, J., Dong, Z.Q., Wang, J., Yu, J., Yan, M.F., Zhang, Y.Q., Wu, G.C. & Wang, Y.Q. A rat model of bone cancer pain induced by intra-tibia inoculation of Walker 256 mammary gland carcinoma cells. *Biochem Biophys Res Commun.* 2006 Jul; 345(4):1292-1298.

Markus, H., Pomeranz, B. & Krushelnycky, D. Spread of saphenous somatotopic projection map in spinal cord and hypersensitivity of the foot after chronic sciatic denervation in adult rat. *Brain Res.* 1984 Mar; 296(1):27-39.

Martenson, M.E., Cetas, J.S. & Heinricher, M.M. A possible neural basis for stress induced hyperalgesia. *Pain.* 2009 Apr; 142(3):236-244.

Martínez-Lavín, M. Is fibromyalgia a generalized reflex sympathetic dystrophy? *Clin Exp Rheumatol.* 2001 Jan-Feb; 19(1):1-3.

Martinez-Lavin, M. Biology and therapy of fibromyalgia. Stress, the stress response system, and fibromyalgia. *Arthritis Res Ther.* 2007; 9(4):216.

Maxwell, D.J., Kerr, R., Rashid, S. & Anderson, E. Characterisation of axon terminals in the rat dorsal horn that are immunoreactive for serotonin 5-HT3A receptor subunits. *Exp Brain Res.* 2003 Mar; 149(1):114-124.

McGaraughty, S. & Heinricher, M.M. Microinjection of morphine into various amygdaloid nuclei differentially affects nociceptive responsiveness and RVM neuronal activity. *Pain.* 2002 Mar; 96(1-2):153-162.

McGaraughty, S., Farr, D.A. & Heinricher, M.M. Lesions of the periaqueductal gray disrupt input to the rostral ventromedial medulla following microinjections of morphine into the medial or basolateral nuclei of the amygdala. *Brain Res.* 2004 May; 1009(1-2):223-227.

Meeus, M. & Nijs, J. Central sensitization: a biopsychosocial explanation for chronic widespread pain in patients with fibromyalgia and chronic fatigue syndrome. *Clin Rheumatol.* 2007 Apr; 26(4):465-473.

Milligan, E.D., Twining, C., Chacur, M., Biedenkapp, J., O'Connor, K., Poole, S., Tracey, K., Martin, D., Maier, S.F. & Watkins, L.R. Spinal glia and proinflammatory cytokines mediate mirror-image neuropathic pain in rats. *J Neurosci.* 2003 Feb; 23(3):1026-1040.

Milligan, E.D., Sloane, E.M., Langer, S.J., Cruz, P.E., Chacur, M., Spataro, L., Wieseler-Frank, J., Hammack, S.E., Maier, S.F., Flotte, T.R., Forsayeth, J.R., Leinwand, L.A., Chavez, R. & Watkins, L.R. Controlling neuropathic pain by adeno-associated virus driven production of the anti-inflammatory cytokine, interleukin-10. *Mol Pain.* 2005 Feb; 1:9.

Mitchell, J.M., Lowe, D. & Fields, H.L. The contribution of the rostral ventromedial medulla to the antinociceptive effects of systemic morphine in restrained and unrestrained rats. *Neuroscience.* 1998 Nov; 87(1):123-133.

Moldofsky H. The significance of the sleeping-waking brain for the understanding of widespread musculoskeletal pain and fatigue in fibromyalgia syndrome and allied syndromes. Joint Bone Spine. 2008 Jul; 75(4):397-402.

Nagakura, Y., Oe, T., Aoki, T. & Matsuoka, N. Biogenic amine depletion causes chronic muscular pain and tactile allodynia accompanied by depression: A putative animal model of fibromyalgia. *Pain.* 2009 Nov; 146(1-2):26-33.

Narita, M., Niikura, K., Nanjo-Niikura, K., Narita, M., Furuya, M., Yamashita, A., Saeki, M., Matsushima, Y., Imai, S., Shimizu, T., Asato, M., Kuzumaki, N., Okutsu, D., Miyoshi, K., Suzuki, M., Tsukiyama, Y., Konno, M., Yomiya, K., Matoba, M. & Suzuki, T. Sleep disturbances in a neuropathic pain-like condition in the mouse are associated with altered GABAergic transmission in the cingulate cortex. *Pain.* 2011 Jun; 152(6):1358-1372.

Nash, J.M. & Thebarge, R.W.. Understanding psychological stress, its biological processes, and impact on primary headache. *Headache.* 2006 Oct; 46(9):1377-1386.

Nasu, T., Taguchi, T. & Mizumura, K. Persistent deep mechanical hyperalgesia induced by repeated cold stress in rats. *Eur J Pain.* 2010 Mar; 14(3):236-244.

Nishiyori, M. & Ueda H. Prolonged gabapentin analgesia in an experimental mouse model of fibromyalgia. Mol Pain. 2008 Nov; 4:52.

O'Brien, E.M., Waxenberg, L.B., Atchison, J.W., Gremillion, H.A., Staud, R.M., McCrae, C.S. & Robinson, M.E. Negative mood mediates the effect of poor sleep on pain among chronic pain patients. *Clin J Pain.* 2010 May; 26(4):310-319.

Okamoto, K., Imbe, H., Morikawa, Y., Itoh, M., Sekimoto, M., Nemoto, K. & Senb,a E. 5-HT2A receptor subtype in the peripheral branch of sensory fibers is involved in the potentiation of inflammatory pain in rats. *Pain.* 2002 Sep; 99(1-2):133-143.

Okamoto, K., Imbe, H., Tashiro, A., Kumabe, S. & Senba, E. Blockade of peripheral 5HT3 receptor attenuates the formalin-induced nocifensive behavior in persistent temporomandibular joint inflammation of rat. *Neurosci Lett.* 2004 Sep; 367(2):259-263.

Okamoto, K., Imbe, H., Tashiro, A., Kimura, A., Donishi, T., Tamai, Y. & Senba, E. The role of peripheral 5HT2A and 5HT1A receptors on the orofacial formalin test in rats with persistent temporomandibular joint inflammation. *Neuroscience.* 2005a; 130(2):465-474.

Okamoto, K., Kimura, A., Donishi, T., Imbe, H., Senba, E. & Tamai, Y. Central serotonin 3 receptors play an important role in the modulation of nociceptive neural activity of trigeminal subnucleus caudalis and nocifensive orofacial behavior in rats with persistent temporomandibular joint inflammation. *Neuroscience.* 2005b; 135(2):569-581.

Okamoto, K., Kimura, A., Donishi, T., Imbe, H., Goda, K., Kawanishi, K., Tamai, Y. & Senba, E. Persistent monoarthritis of the temporomandibular joint region enhances nocifensive behavior and lumbar spinal Fos expression after noxious stimulation to the hindpaw in rats. *Exp Brain Res.* 2006a Apr; 170(3):358-367.

Okamoto, K., Kimura, A., Donishi, T., Imbe, H., Nishie, Y., Matsushita, H., Tamai, Y. & Senba, E. Contribution of peripheral 5-HT2A or 5-HT3 receptors to Fos expression in the trigeminal spinal nucleus produced by acute injury to the masseter muscle during persistent temporomandibular joint inflammation in rats. *Neuroscience.* 2006b Dec; 143(2):597-606.

Okamoto, K., Imbe, H., Kimura, A., Donishi, T., Tamai, Y. & Senba, E. Activation of central 5HT2A receptors reduces the craniofacial nociception of rats. *Neuroscience.* 2007 Jul; 147(4):1090-1102.

Pincus, T., Burton, A.K., Vogel, S. & Field, A.P. A systematic review of psychological factors as predictors of chronicity/disability in prospective cohorts of low back pain. *Spine* (Phila Pa 1976). 2002 Mar; 27(5):E109-120.

Porreca, F., Ossipov, M.H. & Gebhart, G.F. Chronic pain and medullary descending facilitation. *Trends Neurosci.* 2002 Jun; 25(6):319-325.

Quintero, L., Moreno, M., Avila, C., Arcaya, J., Maixner, W. & Suarez-Roca, H. Long-lasting delayed hyperalgesia after subchronic swim stress. *Pharmacol Biochem Behav.* 2000 Nov; 67(3):449-458.

Quintero, L., Cuesta, M.C., Silva, J.A., Arcaya, J.L., Pinerua-Suhaibar, L., Maixner, W. & Suarez-Roca, H. Repeated swim stress increases pain-induced expression of c-Fos in the rat lumbar cord. *Brain Res.* 2003 Mar; 965(1-2):259-268.

Radley, J.J., Sisti, H.M., Hao, J., Rocher, A.B., McCall, T., Hof, P.R., McEwen, B.S. & Morrison, J.H. Chronic behavioral stress induces apical dendritic reorganization in pyramidal neurons of the medial prefrontal cortex. *Neuroscience.* 2004; 125(1):1-6.

Ren, K. & Dubner, R. Descending modulation in persistent pain: an update. *Pain.* 2002 Nov; 100(1-2):1-6.

Rhudy, J.L. & Meagher, M.W. Fear and anxiety: divergent effects on human pain thresholds. *Pain.* 2000 Jan; 84(1):65-75.

Sandkühler, J. Understanding LTP in pain pathways. Mol Pain. 2007 Apr; 3:9.

Sarlani, E. & Greenspan, J.D. Evidence for generalized hyperalgesia in temporomandibular disorders patients. Pain. 2003 Apr; 102(3):221-226.

Satoh, M., Kuraishi, Y. & Kawamura, M. Effects of intrathecal antibodies to substance P, calcitonin gene-related peptide and galanin on repeated cold stress-induced hyperalgesia: comparison with carrageenan-induced hyperalgesia. *Pain.* 1992 May; 49(2):273-278.

Schreiber, K.L., Beitz, A.J. & Wilcox, G.L. Activation of spinal microglia in a murine model of peripheral inflammation-induced, long-lasting contralateral allodynia. *Neurosci Lett.* 2008 Jul; 440(1):63-67.

Schweinhardt, P., Glynn, C., Brooks, J., McQuay, H., Jack, T., Chessell, I., Bountra, C. & Tracey, I. An fMRI study of cerebral processing of brush-evoked allodynia in neuropathic pain patients. *Neuroimage.* 2006 Aug; 32(1):256-265.

Senba, E. & Ueyama, T. Stress-induced expression of immediate early genes in the brain and peripheral organs of the rat. *Neurosci Res.* 1997 Nov; 29(3):183-207.

Shenker, N., Haigh, R., Roberts, E., Mapp, P., Harris, N. & Blake, D. A review of contralateral responses to a unilateral inflammatory lesion. *Rheumatology (Oxford).* 2003 Nov; 42(11):1279-1286.

Shenker, N.G., Haigh, R.C., Mapp, P.I., Harris, N. & Blake, D.R. Contralateral hyperalgesia and allodynia following intradermal capsaicin injection in man. *Rheumatology (Oxford).* 2008 Sep; 47(9):1417-1421.

Sluka, K.A., Kalra, A. & Moore, S.A. Unilateral intramuscular injections of acidic saline produce a bilateral, long-lasting hyperalgesia. *Muscle Nerve.* 2001 Jan; 24(1):37-46.

Staud, R. & Rodriguez, M.E.. Mechanisms of disease: pain in fibromyalgia syndrome. *Nat Clin Pract Rheumatol.* 2006 Feb; 2(2):90-98.

Staud, R., Nagel, S., Robinson, M.E. & Price, D.D. Enhanced central pain processing of fibromyalgia patients is maintained by muscle afferent input: a randomized, double-blind, placebo-controlled study. *Pain.* 2009 Sep; 145(1-2):96-104.

Suarez-Roca, H., Leal, L., Silva, J.A., Pinerua-Shuhaibar, L. & Quintero, L. Reduced GABA neurotransmission underlies hyperalgesia induced by repeated forced swimming stress. Behav Brain Res. 2008 May; 189(1):159-169.

Sugiyo, S., Takemura, M., Dubner, R. & Ren, K. Trigeminal transition zone/rostral ventromedial medulla connections and facilitation of orofacial hyperalgesia after masseter inflammation in rats. J Comp Neurol. 2005 Dec; 493(4):510-523.

Suzuki, R., Morcuende, S., Webber, M., Hunt, S.P. & Dickenson, A.H. Superficial NK1-expressing neurons control spinal excitability through activation of descending pathways. Nat Neurosci. 2002 Dec; 5(12):1319-1326.

Suzuki, R., Rygh, L.J. & Dickenson, A.H. Bad news from the brain: descending 5-HT pathways that control spinal pain processing. *Trends Pharmacol Sci.* 2004 Dec; 25(12):613-617.

Takemura, Y., Yamashita, A., Horiuchi, H., Furuya, M., Yanase, M., Niikura, K., Imai, S., Hatakeyama, N., Kinoshita, H., Tsukiyama, Y., Senba, E., Matoba, M., Kuzumaki, N., Yamazaki, M., Suzuki, T. & Narita, M. Effects of gabapentin on brain hyperactivity related to pain and sleep disturbance under a neuropathic pain-like state using fMRI and brain wave analysis. *Synapse.* 2011 Jul; 65(7):668-676.

Tanabe, M., Takasu, K., Takeuchi, Y. & Ono, H.. Pain relief by gabapentin and pregabalin via supraspinal mechanisms after peripheral nerve injury. *J Neurosci Res.* 2008 Nov; 86(15):3258-3264.

Türp, J.C., Kowalski, C.J., O'Leary, N. & Stohler, C.S. Pain maps from facial pain patients indicate a broad pain geography. *J Dent Res.* 1998 Jun; 77(6):1465-1472.

Umemoto, S., Kawai, Y., Ueyama, T. & Senba E. Chronic glucocorticoid administration as well as repeated stress affects the subsequent acute immobilization stress-induced expression of immediate early genes but not that of NGFI-A. *Neuroscience.* 1997 Oct; 80(3):763-773.

Vera-Portocarrero, L.P., Xie, J.Y., Kowal, J., Ossipov, M.H., King, T. & Porreca F. Descending facilitation from the rostral ventromedial medulla maintains visceral pain in rats with experimental pancreatitis. *Gastroenterology.* 2006 Jun; 130(7):2155-2164.

Verne, G.N. & Price, D.D. Irritable bowel syndrome as a common precipitant of central sensitization. *Curr Rheumatol Rep.* 2002 Aug; 4(4):322-328.

Vidal, C. & Jacob, J. Hyperalgesia induced by non-noxious stress in the rat. *Neurosci Lett.* 1982 Sep; 32(1):75-80.

Vyas, A., Mitra, R., Shankaranarayana, Rao. B.S. & Chattarji, S.. Chronic stress induces contrasting patterns of dendritic remodeling in hippocampal and amygdaloid neurons. *J Neurosci.* 2002 Aug; 22(15):6810-6818.

Wasner, G., Naleschinski, D., Binder, A., Schattschneider, J., McLachlan, E.M. & Baron, R. The effect of menthol on cold allodynia in patients with neuropathic pain. *Pain Med.* 2008 Apr; 9(3):354-358.

Wei, F., Dubner, R., Zou, S., Ren, K., Bai, G., Wei, D. & Guo, W. Molecular depletion of descending serotonin unmasks its novel facilitatory role in the development of persistent pain. *J Neurosci.* 2010 Jun; 30(25):8624-8636.

Wood, P.B. Stress and dopamine: implications for the pathophysiology of chronic widespread pain. *Med Hypotheses.* 2004; 62(3):420-424.

Wood, J.L.& Russo, A.F. Autoregulation of cell-specific MAP kinase control of the tryptophan hydroxylase promoter. *J Biol Chem.* 2001 Jun; 276(24):21262-21271.

Yasuda, T., Miki, S., Yoshinaga, N. & Senba E. Effects of amitriptyline and gabapentin on bilateral hyperalgesia observed in an animal model of unilateral axotomy. *Pain.* 2005 May; 115(1-2):161-170.

Yu, L.C., Hansson, P. & Lundeberg, T. The calcitonin gene-related peptide antagonist CGRP8-37 increases the latency to withdrawal responses bilaterally in rats with unilateral experimental mononeuropathy, an effect reversed by naloxone. *Neuroscience.* 1996 Mar; 71(2):523-531.

Zhuo, M. Cortical excitation and chronic pain. *Trends Neurosci.* 2008 Apr; 31(4):199-207.

Zhuo, M. A synaptic model for pain: long-term potentiation in the anterior cingulate cortex. *Mol Cells.* 2007 Jun; 23(3):259-271.

3

Animal Models of Fibromyalgia

Yukinori Nagakura, Hiroyuki Ito and Yasuaki Shimizu
Drug Discovery Research, Astellas Pharma Inc., Japan

1. Introduction

Fibromyalgia patients are suffering distressing pain-centered symptoms, and therefore the clarification of fibromyalgia pathophysiology and the development of more effective therapeutic options are urgent issues. In the preclinical phase of analgesic drug development, experimental animal pain models have been widely used to investigate analgesic efficacies of candidate agents. In this approach, some kind of manipulations are conducted to induce clinically relevant pain states such as hyperalgesia and allodynia in the experimental animal, and then, pain-associated behaviors are measured as indicators of pain. Efficacy of test drugs on the said pain is finally evaluated. While animal pain models associated with neuropathy or inflammation have been well established and used in drug discovery research (Negus et al., 2006), only a few have been developed for examining fibromyalgia. This chapter focuses on the review of putative animal models of fibromyalgia, which mimic some features of patients with fibromyalgia.

2. Challenges accompanying the development of fibromyalgia animal model

Widespread chronic pain is a core symptom in fibromyalgia patients, and manifestation of this pain is a necessary and extremely important criterion in developing fibromyalgia animal models. Chronic pain can be categorized into three groups according to the mechanism underlying the pain: nociceptive pain, which is caused by tissue damage including inflammation; neuropathic pain, which is caused by a dysfunction in nerve fibers; and central (non-nociceptive) pain, where the central disturbance in pain processing is involved (Clauw, 2009). Animal models for nociceptive and neuropathic pain are well established and have significantly contributed to the development of therapeutic options since relatively early times (Negus et al., 2006), due in part to the relatively easy to mimic etiology of nociceptive and neuropathic pain. For example, controlled damage to the sensory nerve fibers by surgical, infectious, or chemical manipulation can effectively induce neuropathic pain, while damaging the tissues by surgical or inflammatory means can effectively induce nociceptive pain. On the other hand, the development of the animal model of central (non-nociceptive) pain is rather difficult due to the unknown etiology. Namely, it is challenging to specify the relevant trigger of experimental pain due to the poor understanding of the nociceptive source. Consequently, animal models for fibromyalgia, a disease that exhibits typical central (non-nociceptive) pain, are relatively naive.

Also complicating such models is the fact that fibromyalgia is not a simple pain state but rather is accompanied by multiple comorbid symptoms including depression, fatigue, sleep disturbance, irritable bowel syndrome, and irritable bladder syndrome (Clauw, 2009), and patients often experience one or more of these comorbidities in addition to widespread pain. One report plotted the prevalence of fatigue, sleep disturbance, and depression in fibromyalgia patients at approximately 80%, 90%, and 40%, respectively, although the prevalence varies across the population (Sluka, 2009). In 2010, new diagnostic criteria for fibromyalgia were proposed by the American College of Rheumatology (Wolfe et al., 2010). The criteria include widespread pain index, but no longer require tender point examination. Alternatively, they include severity scores of comorbid symptoms such as cognitive symptoms, unrefreshed sleep, and fatigue. Given this correlation between comorbidities and pain in fibromyalgia patients, corresponding animal models should ideally simulate the development of these symptoms in addition to chronic pain; concurrent consideration for depression, fatigue, and visceral organ function is important when investigating the chronic pain state in fibromyalgia animal models.

Given that fibromyalgia predominantly occurs in women, it would be worth to consider the involvement of female animals in the construction of fibromyalgia animal model. However, because estrous cycle influences nociceptive processing in both peripheral and central pathways (Fillingim and Ness, 2000), this cycle could be a confounding bias in the preclinical pain research. Thus, it would be necessary to assess carefully the estrous cycle phase when using female animals in preclinical experiment.

3. Hypotheses for fibromyalgia pathogenesis

Several hypotheses have been proposed regarding fibromyalgia pathogenesis and should be considered when designing a suitable animal model.

Fibromyalgia is generally considered to be a stress-related disorder. Stress plays an exacerbating role in the development of somatic symptoms including chronic pain. There is evidence that disparate stressful events including childhood abuse and physical traumas would trigger the development of fibromyalgia (Clauw, 2009).

Alternatively, an abnormal neuroendocrine system may be involved in the pathogenesis, as stress affects the hypothalamic-pituitary-adrenal axis and the autonomic nervous system. Indeed, altered function of the hypothalamic-pituitary-adrenal axis has been detected in at least some subpopulation of fibromyalgia patients, and aberrations in the autonomic nervous system are also often observed (Bradley, 2009).

Many fibromyalgia patients experience pain in deep tissues such as muscle. Abnormalities in the local muscle tissue may therefore underlie the pathogenesis, as changes of intramuscular microcirculation or muscle energy metabolism have been detected in the musculoskeletal tissue of some fibromyalgia patients. This localized muscle pain may eventually spread throughout the body via a central nervous system-mediated mechanism, thereby leading to nociceptive input from the deep tissue and generating central sensitization in the spinal cord and brain. After this central sensitization is established, a chronic pain state can be maintained by even minimal nociceptive input from the peripheral tissues (Nielsen & Henriksson, 2007). The pain assessment by using spinal nociceptive flexion reflex (NFR), a correlate for the evaluation of central nociceptive pathways, suggests that abnormally processed nociceptive signal input to central nociceptive pathways (i.e., central sensitization) underlies the pathophysiology in fibromyalgia patients (Desmeules et al., 2003). Also, attenuation of inhibitory pain control in the central nervous system seems to be involved in development of fibromyalgia. Ascending nociceptive input to the supraspinal brain is down-regulated by the brainstem descending pain inhibitory pathway, where neurotransmitters associated with the regulation of both pain and mood operate. Dysfunction in this inhibitory pain control has been suggested to play an important role in the pathogenesis of fibromyalgia. Indeed, diffuse noxious inhibitory control (DNIC), an indicator for activity in the central descending pain inhibitory pathway, has been found to be attenuated in many fibromyalgia patients (Clauw, 2009).

4. Approach for developing fibromyalgia animal models

Given the different etiology between fibromyalgia (central pain) and nociceptive/neuropathic pain, animal models more specific to fibromyalgia need to be developed. In this chapter, we discuss several noteworthy animal models that mimic certain major fibromyalgia features based on three types of validation: construct validity ("Is induction of the disease state in accordance with the fibromyalgia etiology?" Animal models should be constructed based on the aforementioned hypotheses), face validity ("Do fibromyalgia-like symptoms successfully manifest?" Animal models should exhibit chronic pain symptoms and accompanying comorbidities), and predictive validity ("How do the symptoms respond to treatments used in the clinical setting?" Although some drugs have been approved for the treatment of fibromyalgia-associated pain, the effect size of these drugs seems not sufficient).

5. Key animal models

The animal models to be reviewed in this chapter are the repeated cold stress model, the unpredictable sound stress model, the intramuscular acidic saline injection model, and the reserpine-induced myalgia model, in which assumed causes of chronic pain are unavoidable stress, stress-induced disruption of the neuroendocrine system, an initial local muscle abnormality that spreads throughout the body, and dysfunctional pain control in the central nervous system, respectively. All four models operate on the common premise that once chronic pain is established, no apparent organic disorder can explain its persistence. Further, the pain symptoms are not localized to the ipsilateral side, instead spreading to the contralateral side of the body as well.

5.1 Repeated cold stress model

5.1.1 Methods

The repeated cold stress model was first introduced as an animal model to study autonomic imbalance in response to sudden changes in environmental temperature (Kita et al., 1975; Ohara et al., 1991). Mice exposed to repeated cold stress develop several abnormal physiologies such as a continuous lowering of blood pressure, sympathicotonic-type electrocardiogram (ECG), depression-like behaviors, and hyperalgesia (Ohara et al., 1991). In this model, animals (usually mice) are kept in a room at a temperature alternating between 24 and 4°C every hour in the day time, and then kept at 4°C overnight. This procedure was repeated for 5 days or more (Kita et al., 1975; Hata et al., 1995).

Nishiyori and Ueda (2008) revisited this stress model to demonstrate its utility as a putative animal model for fibromyalgia. They optimized the repeated cold stress protocol to induce long lasting mechanical and thermal hyperalgesia in mice as follows. On Day 1, the experiment starts with exposing mice to a cold temperature (4°C) overnight until morning the next day. From there, the temperature is then alternated between 24 and 4°C every 30 min in the day time. This is repeated until morning on Day 4. A nociception test is then conducted at room temperature at least 1 h after the end of stress exposure.

5.1.2 Primary features

Nishiyori and Ueda (2008) found that mice exhibited thermal hyperalgesia (reduction in paw withdrawal latency in response to thermal stimulation) and mechanical hyperalgesia (reduction in threshold for paw withdrawal in response to mechanical stimulation by the von Frey apparatus) for more than 12 days following the repeated cold stress. In contrast, mice exposed to constant cold temperature (4°C) stress manifest only short-lasting hyperalgesia. Similarly, the repeated cold stress with the protocol by Ohara et al. (1991) caused mechanical hyperalgesia, which was maintained for 1 week after the start of stress loading, in the tail pressure test in mice.

5.1.3 Constructive validity

Stressors such as physical trauma, catastrophic events, hormonal alterations, and emotional stress play an important role in triggering development of fibromyalgia. Individuals exposed to these stressors have been found to develop chronic widespread pain with a relatively high probability, as indicated by studies investigating the effects of childhood abuse and posttraumatic stress disorder. Similarly, studies using animals have suggested that stressors that can neither be avoided nor controlled are more likely to cause unstable physiological responses and may lead to irreversible health problems (Clauw, 2009). This means that disparate stressors are potential factors in triggering development of fibromyalgia.

Stressors may affect physiological function by altering the hypothalamic-pituitary-adrenal axis and causing a malfunction in autonomic control. Dysfunctional autonomic control may then, in turn, associate with fibromyalgia pathogenesis. For example, the abnormality in heart rate variability, which suggests an autonomic dysfunction, is raised as a risk factor of fibromyalgia (Clauw, 2009).

Based on the suggested involvement of stressors and the dysfunction of neuroendocrine system in the pathogenesis of fibromyalgia, the sudden and frequent change in environmental temperature, which has been known to cause autonomic imbalance and hyperalgesia (Hata et al., 1988), is utilized to cause a long-lasting mechanical and thermal hyperalgesia in animals in the repeated cold stress model. Interestingly, it is not simply constant cold stress but rather repeated cycling of cold stress that causes long-lasting hyperalgesia. This may mean that animals are capable of adapting to cold environmental temperature exposure over a long period, but a sudden and frequent environmental temperature change such as that experienced in the repeated cold stress greatly affects physiological function, including autonomic control. This dysfunction in autonomic control and subsequent development of chronic pain state may be due to changes in certain neurotransmitters, as the repeated cold test has been found to also decrease serotonin and increase noradrenaline and dopamine in the central nervous system (Hata et al., 1987; Hata et al., 1991). The turnover of serotonin in the spinal cord is accordingly reluctant to be activated in mice following exposure to repeated cold stress (Nishiyori et al., 2010).

5.1.4 Face validity

The repeated cold stress test performed by Nishiyori and Ueda (2008) caused long-lasting (12 days or more) mechanical and thermal hyperalgesia in the paws of both male and female mice, suggesting no apparent difference in susceptibility to repeated cold stress between genders. Intriguingly, gonadectomized male mice showed partial reversal of the repeated cold stress-induced mechanical hyperalgesia, whereas gonadectomized females did not. This difference may be related to the predominance of chronic pain occurrence in female, which is a feature of fibromyalgia epidemiology (Clauw, 2009).

In addition to hyperalgesia, the repeated cold stress causes abnormal behaviors in mice, a feature that may be associated with the comorbid symptoms seen in fibromyalgia patients. Mice loaded with repeated cold stress show a shortened immobility time in the forced swim test, a test used to assess the anti-depressive effect of drugs. Although prolonged immobility time is typically associated with depressed condition, the authors suggest that the reduced immobility time may also indicate complex symptoms such as excessive emotion- or anxiety-related depression, as immobility time recovered when the mice were given repeated injections of antidepressants or a single injection of anxiolytics (Hata et al., 1995).

5.1.5 Predictive validity

Gabapentin is structurally related to pregabalin, which has been approved for treatment of fibromyalgia in the USA. It is a ligand for the α2δ1 subunit in voltage-dependent calcium channels as is pregabalin. Gabapentin was approved for the treatment of neuropathic pain such as post herpetic neuralgia, and was statistically more effective than placebo in reducing pain scores in a 12-week randomized controlled trial in fibromyalgia patients (Arnold et al., 2007). Regarding the effect size in this trial, reduction of >or=30% in average pain severity score (range 0-10, where 0=no pain and 10=pain as bad as you can imagine) was 51% for gabapentin and 31% for placebo

In the repeated cold stress model, gabapentin completely recovered both mechanical and thermal hyperalgesia. This efficacy of gabapentin is basically consistent with the study showing that pregabalin significantly attenuates the pain score in phase III clinical trials for fibromyalgia patients (Mease et al., 2008). In mice, the effective dose for relieving hyperalgesia in the repeated cold stress model was over three times smaller than that for reducing nerve injury-induced neuropathic pain, despite the effective dose of pregabalin being similar between fibromyalgia and neuropathic pain patients in a clinical setting (Dworkin et al., 2007; Mease et al., 2008). Further, intracerebroventricular administration of gabapentin significantly attenuated hyperalgesia in the repeated cold stress model but not in the neuropathic pain model, possibly because up-regulation of the α2δ1 subunit may underlie the mechanism for repeated cold stress-induced hyperalgesia (Nishiyori and Ueda, 2008).

No clear evidence has yet been presented that opiate drugs are effective in reducing pain in fibromyalgia patients, as no randomized controlled double blind clinical trial on opiates has been conducted with a sufficient number of fibromyalgia patients (Clauw, 2009). One study using a small number of fibromyalgia patients did not detect any significant benefit when intravenously injecting morphine (Sörensen et al., 1995). In accordance with these findings, neither subcutaneous nor intracerebroventricular injection of morphine has been found to attenuate thermal hyperalgesia in repeated cold stressed mice. Given that serotonin turnover is unlikely to be accelerated by morphine in these animals, a dysfunction in the serotonin-mediated descending inhibitory pathway might explain the absence of morphine analgesia (Nishiyori et al., 2010).

Neurotropin is an extracted substance from the inflamed skin of rabbits inoculated with vaccinia virus. It has been used in the treatment of chronic pain including post-herpetic neuralgia in Japan. The analgesic mechanism of neurotropin has not been fully clarified. However, the enhancement of descending pain inhibitory pathway could be involved in the mechanism and the intraperitoneal injection of neurotropin significantly attenuates the mechanical hyperalgesia in the repeated cold stressed mice (Ohara et al., 1991). There has been no study with a sufficient number of fibromyalgia patients so far, which suggests that neurotropin has beneficial effect in the treatment of fibromyalgia.

5.2 Unpredictable sound stress model

5.2.1 Methods

In this model, animals (usually male rats) are placed in a sound box for 30 min, during which they are randomly exposed every minute to unpredictable sounds ranging in amplitude from 20 to 105 dB and several frequencies for 5 or 10 sec each. Following the test, the animals are returned to their home cage. The sound stress protocol is conducted on Days 1, 3, and 4. Nociceptive and other behavioral experiments are conducted typically 1 or 14 days after the last sound stress delivery (Green et al., 2011; Khasar et al., 2005).

5.2.2 Primary features

Sound-stressed rats exhibit an aggravation of algesic substances (bradykinin or prostaglandin E_2)-induced mechanical hyperalgesia in the paw and muscle pressure tests, although they do not manifest hyperalgesia without the application of an algesic substance. These animals also manifest visceral hypersensitivity based on colorectal distension 14 days after the end of stress, even without application of exogenous algesic substances. Anxiety score as measured by the elevated plus maze test is also increased 14 days after the last stress exposure (Green et al., 2011; Khasar et al., 2005).

5.2.3 Constructive validity

Stress has been shown to be a risk factor for fibromyalgia development, as traumatic events are likely to trigger chronic pain symptoms (Clauw, 2009). Dysfunction in sound stressor processing has been shown to occur in some fibromyalgia patients (Dohrenbusch et al., 1997), and stressors in general have been known to affect physiological functions such as nociception by changing the sympathoadrenal (adrenaline) and hypothalamic-pituitary-adrenal (corticosterone) axes in the neuroendocrine system. Because unpredictable sound stress affects both axes and causes a prolonged elevation of plasma adrenaline levels, the stress-induced release of adrenaline (the product of neuroendocrine system activation) likely plays an important role in developing mechanical hyperalgesia in the unpredictable sound stress model. This working hypothesis is consistent with evidence of elevated plasma adrenaline levels and enhanced bradykinin-induced mechanical hyperalgesia in vagotomized rats (Khasar et al., 1998; Khasar et al., 2003; Levine and Reichling, 2005).

Hypersensitivity in the presence of pronociceptive substances is commonly observed in fibromyalgia patients (Wang et al., 2008), and the requirement of pronociceptive substances for sound-stressed animals to exhibit enhanced somatosensory function may reflect a similar feature. However, the lack of change in the baseline nociceptive threshold in the unpredictable sound stress model is inconsistent with changes noted with other types of stressors such as repeated cold stress (Nishiyori and Ueda, 2008) and repeated swim stress

(Quintero et al., 2000). This difference may be because of the type of stressor or because of the method for measuring nociceptive function. Additionally, visceral hypersensitivity manifests even without the addition of pronociceptive substances, possibly due to the presence of endogenous pronociceptive substances in the gastrointestinal tissues, as these make sound stress-induced hyperalgesia be elicited (Green et al., 2011).

5.2.4 Face validity

In humans, chronic pain symptoms tend to develop on a delay after a traumatic event (Roy-Byrne et al., 2004). In line with this observation, sound-stressed rats develop an enhanced nociceptive response to pronociceptive substances such as bradykinin and prostaglandin E2 in the skin and muscle region not on Day 1 but on Day 14 after the end of stress exposure. Although fibromyalgia patients show somatic hypersensitivities to mechanical stimuli without exogenous application of pronociceptive substances, unpredictable sound stress does not cause significant changes in somatic sensitivity to mechanical stimuli unless pronociceptive substances are applied. Instead, plasma cytokine levels associating with the hypersensitivity seen in fibromyalgia patients (Wang et al., 2008) correlate with repeated sound stress exacerbating the pronociceptive substance-induced nociceptive response. Further, visceral hypersensitivity manifests even without the addition of exogenous pronociceptive substances (Green et al., 2011).
Similar to fibromyalgia patients who show comorbid symptoms such as irritable bowel syndrome, temporomandibular disorder, and anxiety (Clauw, 2009), sound-stressed rats manifest typical comorbid symptoms such as visceral hyperalgesia, temporomandibular hyperalgesia and an elevation of anxiety score on Day 14 after the end of stress exposure (Green et al., 2011). Abnormalities in the hypothalamic-pituitary-adrenal axis and sympathetic nervous system have also been suggested to occur in at least a subpopulation of fibromyalgia patients (Clauw, 2009), and indeed, sound-stressed rats manifest a long-lasting elevation of plasma adrenaline concentration (Khasar et al., 2005).

5.2.5 Predictive validity

Little information is available regarding the pharmacotherapeutic efficacy on nociceptive responses induced by sound stress-induced exacerbation so far. However, because adrenal medullectomy but not sympathectomy inhibits this exacerbation, adrenaline derived from the adrenal medulla appears to play a primary role in the exacerbation mechanism in this model (Khasar et al., 2005). Pharmacotherapy for correcting the neuroendocrine abnormality may improve the enhanced sensitivity.

5.3 Intramuscular acidic saline injection model

5.3.1 Methods

Rats are typically used in this model. Each animal receives two repeated 100-μl injections of acidic saline (adjusted to pH 4.0) in the same unilateral gastrocnemius muscle under inhaled anesthesia, a procedure which is repeated five days later (Sluka et al., 2001).

5.3.2 Primary features

After the second injection of acidic saline, the animal manifests bilateral cutaneous and muscle mechanical hyperalgesia, which is detected as a reduction in the paw withdrawal threshold based on the von Frey filament test and the muscle pressure test in the gastrocnemius muscle. These hyperalgesia symptoms are long-lasting (two weeks or more). However, the animal does not exhibit thermal hyperalgesia, which is measured based on paw withdrawal latency in the radiant heat exposure test. The rat displays no motor impairment, as assessed by the treadmill test, nor any histopathological change specific to the acidic acid saline injection in the affected gastrocnemius muscle (Sluka et al., 2001; Yokoyama et al., 2007).

5.3.3 Constructive validity

In humans, local acidity in the muscle or skin has several causes, including inflammation and ischemia, and can induce pain initially at a local region. A previous study found that muscle pain is caused by an infusion of acid phosphate buffer into the forearm muscle (Issberner et al., 1996). Another study (Graven-Nielsen et al., 1997) showed that repeated infusion of hypertonic saline caused the pain to diffuse from the primary infusion site (tibialis anterior muscle) to other regions like the ankle. The intensity of the diffused pain becomes larger in the short interval of infusion than in the long interval. Taken together, these previous findings suggest that repeated infusions of saline into the muscle can cause pain to spread beyond the original local region. This spreading may be analogous to the widespread musculoskeletal pain that is a main feature in fibromyalgia. Based on the working hypothesis that the spread of initially localized pain underlies the process of fibromyalgia development, the acidic acid injection was used as the trigger of wide spreading pain in the intramuscular acidic saline injection model. In consistent with the clinical features, not single but repeated injections successfully induced the chronic mechanical hyperalgesia in cutaneous and muscle tissues outside the area of injection, both ipsilaterally and contralaterally.

Regarding the mechanism of this spreading in fibromyalgia patients, several studies have suggested involvement of enhanced synaptic activity in the central nervous system, including the spinal cord, and an attenuation of descending inhibitory control from supraspinal centers (Graven-Nielsen and Arendt-Nielsen, 2010). Actually, the nociceptive flexion reflex (NFR), a psychophysiological tool to study spinal nociceptive process (France et al., 2009), suggests that central sensitization is involved in fibromyalgia pain as a key mechanism (Desmeules et al., 2003). Since an injection of lidocaine into the ipsilateral gastrocnemius muscle did not affect the spreading of pain, it is unlikely that the continuous nociceptive signal input from the primary afferent neurons is mainly implicated in the spreading of pain in the intramuscular acidic saline injection model. Alternatively, the development of contralateral hyperalgesia implies the involvement of central mechanism. Actually, intrathecal administration of glutamate receptor antagonists after the second acidic acid saline injection inhibits the spreading of pain (Skyba et al., 2002), and the injection of a local anesthetic or a *N*-methyl-*D*-aspartate (NMDA) receptor antagonist into the rostral ventromedial medulla, which plays a critical role in the descending pain control, also prevents the development of the bilateral hyperalgesia (Tillu et al., 2008; Da Silva et al., 2010). Further, activation of mu- and delta-opioid receptors in the spinal cord attenuates the hyperalgesia (Sluka et al., 2002). These evidences suggest that the sensitization mainly in the spinal cord level rather than the peripheral level and the modulation of descending pain control pathway underlie the pathophysiology in this model in accordance with existing theories for the pathophysiology in fibromyalgia patients (Clauw, 2009).

5.3.4 Face validity

The nociceptive source of fibromyalgia pain has yet to be identified. In general, no apparent inflammation or tissue injury can explain the pain endured by fibromyalgia patients. In line with this, no specific pathological changes specific to acidic saline injection have been detected (Sluka et al., 2001), meaning no organic disorder has been identified that can explain the widespread pain.

Intramuscular acidic saline injection model rats develop long-lasting bilateral muscle hyperalgesia (Yokoyama et al., 2007) in accordance with the deep tissue hyperalgesia and existence of tender points around the body seen in fibromyalgia patients. These animals also exhibit chronic cutaneous tactile allodynia, also consistent with findings in fibromyalgia patients. However, these animals do not develop thermal hyperalgesia (Sluka et al., 2001). Instead, visceral hyperalgesia detected by electromyography occurs concurrently with somatic hyperalgesia, suggesting that somatic afferent input from the gastrocnemius muscle induces visceral hyperalgesia, possibly via viscerosomatic convergence at the level of the lower spinal cord (Miranda et al., 2004). This finding is consistent with those in fibromyalgia patients who frequently show abdominal hypersensitivity, such as irritable bowel syndrome.

5.3.5 Predictive validity

A phase III clinical trial with fibromyalgia patients showed that pregabalin is efficacious and safe for use in treating pain associated with fibromyalgia, although it frequently causes dizziness and somnolence (Mease et al., 2008). A recent systematic review of the effectiveness of antiepileptic drugs for fibromyalgia suggests that pregabalin is modestly effective, although long-term safety and efficacy results are unavailable (Siler et al., 2011). In intramuscular acidic saline injection model, pregabalin significantly recovers the lowered mechanical thresholds at a dose range of 10-100 mg/kg i.p. but significantly reduces motor function outcome, as determined by the rota-rod performance test, at 30 mg/kg i.p. or more (Yokoyama et al., 2007). This result may reflect the clinically narrow margin between analgesic and dizzying doses.

5.4 Reserpine-induced myalgia (RIM) model

5.4.1 Methods

Rats are usually used. They are housed individually in the cage and are subcutaneously administered reserpine at 1 mg/kg once daily for three consecutive days. Hyperalgesia is typically evaluated on Day 5 after the last reserpine injection (Nagakura et al., 2009). Because tenderness to palpation is a diagnostic criterion for fibromyalgia and blunt pressure loading is frequently used to estimate muscular pain thresholds in humans, the muscle pressure test is used (Petzke et al., 2003).

5.4.2 Primary features

In male and female rats, repeated administration of reserpine significantly reduced the muscle pressure threshold based on muscle pressure test results and tactile allodynia based on the von Frey hair test results, lasting for more than one week. This administration decreased the amount of biogenic amines (dopamine, norepinephrine, and serotonin) in the spinal cord, thalamus, and prefrontal cortex, all of which are responsible for pain signal processing in the central nervous system. Further, reserpine significantly increased the immobility time in the forced swim test, which is indicative of depression, a common comorbid symptom in fibromyalgia patients. However, the general condition of the animals did not deteriorate, as both tremor and hypokinesia symptoms disappeared by Day 3 or later after the last reserpine injection. Additionally, no significant difference was noted in either blood pressure or rectal temperature between reserpine- and vehicle-treated groups. Taken together, these findings suggest that reserpine produces chronic pain symptoms and depressive conditions by influencing pathways that regulate nociception and depression (Nagakura et al., 2009).

5.4.3 Constructive validity

Functional magnetic resonance imaging studies have shown augmented central pain processing in fibromyalgia patients. This augmented processing can be thought of as a central nervous system that is hyper-reactive to a wide range of stimuli (Clauw, 2009). While several transmitters have been known to play a role in excitatory and inhibitory pain control in the central nervous system, biogenic amines like dopamine, norepinephrine, and serotonin are particularly important for analgesic control in the descending inhibitory pathway in the brainstem (Mense, 2000; Wood, 2008). Therefore, it is plausible that a dysfunction in biogenic amine-mediated inhibitory pain control underlies the central augmentation and pain-centered complex symptoms seen in fibromyalgia patients.

Concentrations of biogenic amines in cerebrospinal fluid have been reported to be significantly lower in fibromyalgia patients (17 female participants) than non-fibromyalgia subjects (Russell et al., 1992), and a profound disruption of dopamine release in the basal ganglia in response to painful stimulation in the muscle has been observed in fibromyalgia patients (Wood, 2008). Further, interruption of the descending pain inhibitory pathway in the brainstem causes augmented background activity and increased responsiveness to noxious stimulations in spinal nociceptive neurons (Mense, 2000). In addition to their involvement in pain control, biogenic amines in the central nervous system are closely related to the pathophysiology of major depressive disorders (El Mansari et al., 2010). Accordingly, depression is a potentially serious adverse effect when clinically administering reserpine (Leith and Barrett, 1980).

The RIM model uses reserpine to deplete biogenic amines from the nervous system by irreversibly binding to monoamine transporters for storage vesicles. Repeated administration of reserpine in RIM rats causes a chronic decrease in dopamine, norepinephrine, and serotonin in several brain regions and concomitantly induces a decrease in muscle pressure threshold and tactile allodynia for at least one week. The time courses of the changes in biogenic amines concentrations and muscular/tactile response thresholds generally correlate, supporting the assumption that the observed chronic pain symptoms are caused by the attenuation of biogenic amine-mediated central nervous system pain control. These observations are all in accordance with the critical role of biogenic amines in analgesic control. Further, reserpine treatment also induces a depressive condition, consistent with the role biogenic amines have in the pathophysiology of depression. Because a short-term depletion of biogenic amines by tetrabenazine did not significantly affect nociceptive sensitivities, long-term depletion is likely necessary for chronic pain symptoms to develop (Nagakura et al., 2009; Oe et al., 2010).

Intriguingly, in the experiment using single injection of reserpine, hyposensitivity to muscle pressure stimulus occurred in the acute phase (within 24 h) after the reserpine injection, although hypersensitivity developed in the chronic phase, i.e., on Day 2 or later after the injection. The dynamic change of brain biogenic amine tones observed after the single injection of reserpine, which was indicated by the ratio of biogenic amine metabolite/biogenic amine, in the central nervous system seemed to correlate with the determination of nociceptive sensitivity (Oe et al., 2010).

5.4.4 Face validity

Fibromyalgia patients suffer chronic widespread musculoskeletal pain symptoms including tenderness to palpation. In line with this, both male and female RIM rats show a reduced muscle pressure threshold and tactile allodynia for at least one week after injection. Cold and heat hypersensitivities have also been reported to follow reserpine treatment (Oe et al., 2010), paralleling the hypersensitivities to mechanical and thermal stimuli seen in fibromyalgia patients. In addition to chronic widespread pain symptoms, fibromyalgia is also characterized by comorbid affective disorders such as depression. Similarly, reserpine treatment prolongs the immobility time in the forced swim test for up to two weeks in RIM rats (Nagakura et al., 2009).

Disturbances in the central nervous system is assumed to be involved in both the wide spread pain symptoms and the disturbance in the physiology of sleeping-waking (Moldofsky, 2010). Actually, fibromyalgia patients prevalently experience unrefreshed sleep, which is involved as one of factors in the new fibromyalgia criteria (Wolfe et al., 2010). Because dopamine, norepinephrine, and serotonin in brain neurological circuits are considered to play an important role in the regulation of complex sleep-waking cycle (Murillo-Rodríguez et al., 2009), depletion of brain biogenic amines by reserpine may cause a disturbance of sleep-waking cycle in addition to the hyperalgesia. Actually, it has been reported that reserpine treatment significantly affects the sleep cycle including rapid eye movement (REM) sleep and slow-wave sleep (Hoffman & Domino, 1969). It is of interest to investigate the sleep-waking cycle in RIM rats.

5.4.5 Predictive validity

A recent clinical study has suggested that pregabalin significantly relieves fibromyalgia pain symptoms compared to the placebo (Mease et al., 2008). This clinical beneficial effect is consistent with the result in the RIM rat that the muscle mechanical hyperalgesia is attenuated by pharmacotherapy with pregabalin. Other drugs approved for the treatment of fibromyalgia are duloxetine and milnacipran, which enhance norepinephrine and serotonin neurotransmission in the descending analgesic pathway. They significantly reduce the fibromyalgia pain symptoms in the phase III clinical trials (Clauw et al., 2008; Russell et al., 2008). Duloxetine has been shown to significantly recover muscle hyperalgesia but not tactile allodynia in RIM rats. Further, the dopamine receptor agonist pramipexole has been shown to be efficacious in treating fibromyalgia pain symptoms in a phase II clinical study (Holman and Myers, 2005). In line with this, pramipexole attenuates muscle hyperalgesia and tactile allodynia at doses that cause hyper-locomotion in RIM rats. In contrast, diclofenac shows no significant efficacy in RIM rats, a finding consistent with evidence that non-steroidal anti-inflammatory drugs do not improve pain in fibromyalgia patients (Goldenberg et al., 1986; Yunus et al., 1989). The good correlation of analgesic efficacy mentioned above demonstrates the predictive validity of the RIM model (Nagakura et al., 2009).

Model	Constructive validity	Face validity	Predictive validity
Repeated cold stress	• Exposure to difficult-to-adapt stressors • Dysfunction of autonomic control • Decrease in serotonin-mediated central nervous system control	• Long-lasting mechanical and thermal hyperalgesia • Relief from hyperalgesia by gonadectomy in male mice only • Shortened immobility time in forced swimming test and reversal by antidepressants	• Gabapentin (an analogue of pregabalin) is highly effective • Neurotropin (an extracted substance from inflamed skin) is effective • Small efficacy of morphine
Unpredictable sound stress	• Exposure to unpredictable stress • Appearance of hypersensitivity in the presence of pronociceptive substances • Stress-induced release of adrenaline (product of neuroendocrine system activation)	• Enhancement of nociceptive response to pronociceptive substances • Delayed onset of hypersensitivity after stress exposure • Comorbidities including visceral hyperalgesia, temporomandibular hyperalgesia, and anxiety	• Adrenal medullectomy inhibits the occurrence of hypersensitivity
Intramuscular acidic saline injection	• Local acidic condition in the muscle • Ipsilateral and contralateral spreading of hyperalgesia beyond the original local region • Involvement of central nervous system in spreading	• Long-lasting bilateral muscle hyperalgesia • Chronic cutaneous tactile allodynia • Visceral hyperalgesia occurs concurrently with somatic hyperalgesia	• Pregabalin recovers mechanical hyperalgesia but concurrently affects motor function at similar doses
Reserpine-induced myalgia	• Dysfunction of biogenic amine-mediated analgesic control in the central nervous system • Time courses of changes in biogenic amine concentrations and muscular/tactile response thresholds generally correlate	• Long-lasting muscle hyperalgesia • Chronic cutaneous tactile allodynia • Comorbid depression	• Pregabalin attenuates hyperalgesia • Duloxetine (a biogenic amines reuptake inhibitor) and pramipexole (a dopamine receptor agonist) are effective • Diclofenac is not effective

Table 1. Constructive, face, and predictive validities of putative animal models for fibromyalgia.

6. Other relevant animal models

In addition to the above-mentioned four models, several other models mimicking fibromyalgia symptoms have been developed. These models also reveal no apparent organic disorders that can explain the source of chronic pain symptoms, unlike neuropathic and inflammatory pain animal models.

Biochemical and pharmacological studies have found that vagotomy in the stomach (subdiaphragmatic vagotomy) induces cutaneous (Khasar et al., 1998), visceral (Gschossmann et al., 2002) and muscle (Furuta et al., 2009) hyperalgesia in rats in a manner that resembles hyperalgesia in fibromyalgia patients.

Chronic exposure to stressors other than those described above also produces enhanced nociception in animals. The forced swim stress (10-20 min of forced swimming for 3 consecutive days) induces sustained thermal hyperalgesia and increased nociceptive behavior according to the formalin test (Quintero et al., 2000). Chronic restraint stress caused by wrapping with soft wire mesh (Imbe et al., 2004) or restraint in a cylinder (Bardin et al., 2009) also causes thermal hyperalgesia, mechanical allodynia, or cold allodynia in rats. The thermal hyperalgesia induced by wrapping seems to be associated with a significant change in the regulation of extracellular signal-regulated kinase (ERK) in the descending pain control pathway of the central nervous system (Imbe et al., 2004).

Delayed onset of muscle soreness often occurs in humans after exercise. In accordance with this, inducing eccentric contractions in the extensor digitorum longus muscle causes delayed, long-lasting muscle hyperalgesia in rats (Taguchi et al., 2005).

7. Future research

Comparative mechanistic analysis of the animal models introduced in this chapter may clarify common and different pathways underlying those models. For example, it may be of interest to investigate whether there are causal changes in biogenic substances (e.g., hormones and neurotransmitters) common in those models. Investigation of the mechanism responsible for the different phenotypes among those models may be another interest. Such analysis may shed light on the true nature of fibromyalgia pathophysiology.

Different time course of disease development between fibromyalgia patients and animal models need to be taken into consideration. Animals introduced in this chapter develop disease condition with shorter time course than fibromyalgia patients. It may be a future issue to establish and use naturally occurring chronic pain models which develop fibromyalgia-like conditions over slower time course.

To date, studies using the above-mentioned models have mainly focused on the induction of pain symptoms such as mechanical and thermal hyperalgesia. While this is certainly a reasonable approach when developing animal models for fibromyalgia, as a major symptom in fibromyalgia patients is chronic, widespread pain, these patients also frequently show comorbid symptoms such as fatigue, sleep disturbance, cognitive impairment, irritable bowel, depression, and anxiety. Future research should therefore consider how to incorporate such comorbid symptoms into animal models.

Other investigations may consider fibromyalgia biomarkers, as alterations in certain biochemical substances, physiological vital signs, and brain imaging have been reported to occur in fibromyalgia patients (Dadabhoy et al., 2008). These would also be potential biomarkers in animal models. Such investigations would add great value to our limited understanding of fibromyalgia, and support new diagnostics and drug development.

8. Conclusion

As with nociceptive and neuropathic pain, animal models for fibromyalgia are strongly expected to contribute to our understanding of fibromyalgia pathophysiology and lead to more effective therapeutic options. The animal models described in this chapter are relevant for at least some key aspects of fibromyalgia patients. Combination of multiple animal models, each mimicking at least one key feature seen in fibromyalgia patients, may complement the comprehensive understanding of fibromyalgia pathophysiology. These and future animal models will be instrumental for developing more effective treatments for fibromyalgia.

9. Acknowledgment

The authors would like to thank members of Drug Discovery Research, Astellas Pharma Inc. (Tsukuba, Japan) for their insightful discussions.

10. References

Arnold, L.M.; Goldenberg, D.L.; Stanford, S.B.; Lalonde, J.K.; Sandhu, H.S.; Keck, P.E. Jr.; Welge, J.A.; Bishop, F.; Stanford, K.E.; Hess, E.V. & Hudson J.I. (2007) Gabapentin in the treatment of fibromyalgia: a randomized, double-blind, placebo-controlled, multicenter trial. *Arthritis and rheumatism,* Vol.56, No.4, (April 2007) pp. 1336-1344, ISSN 0004-3591

Bardin, L.; Malfetes, N.; Newman-Tancredi, A. & Depoortère, R. (2009) Chronic restraint stress induces mechanical and cold allodynia, and enhances inflammatory pain in rat: Relevance to human stress-associated painful pathologies. *Behavioural brain research,* Vol.205, No.2, (December 2009) pp. 360-366, ISSN:0166-4328

Bradley, L.A. Pathophysiology of fibromyalgia. (2009) *The American journal of medicine,* Vol.122, No.12, (December 2009) pp. S22-30, ISSN 0002-9343

Clauw, D.J. (2009). Fibromyalgia: an overview. *American Journal of Medicine,* Vol.122, No.12, (December 2009) pp. S3-13, ISSN 0002-9343

Clauw, D.J., Mease, P., Palmer, R.H., Gendreau, R.M. & Wang, Y. (2008) Milnacipran for the treatment of fibromyalgia in adults: a 15-week, multicenter, randomized, double-blind, placebo-controlled, multiple-dose clinical trial. *Clinical therapeutics,* Vol.30, No.11, (November 2008) pp. 1988-2004, ISSN 0149-2918

Da Silva, L.F.; Desantana, J.M. & Sluka, K.A. (2010). Activation of NMDA receptors in the brainstem, rostral ventromedial medulla, and nucleus reticularis gigantocellularis mediates mechanical hyperalgesia produced by repeated intramuscular injections of acidic saline in rats. *Journal of Pain,* Vol.11, No.4, (April 2010) pp. 378-387, ISSN 1526-5900

Dadabhoy, D.; Crofford, L.J.; Spaeth, M.; Russell, I.J. & Clauw, D.J. (2008) Biology and therapy of fibromyalgia. Evidence-based biomarkers for fibromyalgia syndrome. *Arthritis Research & Therapy,* Vol.10, No.4, (August 2008) pp. 211, ISSN 1478-6354Desmeules, J.A.; Cedraschi, C.; Rapiti, E.; Baumgartner, E.; Finckh, A.; Cohen, P.; Dayer, P. & Vischer, T.L. (2003) Neurophysiologic evidence for a central sensitization in patients with fibromyalgia. *Arthritis and rheumatism,* Vol.48, No.5, (May 2003) pp.1420-1429, ISSN 0004-3591

Dohrenbusch, R.; Sodhi, H.; Lamprecht, J. & Genth, E. (1997). Fibromyalgia as a disorder of perceptual organization? An analysis of acoustic stimulus processing in patients with widespread pain. *Zeitschrift fur Rheumatologie,* Vol.56, No.6, (December 1997) pp. 334-341, ISSN 0340-1855

Dworkin, R.H.; O'Connor, A.B.; Backonja, M.; Farrar, J.T.; Finnerup, N.B.; Jensen, T.S.; Kalso, E.A.; Loeser, J.D.; Miaskowski, C.; Nurmikko, T.J.; Portenoy, R.K.; Rice, A.S.; Stacey, B.R.; Treede, R.D.; Turk, D.C. & Wallace, M.S. (2007). Pharmacologic management of neuropathic pain: evidence-based recommendations. *Pain,* Vol.132, No.3, (December 2007) pp. 237-251, ISSN 0304-3959

El Mansari, M.; Guiard, B.P.; Chernoloz, O.; Ghanbari, R.; Katz, N. & Blier, P. (2010). Relevance of norepinephrine-dopamine interactions in the treatment of major depressive disorder. CNS *Neuroscience & Therapeutics,* Vol.16, No.3, (June 2010) pp. e1-17, ISSN 1755-5930

Fillingim, R.B. & Ness, T.J. (2000). Sex-related hormonal influences on pain and analgesic responses, Neuroscience and Biobehavioral Reviews Vol.24, No.4, (June 2000) pp. 485–501, ISSN 0149-7634

France, C.R., Rhudy, J.L. & McGlone, S. (2009) Using normalized EMG to define the nociceptive flexion reflex (NFR) threshold: further evaluation of standardized NFR scoring criteria. *Pain,* Vol.145, No.1-2, (September 2009) pp. 211-218, ISSN 0304-3959

Furuta, S.; Shimizu, T.; Narita, M.; Matsumoto, K.; Kuzumaki, N.; Horie, S.; Suzuki, T. & Narita, M. (2009) Subdiaphragmatic vagotomy promotes nociceptive sensitivity of deep tissue in rats. *Neuroscience,* Vol.164, No.3, (December 2009) pp. 1252-1262, ISSN 0306-4522

Goldenberg, D.L.; Felson, D.T. & Dinerman, H. (1986). A randomized, controlled trial of amitriptyline and naproxen in the treatment of patients with fibromyalgia. *Arthritis and rheumatism,* Vol.29, No.11, (November 1986) pp. 1371–1377. ISSN 0004-3591

Graven-Nielsen, T. & Arendt-Nielsen, L. (2010). Assessment of mechanisms in localized and widespread musculoskeletal pain. *Nature Reviews Rheumatology,* Vol.6, No.10, (October 2010) pp. 599-606, ISSN 1759-4790

Graven-Nielsen, T.; Arendt-Nielsen, L.; Svensson, P. & Jensen, T.S. (1997). Stimulus-response functions in areas with experimentally induced referred muscle pain--a psychophysical study. *Brain Research,* Vol.744, No.1, (January 1997) pp. 121-128, ISSN 0006-8993

Green, P.G.; Alvarez, P.; Gear, R.W.; Mendoza, D. & Levine, J.D. (2011). Further validation of a model of fibromyalgia syndrome in the rat. *J Pain,* Vol.12, No.7, (July 2011) pp. 811-818, ISSN 1526-5900

Gschossmann, J.M.; Mayer, E.A.; Miller, J.C. & Raybould, H.E. (2002) Subdiaphragmatic vagal afferent innervation in activation of an opioidergic antinociceptive system in response to colorectal distension in rats. *Neurogastroenterology and motility,* Vol.14, No.4, (August 2002) pp. 403-408, ISSN 1350-1925

Hata, T.; Itoh, E. & Kawabata, A. (1991) Changes in CNS levels of serotonin and its metabolite in SART-stressed (repeatedly cold-stressed) rats. *Japanese Journal of Pharmacology,* Vol.56, No.1, (May 1991) pp. 101-104, ISSN 0021-5198

Hata, T.; Itoh, E. & Nishikawa, H. (1995). Behavioral characteristics of SART-stressed mice in the forced swim test and drug action. *Pharmacology, Biochemistry & Behavior,* Vol.51, No.4, (August 1995) pp. 849-853, ISSN 0091-3057

Hata, T.; Kita, T.; Itoh, E. & Kawabata, A. (1988). The relationship of hyperalgesia in SART (repeated cold)-stressed animals to the autonomic nervous system. *Journal of Autonomic Pharmacology,* Vol.8, No.1, (March 1988) pp. 45-52, ISSN 0144-1795

Hata, T.; Kita, T.; Kamanaka, Y.; Honda, S.; Kakehi, K.; Kawabata, A. & Itoh, E. (1987) Catecholamine levels in the brain of SART (repeated cold)-stressed rats. *Journal of Autonomic Pharmacology,* Vol.7, No.3, (September 1987) pp. 257-266, ISSN 0144-1795

Holman, A.J. & Myers, R.R. (2005) A randomized, double-blind, placebo-controlled trial of pramipexole, a dopamine agonist, in patients with fibromyalgia receiving concomitant medications. *Arthritis and rheumatism,* Vol.52, No.8, (August 2005) pp. 2495-505, ISSN 0004-3591

Imbe, H.; Murakami, S.; Okamoto, K.; Iwai-Liao, Y. & Senba, E. (2004) The effects of acute and chronic restraint stress on activation of ERK in the rostral ventromedial medulla and locus coeruleus. *Pain,* Vol.112, No.3, (December 2004) pp. 361-371, ISSN 0304-3959

Issberner, U.; Reeh, P.W. & Steen, K.H. (1996). Pain due to tissue acidosis: a mechanism for inflammatory and ischemic myalgia? *Neuroscience Letters,* Vol.208, No.3, (April 1996) pp. 191-194, ISSN 0304-3940

Khasar, S.G.; Green, P.G. & Levine, J.D. (2005) Repeated sound stress enhances inflammatory pain in the rat. *Pain,* Vol.116, No.1-2, (July 2005) pp. 79-86, ISSN 0304-3959

Khasar, S.G.; Green, P.G.; Miao, F.J. & Levine, J.D. (2003). Vagal modulation of nociception is mediated by adrenomedullary epinephrine in the rat. *European Journal of Neuroscience,* Vol.17, No.4, (February 2003) pp. 909-915, ISSN 0953-816X

Khasar, S.G.; Miao, F.J.; Jänig, W. & Levine, J.D. (1998). Vagotomy-induced enhancement of mechanical hyperalgesia in the rat is sympathoadrenal-mediated. *Journal of Neuroscience,* Vol.18, No.8, (April 1998) pp. 3043-3049, ISSN 0270-6474

Kita, T.; Hata, T.; Yoneda, R. & Okage, T. (1975). Stress state caused by alternation of rhythm in environmental temperature, and the functional disorders in mice and rats. *Folia Pharmacologica Japonica,* Vol.71, No.2, (March 1975) pp. 195–210, ISSN 0015-5691

Leith, N.J. & Barrett, R.J. (1980). Effects of chronic amphetamine or reserpine on self-stimulation responding: animal model of depression? *Psychopharmacology (Berlin),* Vol.72, No.1, (January 1980) pp. 9–15, ISSN 0033-3158

Levine, J.D. & Reichling, D.B. (2005). Fibromyalgia: the nerve of that disease. *The Journal of rheumatology Supplement* Vol.75 (August, 2005) pp. 29-37, ISSN 0380-0903

Mease, P.J.; Russell, I.J.; Arnold, L.M.; Florian, H.; Young, J.P. Jr; Martin, S.A. & Sharma, U. (2008). A randomized, double-blind, placebo-controlled, phase III trial of pregabalin in the treatment of patients with fibromyalgia. *Journal of Rheumatology,* Vol.35, No.3 (March 2008) pp. 502-514, ISSN 0315-162X

Mense, S. (2000). Neurobiological concepts of fibromyalgia – the possible role of descending spinal tracts, *Scandinavian Journal of Rheumatology Supplement,* Vol.113, pp. 24–29, ISSN 0301-3847

Miranda, A.; Peles, S.; Rudolph, C.; Shaker, R. & Sengupta, J.N. (2004). Altered visceral sensation in response to somatic pain in the rat. *Gastroenterology,* Vol.126, No.4, (April 2004) pp. 1082-1089, ISSN 0016-5085Moldofsky, H. (2010) Rheumatic manifestations of sleep disorders. *Current opinion in rheumatology,* Vol.22, No.1 (January 2010) pp.59-63, ISSN:1040-8711

Murillo-Rodríguez, E.; Arias-Carrión, O.; Sanguino-Rodríguez, K.; González-Arias, M. & Haro, R. (2009) Mechanisms of sleep-wake cycle modulation. *CNS & neurological disorders drug targets*, Vol.8, No.4, (August 2009) pp. 245-253, ISSN 1871-5273

Nagakura, Y.; Oe, T.; Aoki, T. & Matsuoka, N. (2009). Biogenic amine depletion causes chronic muscular pain and tactile allodynia accompanied by depression: A putative animal model of fibromyalgia. *Pain*, Vol.146, No.1-2, (November 2009) pp. 26-33, ISSN 0304-3959

Nielsen, L.A. & Henriksson, K.G. (2007) Pathophysiological mechanisms in chronic musculoskeletal pain (fibromyalgia): the role of central and peripheral sensitization and pain disinhibition. *Best practice & research. Clinical rheumatology*, Vol.21, No.3, (June 2007) pp. 465-480, ISSN 1521-6942

Negus, S.S.; Vanderah, T.W.; Brandt, M.R.; Bilsky, E.J.; Becerra, L. & Borsook, D. (2006) Preclinical assessment of candidate analgesic drugs: recent advances and future challenges. *The Journal of pharmacology and experimental therapeutics*, Vol.319, No.2, (November 2006) pp. 507-514, ISSN 0022-3565

Nishiyori, M.; Nagai, J.; Nakazawa, T. & Ueda, H. (2010) Absence of morphine analgesia and its underlying descending serotonergic activation in an experimental mouse model of fibromyalgia. *Neuroscience letters*, Vol.472, No.3, (March 2010) pp.184-187, ISSN:0304-3940

Nishiyori, M. & Ueda, H. (2008). Prolonged gabapentin analgesia in an experimental mouse model of fibromyalgia. *Molecular Pain*, Vol.4, (November 2008) pp. 52, ISSN 1744-8069

Oe, T.; Tsukamoto, M. & Nagakura Y. (2010) Reserpine causes biphasic nociceptive sensitivity alteration in conjunction with brain biogenic amine tones in rats. *Neuroscience*, Vol.169, No.4, (September 2010) pp. 1860-1871, ISSN 0306-4522

Ohara, H.; Kawamura, M.; Namimatsu, A.; Miura, T.; Yoneda, R. & Hata, T. (1991). Mechanism of hyperalgesia in SART stressed (repeated cold stress) mice: antinociceptive effect of neurotropin. *Japanese Journal of Pharmacology*, Vol.57, No.2, (October 1991) pp. 243-250, ISSN 0021-5198

Petzke, F.; Clauw, D.J.; Ambrose, K.; Khine, A. & Gracely, R.H. (2003) Increased pain sensitivity in fibromyalgia: effects of stimulus type and mode of presentation. *Pain*, Vol.105, No.3, (October 2003) pp. 403-413, ISSN 0304-3959Quintero, L.; Moreno, M.; Avila, C.; Arcaya, J.; Maixner W. & Suarez-Roca, H. (2000) Long-lasting delayed hyperalgesia after subchronic swim stress. *Pharmacology, biochemistry & behavior*, Vol.67, No.3, (November 2000) pp. 449-458, ISSN 0091-3057

Roy-Byrne, P.; Smith, W.R.; Goldberg, J.; Afari, N. & Buchwald, D. (2004). Post-traumatic stress disorder among patients with chronic pain and chronic fatigue. *Psychological Medicine*, Vol.34, No.2, (February 2004) pp. 363-368, ISSN 0033-2917

Russell, I.J.; Vaeroy, H.; Javors, M. & Nyberg, F. (1992). Cerebrospinal fluid biogenic amine metabolites in fibromyalgia/fibrositis syndrome and rheumatoid arthritis. *Arthritis and rheumatism*, Vol.35, No.5, (May 1992) pp. 550-556, ISSN 0004-3591

Russell, I.J., Mease, P.J., Smith, T.R., Kajdasz, D.K., Wohlreich, M.M., Detke, M.J., Walker, D.J., Chappell, A.S. & Arnold, L.M. (2008) Efficacy and safety of duloxetine for treatment of fibromyalgia in patients with or without major depressive disorder: Results from a 6-month, randomized, double-blind, placebo-controlled, fixed-dose trial. *Pain*, Vol.136, No.3, (June 2008) pp. 432-444, ISSN 0304-3959

Siler, A.C.; Gardner, H.; Yanit, K.; Cushman, T. & McDonagh, M. (2011). Systematic review of the comparative effectiveness of antiepileptic drugs for fibromyalgia. *Journal of Pain,* Vol.12, No.4, (April 2011) pp. 407-415, ISSN 1526-5900

Skyba, D.A.; King, E.W. & Sluka, K.A. (2002). Effects of NMDA and non-NMDA ionotropic glutamate receptor antagonists on the development and maintenance of hyperalgesia induced by repeated intramuscular injection of acidic saline. *Pain,* Vol.98, No.1-2, (July 2002) pp. 69-78, ISSN 0304-3959

Sluka, K.A. (2009). Pain Syndromes: Myofascial pain and Fibromyaliga. In: *Mechanisms and management of pain for the physical therapist,* K.A. Sluka KA, (Ed), 277-295, *IASP Press,* ISBN 13-978-0-931092-77-0, Seatlle, USA

Sluka, K.A.; Kalra, A. & Moore, S.A. (2001). Unilateral intramuscular injections of acidic saline produce a bilateral, long-lasting hyperalgesia. *Muscle & nerve,* Vol.24, No.1, (January 2001) pp. 37-46, ISSN 0148-639X

Sluka, K.A.; Rohlwing, J.J.; Bussey, R.A.; Eikenberry, S.A. & Wilken, J.M. (2002). Chronic muscle pain induced by repeated acid Injection is reversed by spinally administered mu- and delta-, but not kappa-, opioid receptor agonists. *Journal of Pharmacology And Experimental Therapeutics,* Vol.302, No.3, (September 2002) pp. 1146-1150, ISSN 0022-3565

Sörensen, J.; Bengtsson, A.; Bäckman, E.; Henriksson, K.G. & Bengtsson, M. (1995). Pain analysis in patients with fibromyalgia. Effects of intravenous morphine, lidocaine, and ketamine. *Scandinavian Journal of Rheumatology,* Vol.24, No.6, pp. 360-365, ISSN 0300-9742

Taguchi, T.; Matsuda, T.; Tamura, R.; Sato, J. & Mizumura, K. (2005) Muscular mechanical hyperalgesia revealed by behavioural pain test and c-Fos expression in the spinal dorsal horn after eccentric contraction in rats. *The Journal of physiology,* Vol.564, No. 1, (April 2005) pp. 259-268, ISSN 0022-3751

Tillu, D.V.; Gebhart, G.F. & Sluka, K.A. (2008). Descending facilitatory pathways from the RVM initiate and maintain bilateral hyperalgesia after muscle insult. *Pain,* Vol.136, No.3, (June 2008) pp. 331-339, ISSN 0304-3959

Wang, H.; Moser, M.; Schiltenwolf, M. & Buchner, M. (2008). Circulating cytokine levels compared to pain in patients with fibromyalgia – a prospective longitudinal study over 6 months. *Journal of Rheumatology,* Vol.35, No.7, (July 2008) pp. 1366-1370, ISSN 0315-162X

Wolfe, F.; Clauw, D.J.; Fitzcharles, M.A.; Goldenberg, D.L.; Katz, R.S.; Mease, P.; Russell, A.S.; Russell, I.J.; Winfield, J.B. & Yunus, M.B. (2010) The American College of Rheumatology preliminary diagnostic criteria for fibromyalgia and measurement of symptom severity. *Arthritis care & research,* Vol.62, No.5, (May 2010) pp. 600-610, ISSN 2151-464X

Wood, P.B. (2008). Role of central dopamine in pain and analgesia, *Expert Review of Neurotherapeutics,* Vol.8, No.5, (May 2008) pp. 781–797, ISSN 1473-7175

Yokoyama, T.; Maeda, Y.; Audette, K.M. & Sluka, K.A. (2007). Pregabalin reduces muscle and cutaneous hyperalgesia in two models of chronic muscle pain in rats. *Journal of Pain,* Vol.8, No.5, (May 2007) pp. 422-429, ISSN 1526-5900

Yunus, M.B.; Masi, A.T. & Aldag, J.C. (1989). Short term effects of ibuprofen in primary fibromyalgia syndrome: a double blind, placebo controlled trial, *Journal of Rheumatology,* Vol.16, No.6, (April 1989) pp. 527–532, ISSN 0315-162X

4

Psychosocial Factors in Fibromyalgia: A Qualitative Study on Life Stories and Meanings of Living with Fibromyalgia

Paula J. Oliveira and Maria Emília Costa
Psychology Centre of University of Porto, Faculty of Psychology and Educational Sciences, University of Porto, Portugal

1. Introduction

Fibromyalgia (FM) is defined by the American College of Rheumatology (ACR) 1990 classification criteria as a syndrome in which an individual is required to have both a history of chronic widespread pain and the presence of, at least, 11 of 18 tender points (Wolfe et al., 1990). However, additional symptoms such as sleep disorders, fatigue, and psychological distress are also common in clinical practice (Blotman & Branco, 2007; Wilke, 2009) which has contributed over time to questioning if tender points would be the most appropriate measure to capture the essence of FM (Wilke, 2009). In 2010, Wolfe et al., proposed the ACR preliminary diagnostic criteria for FM including two variables that best defined FM and its symptom spectrum: the widespread pain index (WPI) and the symptom severity scale (SS scale). The WPI is a measure of the number of painful body regions and the SS scale assesses cognitive problems, unrefreshed sleep, fatigue, and somatic symptoms. The authors combined the WPI and the SS scale in order to recommend a new case definition of FM: (WPI≥7 AND SS ≥5) OR (WPI 3-6 AND SS≥9); moreover, the symptoms have to be present at a similar level for at least 3 months and the patient does not have a disorder that explain the pain (Wolfe et al., 2010). In terms of gender prevalence, FM is more common among women between 20 and 50 years (Blotman & Branco, 2007).

Although the exact etiology and pathogenesis of FM are still unknown there is evidence that psychosocial factors could play an important role. In fact, FM has been conceptualized within biopsychosocial perspectives, in which physiological, psychological, and social factors are considered as interacting in different ways and at different stages. In a review article, Eich et al. (2000) evaluated the role of psychosocial factors in the development of FM supporting that psychosocial factors can be relevant at different etiological levels and can be classified into predisposing, triggering, and stabilising/"chronifying". On the other hand, Van Houdenhove and Egle (2004) conceptualized stress as playing a key role in the pathogenesis of FM, placing emphasis on the relationships among adverse life experiences, stress regulation, and pain-processing mechanisms. One year later, the authors highlighted, from an etiologic point of view, studies concerning the role of adverse life events, personality and lifestyle factors, post-traumatic stress, and negative childhood experiences (Van Houdenhove et al., 2005). Thus, the proposed integrative biopsychosocial models consider the role of the individuals' life events in the development of the syndrome. In a related conceptualization, Quartilho (2004) proposed that FM should be understood both in a cross-sectional and a longitudinal perspective. In a retrospective qualitative study with FM patients, the author has found empirical support for the classification of predisposing, triggering, and perpetuating factors, identifying specific categories within each one. Therefore, previous theoretical and empirical contributions highlight that psychosocial factors in FM could be classified as predisposing, triggering, and perpetuating; moreover this classification requires both a retrospective and a cross-sectional perspective in order to understand it, given the distinct periods of time wherein vulnerabilities can be identified. More specifically, predisposing factors include adverse events during life-time contributing to individual vulnerability due to their long-lasting psychological effects. These factors are not necessarily causal and include previous adverse life events, in general, and physical and emotional traumas, in particular. These conditions can contribute to further manifestations of low self-esteem, low self-efficacy, and negative affect which increases the risk of dysfunctional life styles and unsatisfying relationships (Eich et al., 2000; Van Houdenhouve & Egle, 2004). On the other hand, triggering factors are those preceding the pain onset and causing it directly (Eich et al., 2000), such as loss of meaningful relationships, critical changes in life conditions, and severe physical diseases. Although triggering factors are usually well-described in terms of a specific time and situation, they can also comprise multiple situations reflecting an extended period of physical and psychosocial stress (Van Houdenhove & Egle, 2004). Lastly, perpetuating factors can help to explain the maintenance of FM, once arisen; these factors can work against the natural remission or can lead to a chronic state of illness by increasing the frequency or intensity of the symptoms. Possible perpetuating factors in FM can be depression, anxiety, worrying, catastrophic thinking, and dysfunctional health-care seeking behaviour. Thus, FM is conceptualized as the end stage of an accumulation of biological and psychosocial vulnerability factors, over time. The successive emergence of those factors promoting the individual's vulnerability ends at the appearance of the first symptoms establishing an irreversible stage (Eich et al., 2000; Quartilho, 2004; Van Houdenhove & Egle, 2004; Van Houdenhove et al., 2005). Therefore we stress the relevance of considering life stories of FM individuals in order to identify key moments of biological and psychosocial deregulation which can promote the further development of the syndrome and its maintenance once arisen. Moreover, the lack of empirical studies performed in this domain (Quartilho, 2004) support the relevance of finding empirical support for theoretical conceptualizations (Eich et al., 2000; Van Houdenhove & Egle, 2004), namely concerning identification of specific dimensions inside each of the major categories: predisposing, triggering, and perpetuating factors.

On the other hand, and based in Quartilho (2004), we support the relevance of proposing an integrated comprehension of FM experience considering both a longitudinal (life stories) and a cross-sectional perspective (meanings of living with FM in the present time). This combined perspective can be achieved through patients' narratives (Quartilho, 2001a, 2001b; Yardley, 1997). In fact, emphasizing a meaning-based perspective for comprehension of FM, various qualitative studies have been useful in order to understand the psychological representations of patients concerning their condition, as narrated by their own words (e.g. Sim & Madden, 2008). One of the particularities that can be found in FM patients is the narrated importance of obtaining a clinical diagnosis and, thus, an external validation of their suffering given the absence of "visible" symptoms to others (e.g. Cunningham & Jillings, 2006; Hellström et al., 1999; Schaefer, 1995; Undeland & Malterud, 2007). Moreover,

several studies have shown that FM has impact on different life contexts of individuals living with such a chronic condition (e.g., Aïni et al., 2010; Arnold et al., 2008; Asbring, 2001; Cunningham & Jillings, 2006; Söderberg & Lundman, 2001), and some of them have stressed specific dimensions, such as professional life (Gauer, 2009; Liedberg & Henriksson, 2002), couple relationship (e.g. Schmidt, 2008), and sexual experience (e.g. Garcia-Campayo & Alda, 2004). Furthermore, the finding and evolution of adaptive coping strategies in the context of FM tend to assume an important role in living with FM (e.g. Cunningham & Jillings, 2006; Hellström et al., 1999; Schaefer, 1995, 1997; Söderberg & Lundman, 2001). Another meaning-making process identified in FM patients' narratives is related to the perceived control over symptoms (e.g., Asbring, 2001; Hellström et al., 1999), which can range from perceived absence of control to a perceived total control over symptoms. The perceived control dimension has been highlighted as being associated to positive physical and emotional health outcomes (e.g. Hagger & Orbell, 2003; Scharloo & Kaptein, 1997) and more adaptive coping strategies (Joyce-Moniz & Barros, 2005) in chronic patients. However, as far as we know, there are no qualitative studies which have assessed the link between perceived control meanings and narrated life stories. Thus, one of the innovations of the present study will be to answer the following research question: Are the perceived control meanings associated to patients' life stories and coping strategies, as well?
Considering the previous theoretical and empirical contributions the aims of the present study are:

a. To access the life stories of patients with FM in order to identify specific psychosocial factors which are considered to have a predisposing, triggering, and perpetuating role in FM;
b. To identify the meanings associated to living with FM at the present time;
c. To explore relationships among perceived control meanings and patients' life stories as well as coping strategies.

2. Method

2.1 Sample and study design

Twenty female who met the ACR criteria for the classification of FM (Wolfe et al., 1990) were recruited through Myos - Portuguese National Association against Fibromyalgia and Chronic Fatigue Syndrome. Participants were excluded if they had significant physical or psychiatric comorbidity (namely severe pain not related to FM and psychotic disorders) that might have interfered with their experience of, or ability to talk about, FM. Purposive sampling ensured the inclusion of individuals differing in various sociodemographic variables in order to maximize the range of the narrated experiences. The sample predominantly consisted of "married" individuals (80%), but included "single" and "divorced" persons, as well. In terms of education "elementary school" and "graduate level" were the most common levels, each one with a sample representation of 30%. Regarding the employment status, 55% of the sample were "currently employed", followed by "retired" (25%) and "unemployed" (20%) persons. Mean symptom duration was 16 years and the mean time elapsed since diagnosis was 4 years.

2.2 Procedure

First, approval of the study protocol and procedures was obtained from Direction of *Myos*. Then, participants gave their written informed consent after having received detailed information about the study and having being assured of anonymity and confidentiality of their interview responses. Next, personal tape-recorded interviews were conducted with each participant. The interviews were about 60 to 150 minutes in length and took place either in the researcher's office or in the participant's home. The data of the current study were collected within the scope of the PhD. dissertation of the first author (Oliveira, 2008).

2.3 Measure (semi-structured interview guide)

A semi-structured interview guide was used in order to collect information concerning life stories of FM patients, as well as meanings associated to the experience of living with FM. The first section of the guide aimed to explore participants' life stories according to their own personal narrative. The second section of the guide included questions directed to (i) the experience of having the first symptoms until the moment when participants received a diagnosis of FM; (ii) the impact of having FM in the current time; (iii) used ways of dealing with FM; (iv) the meaning-making processes underlying: the believed causes that participants felt to contribute to the predisposition, triggering, and perpetuating of FM; the perceived control over FM; and the future perspectives. Questions were initially open-ended to guarantee participants were not biased by the topics of interest; direct questions were only used when relevant topics did not arise in response to the open-ended questions. This methodology has given participants the opportunity to mention spontaneously any relevant domains not previously included in the interview guide.

2.4 Data analysis

Interviews were audio-taped and transcribed with all participant identifiers removed during transcription. Accordingly to the theoretical principles of grounded theory a data analysis was performed (Strauss & Corbin, 1998), using the QSR NVivo7 software. Each interview was read through several times in order to get a sense of their content. Then, coding of quotes involved assigning appropriate codes to patient statements determined by the underlying concept. That is, textual units were identified and categorized with the aim of constructing a list of categories for each interview. The emerging categories were compared to identify those commonly occurring among the interviews. This allowed us to develop an iterative coding scheme, driven by the participants' experience. After that, a final list of categories and subcategories were constructed. Trustworthiness of the data was addressed through close adherence to the method of grounded theory. Consensus about the process of analyzing data was reached through extensive discussion between the researchers.

3. Results and discussion

3.1 Life stories of individuals living with FM

Findings from our thematic analysis of the life stories of individuals with FM support previous conceptualizations of predisposing, triggering, and perpetuating factors in origin and maintenance of FM (Eich et al., 2000; Quartilho, 2004; Van Houdenhove & Egle, 2004; Van Houdenhove et al., 2005). Moreover, our results advance knowledge in this domain by identifying specific factors within these major categories. Table 1 identifies the categories and respective subcategories that have emerged from the life stories, as well as their prevalence in our sample.

Predisposing factors	Triggering factors	Perpetuating factors
Adverse life events (75%) Significant losses (65%) Personality traits (45%) Physical and/or verbal violence (45%) Family conflicts (45%) Excessive striving for high achievement at work (35%) Emotional neglect (35%) Nuclear family psychosocial problems (35%) Job dissatisfaction (30%) Early assumption of responsibility (30%) Unsatisfying relationships (30%) History of sexual abuse (10%) Compensatory overactive lifestyles (10%)	Not identified (40%) Extended period of physical and/or psychosocial stress (35%) Traumatic physical experience (20%) Traumatic emotional experience (20%) Painful injury (15%)	Family relationships (45%) Sleep disturbances (35%) Catastrophic thoughts (30%) Depressive/anxious symptomatology (30%) Social withdrawal (30%) Low illness acceptance (25%) Dysfunctional health care-seeking behaviour (25%) Extended period of physical/psychosocial stress (15%) Somatic vigilance (15%)

Table 1. Predisposing, Triggering, and Perpetuating factors identified in life stories of FM patients and prevalence of each one in the sample.

In the "**predisposing factors**" category, 13 subcategories were identified reflecting physical, psychological, and social aspects which can promote individual vulnerability across the time. It is worth noting that in the entire sample (20 participants) it was possible to identify the presence of several specific predisposing factors whose prevalence varied considerably among the participants (2 to 13). This result support the assumption that not all FM patients present extremely adverse life stories and the presence of protective factors can be useful to further individual well-being (Van Houdenhove & Egle, 2004). More specifically, 75% of the sample narrated "adverse life events" whose recall triggered, even at the present time, powerful negative emotions. Moreover, a diversity of subcategories was identified as potentially promoting individual vulnerability due to their long-lasting psychological effects (Eich et al., 2000; Quartilho, 2004; Van Houdenhove & Egle, 2004; Van Houdenhove et al., 2005). More specifically, some of those subcategories reflect a more intrapersonal level, such as "significant losses", "personality traits", "excessive striving for high achievement at work", "job dissatisfaction", "early assumption of responsibility", "compensatory overactive lifestyles"; while others assume a more clear position in interpersonal context, namely "physical and/or verbal violence", "family conflicts", "emotional neglect", "nuclear family psychosocial problems", "unsatisfying relationships", and "history of sexual abuse".

Regarding **"triggering factors"**, results moderately supported its role in the appearance of the first FM symptoms. In fact, in 40% of our sample any event associated to the appearance of the symptoms was identified. However, a considerable percentage of the sample (35%) narrated an "extended period of physical and/or psychosocial stress" before the manifestation of first FM symptoms. Factors associated to biological deregulation seemed to assume an important role as well, such as, the occurrence of a "traumatic physical experience" (20%) and "painful injury" (15%). And, to a lesser degree, "traumatic emotional experiences" were also mentioned. These results support theoretical conceptualizations of the role of the stress, in it physical and psychological dimensions, as trigger of FM syndrome (Van Houdenhove & Egle, 2004).

In **"perpetuating factors"** category, it is worth noting that the majority of sample presented a wide variety of biological, psychological, and social factors negatively associated with health-related quality of life, thus potentially intensifying the experience of FM (Eich et al., 2000; Quartilho, 2004). More specifically, "family relationships" (45% of the participants are included here) were identified as likely having a deleterious effect on perceived health status. Family conflicts, lack of affectionate relationships, dysfunctional couple relationship, negative manifestations of spouse, and conflicts are perceived as being associated to severe depressive symptoms by participants. This result supports the importance of family relationships on individual well-being (Quartilho, 2001b). At an intrapersonal level, dimensions as "sleep disturbances", "catastrophic thoughts", "depressive/anxious symptomatology", "low illness acceptance" and "somatic vigilance" emerge as potential factors of aggravating FM symptoms, which gives empirical support to previous theoretical conceptualizations (Eich et al., 2000; Quartilho, 2004; Van Houdenhove & Egle, 2004; Van Houdenhove et al., 2005). Furthermore, "social withdrawal", "dysfunctional health care-seeking behaviour" leads us to the inclusion of individual in their social context. Consistent with the theoretical model of Van Houdenhove and Egle (2004), an "extended period of physical/psychosocial stress" was also present.

Our results are thus compatible with perspectives that development and maintenance of FM can be associated to the role of the stress; a biological predisposition interacting with an unfavourable psychosocial context could create conditions to the appearance and perpetuation of FM in individuals' life.

3.2 Meanings of individuals living with FM

Concerning the meanings of individuals living with FM in the present time, four categories have emerged. More specifically, the patients were involved in great efforts to get their illness confirmed until FM had been diagnosed (**Finding a diagnosis: fibromyalgia**) and, presently living with FM is experienced as bringing several changes to the participants' life (**A changed life**); these patients seem to make use of different strategies to cope with those changes (**Coping with fibromyalgia**), as well as develop new perspectives about themselves and surrounding contexts (**Perspectives about a new reality**).

Regarding the first category "**Finding a diagnosis: fibromyalgia"** it can be noted that this was a lengthy process for most participants (the average time for obtaining a diagnosis after the appearance of the first symptoms was nearly 12 years). In the early (pre-diagnosis) stage, when their "perceived first symptoms" are still unknown, patients tended to attribute them different degrees of severity. This experience led to "health care seeking behaviours" in order to identify what is happening to them. However, the successive consultations, normal results of various medical tests, and the absence of a diagnosis tended to be accompanied by "negative feelings" such as sadness, anger, and questioning of their own mental health. As patients do not have a label for their situation, "family and work colleagues" tend to be suspicious of their complaints. For all these reasons, people tended to describe the moment they "get the diagnosis" with feelings of relief. This entire process can be, at least partially, theoretically explained by self-regulation model of Howard Leventhal (1980, 1997). In fact, the dissonance between symptom perception and social/medical messages, added to incomprehension of significant ones, promote the individuals' difficulty on the process of creating cognitive representations which are needed for the self-regulation processes in illness. Therefore, it is understandable the occurrence of negative emotional responses during this process and positive ones (mainly, relief) when obtaining a clinical diagnosis. Thus, the diagnosis can help to eliminate the initial dissonance between symptom perception and social messages.

Concerning the category **"A changed life",** it is related to the perception of participants in the impact of FM on their lives, particularly in terms of "disturbing symptoms", "daily activities", "marital relationship", "sexual life", "family life", "professional life", and "social life" (Table 2). As stated in previous studies, a wide range of life contexts seems to have been affected by the presence of FM (Aïni et al., 2010; Arnold et al., 2008; Asbring, 2001; Cunningham & Jillings, 2006; Söderberg & Lundman, 2001). Considering the high impact of FM in patients' lives, these context-related impacts would presumably exert a negative influence on the health-related quality of life; however, the narrated impact of FM in most situations was associated to a movement towards higher levels of adaptation.

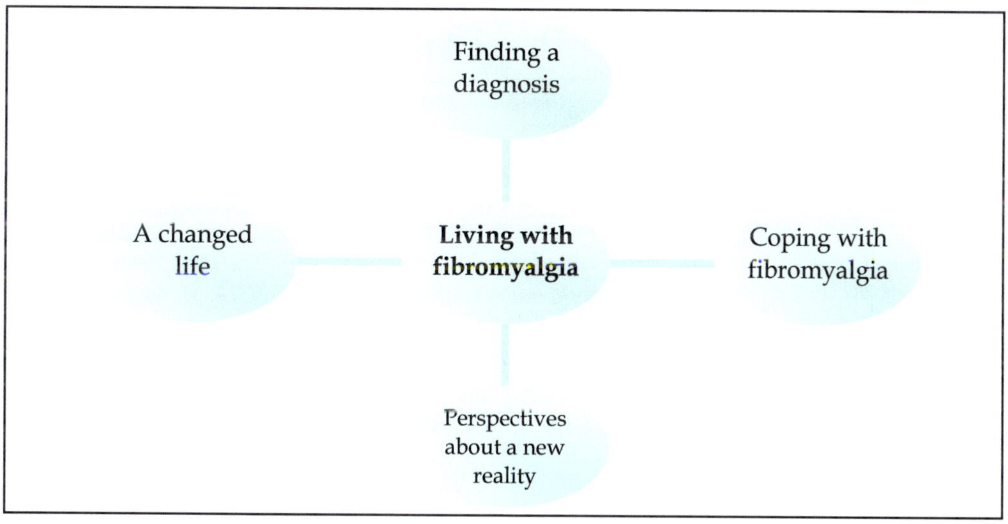

Fig. 1. Meanings of living with fibromyalgia

"A CHANGED LIFE"
Disturbing symptoms
"The most debilitating were, undoubtedly, the cognitive symptoms. Considering the personal, social, and professional areas I think that was the fatigue; not being able to stand up, having to lie down a couple of times a day in order to be minimally lucid and be able to stand on my feet. And, then, the pain." (I1, 40 years old)
Daily activities
"...at home I don't have that vitality, that strength I had some time ago; now, I don't have it. I say this with great sadness and sometimes tears run down my face because I'm not anymore the person used to be." (I6, 51 years old)
Marital relationship
"... each day we could move forward, although we permanently felt this weight, we were closer to each other. Step by step, closer. After these years, after the diagnosis, and having realized that all this happened, we had also gone through a process of knowing each other. " (I3, 32 years old)
Sexual life
"My husband understands me... in the area of sexuality (FM) affected me a lot, but now it does not affect me so much... when I'm not able, I am not, but when I am, I express myself and my husband knows" (I10, 51 years old)
Family life
"I've always felt their support; with them I was able to put up with my pain. Last year when I was very ill they gave me a tremendous strength, really, and when I feel worse they support me and help me several times ... "(I17, 45 years old)
Professional life
"... when I am at work the first two hours go well, but then I start to get tired, sweating, having body aches all over; so I get very anxious and I feel that I can no longer work, but I have to work...I work four hours each day and I have to work, but sometimes I get angry. "(I14, 48 years old)
Social life
"In the past, I was able to find time for having fun, going to the movies, going for walks with friends, going to parties; now I'm not capable of it. I either do something or something else. "(I7, 46 years old)

Table 2. Subcategories of category "A changed life" reflecting the impact of FM in patients lives.

Concerning "disturbing symptoms", although pain was the most mentioned, other symptoms were also stated, such as cognitive disturbance, sleep disorders, and depressive symptoms. This result is in accordance to previous studies developed with rheumatic patients (e.g. Pimm, 1997) and reflects the range of symptoms that can be presented by FM patients (Blotman & Branco, 2007; Wilke, 2009; Wolfe et al., 2010). In terms of "daily activities", Söderberg and Lundman (2001) have shown that patients reported a decrease in level of activities (after the occurrence of FM) being associated to adjustment and emotional problems (Asbring, 2001; Söderberg & Lundman, 2001) which is supported by the current data. However, individuals in the present study also have shown ways of facing difficulties revealing adjustment to felt limitations. Regarding "marital relationship", as supported by previous studies (Asbring, 2001; Schmidt, 2008; Söderberg et al., 2003), FM can be associated with either the weakening or strengthening of the relationship. Our results have supported the role of social support, comprehension, trust, and conflict management as important variables in this process. Also in the context of the romantic relationship, the "sexual life" was considerably impacted by FM as supported by previous literature (Garcia-Campayo & Alda, 2004; Söderberg & Lundman, 2001; Söderberg et al., 2003). Our results have highlighted pain as one of the most disabling symptoms in sexual activity, affecting both sexual desire and satisfaction. Although the avoidance of sexual activity due to FM was found, it was also noted that communication with the partner and coping strategies directed to sexual dimension were crucial in the process of dealing with felt difficulties. Turning now to the category of "family life", it is worth noting that our results have shown that social support, comprehension, and trust tend to be magnified or decreased due to FM. Moreover, the redefinition of roles inside the family is common and the family rituals are modified in order to respond better to patients' needs. Söderberg and Lundman (2001), as well as Söderberg et al. (2003), have found similar results in this dimension. Concerning "professional life", our results have shown associations between FM and lows levels of productivity and difficulties in following rigid schedules. Moreover, when FM contributed to abandoning active life this situation was experienced as emotionally disturbing (Asbring, 2001; Liedberg & Henriksson, 2002). As stated by Liedberg and Henriksson (2002), and supported by our results, the matching of workplace conditions to individuals' needs can promote the maintenance of the posts for them. It was also found that the attitude of colleagues and employers can vary considerably from accepting to rejecting. Regarding "social life", results have shown a decrease of social contacts and a low willingness to have them. However, distrust and lack of understanding by friends was felt only in a small percentage of our sample; previous studies have shown a different pattern wherein unfavourable attitudes from friends were frequently reported (Söderberg & Lundman, 2001; Söderberg et al., 2003). Thus, our results suggest a previous higher quality relationship with friends in our sample.

Regarding "**Coping with fibromyalgia**", we have found a common adoption of "traditional therapeutic strategies" used in the treatment of FM (specifically, pharmacological treatment, psychotherapy, physiotherapy, and physical exercise) as well as "complementary/alternative therapeutic strategies" (namely, acupuncture, reiki, and shiatsu). Finally, it is also usual for patients to adopt "individual strategies of coping" (cognitive and behavioural) in order to handle the situation in their daily lives. Concerning traditional therapeutic strategies, pharmacological treatment is the most used by our participants with a perceived moderate degree of efficacy; moreover, complementary therapeutic strategies are also used (Cunningham & Jillings, 2006; Schaefer, 1997). In our sample, participants tend to use various strategies simultaneously, but each one use a specific combination pattern. This aspect gives support to the importance of a multidisciplinary and individualized treatment of FM (Blotman & Branco, 2007). However, monetary constraints were reported in our study and can be associated with failures to follow a specific treatment regimen. Concerning the "individual strategies of coping" (cognitive and behavioural) these are useful in order to achieve a higher level of autonomy in the process of dealing with FM (Cunningham & Jillings, 2006; Schaefer, 1997) and, in our sample, each of the participants have mentioned particular coping strategies they believe to be effective in the process of coping with FM.

The category "**Perspectives about a new reality**" included the following subcategories: "causal attributions", "perception of control", and "future perspectives". The "*causal attributions*" encompass the meanings built by participants about factors that may have contributed to the predisposition, triggering, and perpetuating of the syndrome. The process of meaning-making "causal attributions" is related to the human tendency to assess the world as predictable and controllable, namely in domain of health and illness (Ogden, 2004) and the absence of such attributions can be related to lower physical and psychosocial adjustment (Pimm, 1997). In our sample, all the participants formed some kind of attributions concerning predisposing, triggering, or perpetuating factors which support the relevance of giving a coherent meaning to their lives. Specifically, attributions acknowledging the psychosocial nature of predisposing factors of FM were commonly found which support previous studies (e.g. Hellström et al., 1999). That is, participants believe that adversity in their lives has contributed to their illness. Concerning triggering factors, the psychosocial attributions were also common, but in some cases there were no attributions made for the appearance of the first symptoms; physiological and mixed (physiological/psychosocial) were found as well, which partially support results found by Hellström et al. (1999). Regarding perpetuating factors, the predominance of psychosocial attributions was again found, followed by physical efforts, climate conditions and only one participant had not identified aggravating factors of FM. These results support those found by Schaeffer (1997), unlike those found by Hellström et al. (1999) who reported that FM patients considered their symptoms as unpredictable.

The category of "*perception of control*" represents participants' perceptions of their level of control over their illness. The narratives of participants have assumed various degrees of perceived control which can be placed in three mutually exclusive subcategories: "total perceived control" (7 participants), "intermediate perceived control" (7 participants), and "absence of perceived control" (6 participants) (table 3).

Joyce-Moniz and Barros (2005) stated that the individual position over a continuum of perceived control can have different consequences in terms of physical and emotional well-being. Concerning a perception of "total control", this is not always the most desirable representation, given that in some circumstances a perception of low control can be more adaptive, such as in those situations which are not objectively controllable (Ogden, 2004). Specifically with FM patients, the helplessness construct can be mentioned as conceptually close to "absence of perceived control" given that the former reflects the degree to which participants feel helpless in controlling pain and the course of FM (Palomino et al., 2007). In fact, helplessness, as the cognitive meaning of having FM, was found as playing a more central role in predicting depressive symptomatology than illness-related stressors, such as pain or disability (Palomino et al., 2007).

"PERCEPTION OF CONTROL"
Total perceived control
"I control my illness. Undoubtedly. I control it in all senses. When I feel that intense pain I try to control, to move forward. I can feel the worst pain but if someone knocks on my door I welcome that person with a smile. Nobody realizes." (I8, 67 years old)
Intermediate perceived control
"For me fibromyalgia is something that I have to work on in order to deal with felt difficulties; when I'm not able I try to find support and better days will come. I know I'll have bad days, but those are only that: bad days." (I3, 32 years old)
Absence of perceived control
"I want to make certain activities and I am manipulated. I have someone who controls me…is fibromyalgia. I consider it as a living being. In my professional life I wish to hurry up and I'm not capable. 'She' commands more than me." (I19, 35 years old)

Table 3. Perception of control over FM.

Concerning the "*future perspectives*", this subcategory includes feelings and thoughts that people have about what will happen to them in relation to FM and, as the previous subcategory, is common to find participants discourses ranging from a very positive perspective to a very negative one. More specifically, we considered the following subcategories: positive perspective (3 participants), mixed perspective (5 participants), absence of future perspective (5 participants) and negative perspective (7 participants). Our results partially support those found by Hellstrom et al. (1999) and Söderberg and Lundman (2001). The first ones have stated mainly the prevalence of absence of future perspectives and the second ones have highlighted the positive perspectives despite the difficulties that need to be faced. These differences can be related to distinct levels of acceptance of FM in the various samples.

3.3 Are the meanings of perceived control associated to patients' life stories and coping strategies, as well?

In order to respond to the research question: "Are the meanings of perceived control associated to patients' life stories and coping strategies, as well?" four matrixes of intersecting subcategories were developed. Firstly, participants were grouped in three mutually exclusive subcategories of "perceived control" over FM considering their narratives on this topic. An even distribution was found across the subcategories "total perceived control" (7 participants), "intermediate perceived control" (7 participants), and "absence of perceived control" (6 participants). This result is consistent with previous literature (Joyce-Moniz & Barros, 2005). Once subgroups were formed we were able to perform four intersection matrixes in order to respond our research question, namely: "Perceived control and Predisposing factors"; "Perceived control and Triggering factors"; "Perceived control and Perpetuating Factors"; and "Perceived control and Coping with fibromyalgia".

Considering the intersection matrix of "perceived control and predisposing factors" (Table 4), participants who experience absence of control tend to present the highest prevalence of predisposing factors in their life stories, compared to participants with intermediate control, although this difference is not very accentuated. However, individuals in the subgroup "total control" have shown, clearly, the lowest prevalence of predisposing factors in their life stories.

Subcategories	Absence of Control		Intermediate Control		Total Control	
	Part.	T.U.	Part.	T.U.	Part.	T.U.
Adverse life events	**5**	30	**5**	34	**5**	8
Significant losses	**3**	7	**5**	20	**5**	8
Personality traits	**3**	13	**4**	18	**2**	10
Physical and/or verbal violence	**4**	12	**2**	8	**3**	10
Family conflicts	**4**	7	**3**	10	**2**	9
Excessive striving for achievement at work	**3**	6	**3**	7	**1**	2
Emotional neglect	**5**	7	**2**	2	**0**	0
Nuclear family psychosocial problems	**3**	10	**3**	3	**1**	1
Job dissatisfaction	**2**	3	**2**	8	**2**	4
Early assumption of responsibility	**1**	1	**2**	5	**3**	11
Unsatisfying relationships	**3**	33	**3**	12	**0**	0
History of sexual abuse	**1**	3	**0**	0	**1**	4
Compensatory overactive lifestyles	**1**	5	**1**	1	**0**	0

N = 20. Part. (number of participants); T.U. (number of text units).

Table 4. Intersection matrix "Perceived Control and Predisposing Factors".

Regarding the identification of triggering factors, based on classification of perceived control, it must be noted that in only 12 of 20 participants (60%) those factors were identified; although the smaller number of individuals, the lowest prevalence of triggering factors were found in the subgroup of "perceived total control" (Table 5).

Subcategories	Absence of Control		Intermediate Control		Total Control	
	Part.	T.U.	Part.	T.U.	Part.	T.U.
Not identified	**2**	-	**2**	-	**4**	-
Extended period of physical and/or psychosocial stress	**1**	1	**3**	3	**3**	10
Traumatic physical experience	**1**	1	**3**	4	**0**	0
Painful injury	**1**	1	**2**	3	**0**	0
Traumatic emotional experience	**2**	3	**0**	0	**0**	0

N = 20. Part. (number of participants); T.U. (number of text units).

Table 5. Intersection matrix "Perceived Control and Triggering Factors".

In what concerns to perpetuating factors the results pattern was very similar to those obtained in predisposing factors. The participants with "total control" have shown the lowest prevalence of perpetuating factors in their life stories, followed by the "intermediate control" participants and, at last, the prevalence was the highest among participants with "absence of control" (Table 6).

Subcategories	Absence of Control		Intermediate Control		Total Control	
	Part.	T.U.	Part.	T.U.	Part.	T.U.
Family relationships	**3**	16	**3**	9	**3**	4
Sleep disturbances	**2**	5	**2**	2	**3**	7
Catastrophic thoughts	**4**	7	**1**	3	**1**	3
Depressive/anxious symptomatology	**3**	13	**2**	2	**1**	9
Social withdrawal	**2**	2	**3**	4	**1**	4
Low illness acceptance	**3**	3	**2**	5	**0**	0
Dysfunctional health care-seeking behaviour	**2**	2	**2**	5	**1**	7
Somatic vigilance	**1**	1	**2**	6	**0**	0
Physical/psychosocial stress	**1**	3	**1**	1	**1**	3

N = 20. Part. (number of participants); T.U. (number of text units).

Table 6. Intersection matrix "Perceived Control and Perpetuating Factors".

On regard to the intersection matrix of coping with fibromyalgia and perceived control, participants with "absence of control" were those who less use therapeutic strategies and individual coping. In the groups of "intermediate control" and "total control" a similar use was found between them and contrasting with the former group. However, in the group "total control" more references to therapeutic strategies (T.U.) were found, which can suggest a higher involvement with that kind of strategies (Table 7).

	Subcategories	Absence of Control		Intermediate Control		Total Control	
		Part.	T.U.	Part.	T.U.	Part.	T.U.
Traditional therapeutic strategies	Pharmacological treatment	**5**	10	**6**	6	**4**	7
	Psychotherapy	**1**	2	**5**	7	**2**	2
	Physiotherapy	**5**	7	**2**	3	**1**	1
	Physical exercise	**2**	2	**1**	1	**5**	13
	Complementary therapies	**2**	2	**2**	4	**4**	13
Individual coping	Cognitive coping	**1**	1	**6**	14	**5**	10
	Behavioural coping	**1**	5	**4**	10	**5**	12

N = 20. Part. (number of participants); T.U. (number of text units).

Table 7. Intersection matrix "Perceived Control and Coping with Fibromyalgia"

In general, results suggest that individuals with the lowest perceived control on FM tend to report life stories with a higher prevalence of predisposing, triggering, and perpetuating factors and lower use of coping strategies with FM (use of therapies and individual coping), as well. One of the possible explanations can be that life stories with a large number of adverse events could have contributed to a sense of victimization and helplessness, promoting the development of low perceived control meanings facing life experiences in general, and the experience of FM, in particular (Quartilho, 2001a). It must be noted that the association between lower perceived control and higher prevalence of perpetuating factors suggests that these participants are likely to experience worse levels of health-related quality of life. However, additional studies are needed in order to confirm this. The use of self-report measures could be useful in order to identify these aspects, such as the Rheumatology Attitudes Index (RAI) – 5-item Helplessness Subscale (DeVellis & Callahan, 1993; cit. in Palomino et al., 2007) which assesses helplessness in controlling pain and the course of FM, simultaneously with measures of FM severity (Wilke, 2009; Wolfe et al., 2010).

Finally, perceived control was associated to prevalence of strategies to cope with FM (therapeutic and individual coping). As stated by Ogden (2004), the belief of being able to control the illness tends to be associated to the maintenance of healthy behaviours and elimination of harmful ones.

4. Conclusions

The results have supported the categorization of three types of psychosocial factors proposed in literature as having a role in origin and maintenance of FM syndrome. Moreover, specific themes within each one of categories could be identified. As shown in patients' life stories the susceptibility to FM may be based in developmental factors which constitute predisposing factors to it. Moreover, FM symptoms often begin after a period of physical or psychosocial adversity. Finally, some factors may contribute to the perpetuation of symptoms and disability in FM. The clinical implications of these data refer to the importance of promoting changes on psychosocial perpetuating factors in order to improve health-related quality of life in these patients. Furthermore, the differentiation of FM patients in subgroups based on the relative weight of predisposing, triggering, and perpetuating factors could be useful in tailoring interventions adapted to particular needs of each subgroup. On the other hand, the meanings of patients living with FM shed light on their inner experiences which can lead health-care providers in the process of communicating more efficiently with these patients. We focus on the relevance of narrative therapy with FM patients in order to promote an integration of life stories with current experiences allowing a more complex cognitive and emotional perspective on their lives. Future research should focus on life stories in patients suffering from other chronic medical conditions, and even healthy individuals, in order to better understand similarities and differences with narratives of FM patients. On the whole, our results support the relevance of a biopsychosocial approach in assessment and intervention with FM patients, given the multifactorial genesis and maintenance of FM syndrome, which require an articulation of Psychology contributions with other health-care disciplines.

5. References

Aïni, K., Curelli-Chereau, A., & Antoine, P. (2010). L'expérience subjective de patients avec une fibromyalgie: analyse qualitative [The subjective experience of patients living with fibromyalgia: A qualitative analysis]. *Annales médico-psychologiques,* Vol.168, No.4, pp. 255-262.

Arnold, L. M., Crofford, L. J., Mease, P. J., Burgess, S. M., Palmer, S. C., Abetz, L., et al (2008). Patient perspectives on the impact of fibromyalgia. *Patient Education & Counseling,* Vol.73, No.1, pp. 114-120.

Asbring, P. (2001). Chronic illness – a disruption in life: Identity-transformation among women with chronic fatigue syndrome and fibromyalgia. *Journal of Advanced Nursing,* Vol.34, No.3, pp. 312-319.

Blotman, F., & Branco, J. (2007). *Fibromyalgia: Daily aches and pain,* Éditions Privat, Toulouse.

Cunningham, M., & Jillings, C. (2006). Individuals' descriptions of living with fibromyalgia. *Clinical Nursing Research,* Vol.15, No.4, pp. 258-273.

Eich, W., Hartmann M., Muller, A., & Fischer, H. (2000). The role of psychosocial factors in fibromyalgia syndrome. *Scandinavian Journal of Rheumatology,* Suppl. 113, pp. 30-31.

Garcia-Campayo, J., & Alda, M. (2004). La vivencia de la sexualidad en pacientes con fibromialgia: Un estudio cualitativo [Experience of sexuality in fibromyalgia patients: A qualitative study]. *Archivos de Psiquiatria,* Vol.67, No.3, pp. 157-168.

Gauer, J. S. (2009). *Fibromyalgia in the workplace: Exploring the impact of chronic pain upon white-collar workers.* Unpublished doctoral dissertation. Capella University.

Joyce-Moniz, L., & Barros, L. (2005). *Psicologia da doença para cuidados de saúde [Psycholgy of illness for health care practice],* Edições Asa, Porto.

Leventhal, H., Benyamini, Y., Brownlee, S., Diefenbach, M., Leventhal, E., Patrick-Miller, L. et al. (1997). Illness representations: Theoretical foundations. In: *Perceptions of health and illness: Current research and applications,* K. Petrie, & J. Weinman (Eds.), pp. 19-46, Harwood Academic Publishers, Amsterdam.

Leventhal, H., Meyer, D., & Nerenz, D. (1980). The common sense representation of illness danger. In: Medical Psychology: Vol.2, S. Rachman (Ed.), pp. 7-30, Pergamon: New York.

Liedberg, G. M., & Henriksson, C. M. (2002). Factors of importance for work disability in women with fibromyalgia: an interview study. *Arthritis and Rheumatism,* Vol.47, No.3, pp. 266-274.

Hagger, M., & Orbell, S. (2003). A meta-analytic review of the common-sense model of illness representations. *Psychology and Health,* Vol.18, No.2, pp. 141-184.

Hellström, O., Bullington, J., Karlsson, G., Lindqvist, P., & Mattsson, B. (1999). A phenomenological study of fibromyalgia. Patient perspectives. *Scandinavian Journal of Primary Health Care,* Vol.17, pp. 11-16.

Ogden, J. (2004). *Psicologia da Saúde [Health Psychology]* (2ª ed.), Climepsi Editores, Lisboa.

Oliveira, P. (2008). *Variáveis psicossociais e fibromialgia [Psychosocial variables and fibromaylgia].* Unpublished doctoral dissertation. Faculty of Psychology and Educational Sciences of University of Porto, Portugal.

Palomino, R., Nicassio, P., Greenberg, M., & Medina, E. (2007). Helplessness and loss as mediators between pain and depressive symptoms in fibromyalgia. *Pain,* Vol.129, pp. 185-194.

Pimm, T. (1997). Self regulation and psycho-educational interventions for rheumatic disease. In: *Perceptions of health and illness: Current research and applications,* K. Petrie, & J. Weinman (Eds.), pp. 349-378, Harwood Academic Publishers: Amsterdam.

Quartilho, M. (2001a). *Cultura, medicina e psiquiatria: Do sintoma à experiência [Culture, medicine, and psychiatry: From symptom to experience],* Quarteto Editora, Coimbra.

Quartilho, M. (2001b). Dor crónica: Aspectos psicológicos, sociais e culturais [Chronic pain: Psyhcological, social and cultural dimensions]. *Acta Reumatológica Portuguesa,* Vol.26, pp. 255-262.

Quartilho, M. (2004). Fibromialgia: Consenso e controvérsia [Fibromyalgia: Consensus and controversy]. *Acta Reumatológica Portuguesa,* Vol.29, pp. 111-129.

Schaefer, K. (1995). Struggling to maintain balance: A study of women living with fibromyalgia. *Journal of Advanced Nursing,* Vol.21, pp. 95-102.

Schaefer, K. (1997). Health patterns of women with fibromyalgia. *Journal of Advanced Nursing,* Vol.26, pp. 565-571.

Scharloo, M., & Kaptein, A. (1997). Measurement of illness perceptions in patients with chronic somatic illness: A review. In: *Perceptions of health and illness: Current research and applications,* K. Petrie, & J. Weinman (Eds.), pp. 103-154, Harwood Academic Publishers, Amsterdam.

Schmidt, K. M. (2008). *The couple as a pair bond and the lived experience of dyadic coping with fibromyalgia: A phenomenological study.* Unpublished doctoral dissertation. Capella University.

Sim, J., & Madden, S. (2008). Illness experience in fibromyalgia syndrome: A metasynthesis of qualitative studies. *Social Science & Medicine,* Vol.67, No.1, pp. 57-67.

Söderberg, S., & Lundman, B. (2001). Transitions experienced by women with fibromyalgia. *Health Care for Women International,* Vol.22, pp. 617-631.

Söderberg, S., Strand, M., Haapala, M., & Lundman, B. (2003). Living with a woman with fibromyalgia from de perspective of the husband. *Journal of Advanced Nursing,* Vol.42, No.2, pp. 143-150.

Strauss, A.L., & Corbin, J.M. (1998). *Basics of qualitative research: Techniques and procedures for developing grounded theory* (2nd ed.), Sage, Thousand Oaks, CA.

Undeland, M., & Malterud, K. (2007). The fibromyalgia diagnosis - hardly helpful for the patients? *Scandinavian Journal of Primary Health Care,* Vol.25, No.4, pp. 250-255.

Van Houdenhove, B., & Egle, U. (2004). Fibromyalgia: A stress disorder? Piecing the biopsychosocial puzzle together. *Psychotherapy and Psychosomatics,* Vol.73, No.5, pp. 267-275.

Van Houdenhove, B., Egle, U., & Luyten, P. (2005). The role of life stress in fibromyalgia. *Current Rheumatology Reports,* Vol.7, pp. 365–370.

Wilke, W. (2009). New developments in the diagnosis of fibromyalgia syndrome: Say goodbye to tender points? *Cleveland Clinic Journal of Medicine,* Vol.76, No.6, pp. 345-352.

Wolfe, F., Claw, D., Fitzcharles, M., Goldenberg, D., Katz, R., Mease, P. et al. (2010). The American College of Rheumatology preliminary diagnostic criteria for fibromyalgia and measurement of symptom severity. *Arthritis Care & Research,* Vol.62, No.5, pp. 600-610.

Wolfe, F., Smythe, H., Yunus, M., Bennett, R., Bombardier, C., Goldenberg, D. et al. (1990). The American College of Rheumatology 1990 criteria for the classification of fibromyalgia: Report of the multicenter criteria committee. *Arthritis and Rheumatism,* Vol.33, No.2, pp. 160–172.

Yardley, L. (1997). Introducing material-discursive approaches to health and illness, In: *Material discourses of health and illness,* L. Yardley (Ed.), pp. 1-24, Routledge, London.

5

The Role of Oxidative Stress and Mitochondrial Dysfunction in the Pathogenesis of Fibromyalgia

Mario D. Cordero[1], Manuel de Miguel[2] and José Antonio Sánchez Alcázar[1]
[1]*Centro Andaluz de Biología del Desarrollo (CABD-CSIC), Universidad Pablo de Olavide and Centro de Investigación Biomédica en Red de Enfermedades Raras (CIBERER), ISCIII, Sevilla,*
[2]*Departamento de Citología e Histología Normal y Patológica, Facultad de Medicina, Universidad de Sevilla, Sevilla, Spain*

1. Introduction

Fibromyalgia (FM) is a common pain syndrome accompanied by other symptoms such as tender spots, decreased pain threshold, fatigue, headache, sleep disturbances, and depression. It is a chronic condition characterized by a pattern of vague symptoms that are difficult to diagnose and treat. FM is diagnosed according to the classification criteria established by the American College of Rheumatology (ACR) (Wolfe et al., 1990) and routine laboratory investigations usually yield normal results (Yunus et al., 1981). The prevalence of FM in industrialized countries ranges from 0,4% to 4% (it affects at least 5 million individuals in the United States and 800.000 in Spain) in the population being 11 time more frequent in women than in men (Lawrence et el., 2008). Its high prevalence makes fibromyalgia a major problem in developed countries in the recent years. FM causes work absenteeism and has been associated with high medical services utilization cost and considerable disability. Furthermore, the use of medications and medical necessities increased markedly across many measures once diagnosis was made. It has been estimated that annual health service cost of FM patients was twice that of patients with chronic widespread pain and pain-free controls. The fact that its diagnostic criteria are only clinical, and that its etiopathogenesis has not yet been clarified makes very difficult the study and therapeutical approach of the disease. Although the etiology of FM remains unclear, evidence suggests that biologic, genetic, and environmental factors are involved.It is considered that the changes in the neuronal activity in the central nervous system, abnormal metabolism of biogenic amines, immunological disorders and oxidative stress may among others factors contribute to the development of the disease. For all these reasons is urgent to do more research in the diagnosis, pathophysiology and therapy of FM.

Fibromyalgia syndrome has been related to disturbances of hypothalamic–pituitary axis together with neurotransmission imbalance, involving excitatory amino acids, catecholamines, substance P and serotonin (5-HT) (Russell et al, 1994; Crofford et al, 1996; Neeck, 2002). Patient's symptoms may derive from poor stressor modulation, sensitization of specific nociceptor neurons and pain threshold diminution in response to multiple environmental factors, such as mechanical or emotional trauma, chronic stress or even infections. In recent years, new information to our understanding of FM pathophysiology has emerged. Some genetic polymorphisms and antibodies have been associated with FM, as the serotoninergic system genotype of 5–HTT (Bazzichi et al., 2006a; Tander et al., 2008), catechol-O-methytransferase gene polymorphism (Gursoy et al., 2004), D4 dopamine receptor exon II repeat polymorphism (Buskila et al., 2004), and antibodies against serotonin (Klein et al., 1992; Werle et al., 2001). It has also been postulated alterations in the metabolism, transport and reuptake of serotonin (Alnigenis & Barland, 2001; Schwarz et al., 2002) and substance P (Staud & Spaeth, 2008). Moreover, cytokines homeostasis has been considered to play a role in the pathogenesis of FM (Wallace, 2001; Wallace et al., 2006). Conversely, several studies have shown mitochondrial dysfunction and high levels of oxidative stress markers in FM patients, suggesting that this process may contribute to the pathophysiology of this disease. However, whether oxidative stress is the cause or the effect in FM is controversial (Ozgocmen et al, 2006).

2. Mitochondrial dysfunction in disease and FM

2.1 About mitochondria

Mitochondria are dynamic organelles that play a central role in many cellular functions including the generation of chemical energy (adenosine triphosphate, ATP), heat, and intracellular calcium homeostasis. They are also responsible for the formation of reactive oxygen species (ROS) and for triggering the programmed cell death or apoptosis (Turrens, 2003). The primary metabolic function of mitochondria is oxidative phosphorylation, an energy-generating process that couples oxidation of respiratory substrates to the synthesis of ATP (Pieczenik & Neustadt, 2007). The mitochondrial respiratory chain (MRC) is composed of five multisubunit enzyme complexes. Both the mitochondrial DNA (mtDNA) and the nuclear DNA (nDNA) encode for polypeptide components of these complexes. Electron transport between MRC complexes I–IV is coupled to the extrusion of protons across the inner mitochondrial membrane by proton pump components of the respiratory chain. This movement of protons creates an electrochemical gradient ($\Delta\psi m$) across the inner mitochondrial membrane. Protons return to the mitochondrial matrix by flowing through ATP synthase (complex V), which utilizes the energy thus produced to synthesize ATP from adenosine diphosphate (ADP) and inorganic phosphate (Pi). Both the mtDNA and the nDNA encode for polypeptide components of these complexes. As a consequence, mutations in either genome can cause MRC dysfunction that impairs transport of electrons and/or protons and decreases ATP synthesis. Primary or secondary genetic diseases affecting MRC or secondary mitochondrial dysfunctions usually affect brain and skeletal muscle because of their energy requirements. Besides MRC enzyme complexes, two electron carriers, coenzyme Q_{10} (CoQ) and cytochrome c, are essential for mitochondrial synthesis of ATP. CoQ transports electrons from complexes I and II to complex III and is essential for the stability of complex III. CoQ is a lipid-soluble component of virtually all cell membranes. It is composed of a benzoquinone ring with a polyprenyl side-chain. The number of isoprene units is specie specific, e.g. 10 in humans (CoQ_{10}). CoQ also functions as an antioxidant that protects cells both by direct ROS scavenging and by regenerating other antioxidants such as vitamins C and E (Turunen et al., 2004). CoQ deficiency impairs oxidative phosphorylation and causes clinically heterogeneous mitochondrial diseases named CoQ deficiency syndrome. An increasing number of patients with primary inherited CoQ deficiencies are being identified (Littarru & Tiano, 2010). These forms are transmitted as autosomal recessive traits and respond to CoQ supplementation, making accurate diagnosis of great practical importance. CoQ deficiency can be also a secondary consequence of different diseases or by treatment with drugs such as statins. Given the critical role of CoQ in mitochondria function, it has been suggested that CoQ levels could be a useful biological marker of mitochondrial function (Haas et al., 2008). CoQ deficiency induces decreased mitochondrial respiratory enzymes activity, reduced expression of mitochondrial proteins involved in oxidative phosphorylation, decreased mitochondrial membrane potential, increased production of ROS, mitochondrial permeabilization, mitophagy of dysfunctional mitochondria, reduced growth rates and cell death (Quinzii et al., 2008; Rodriguez-Hernandez et al., 2009, Cotan et al, 2011).

2.1.1 Reactive Oxygen Species (ROS)

In addition to energy, mitochondrial oxidative phosphorylation also generates ROS. When the MRC becomes highly reduced, the excess electrons from complex I or complex III can be passed directly to O_2 to generate superoxide anion (O_2^-). Superoxide is transformed to hydrogen peroxide (H_2O_2) by the detoxification enzymes manganese superoxide dismutase (MnSOD) or copper/zinc superoxide dismutase (Cu/Zn SOD), and then to water by catalase, glutathione peroxidase (GPX) or peroxidredoxin III (PRX III). However, when these enzymes cannot convert ROS such as the superoxide radical to water fast enough, oxidative damage occurs and accumulates in the mitochondria. If H_2O_2 encounters a reduced transition metal or is mixed with O_2^-, the H_2O_2 can be further reduced to hydroxyl radical ($OH^•$), the most potent oxidizing agent among ROS. Additionally, nitric oxide (NO) is produced within the mitochondria by mitochondrial nitric oxide synthase (mtNOS) and also freely diffuses into the mitochondria from the cytosol. NO reacts with O_2^- to produce peroxynitrite ($ONOO^-$). Together, these two radicals as well as others can do great damage to mitochondria and other cellular components (Turrens, 2003).

Under normal physiological conditions, ROS production is highly regulated. However, if the respiratory chain is inhibited, or key mitochondrial components, such as CoQ, are deficient, then, electrons accumulate on the MRC carriers, greatly increasing the rate of a single electron being transferred to O_2 to generate O_2^-. An excessive mitochondrial ROS production can exceed the cellular antioxidant defense and the cumulative damage can ultimately destroy the cell by necrosis or apoptosis.

2.1.2 Selective degradation of mitochondria: Mitophagy

Degradation of excess or dysfunctional organelles is one of the major problems that eukaryotes face to maintain cell integrity and to adapt cellular activities to environmental changes. To solve this fundamental issue, cells utilize autophagy, which is a self-eating system that generates double-membrane vesicles called autophagosomes, sequesters cytoplasmic components as cargoes, and transports them to lysosomes for degradation (Klionsky, 2005). In the past decade, more than 30 autophagy-related genes (ATG) required for selective and/or nonselective autophagic functions have been identified. Selective autophagy contributes to the

control of both quality and quantity of organelles. It is conceivable that mitochondria are the primary targets of selective autophagy, because they accumulate oxidative damage due to their own by-products, ROS (Bhatia-Kiššová I & Camougrand, 2010).
Consistent with this idea, autophagy-dependent clearance of dysfunctional mitochondria is important for organelle quality control. The term mitophagy refers to the selective removal of mitochondria by autophagy. It has been proposed that ROS damage can induce mitochondria permeabilization by the opening of permeability transition pores in the mitochondrial inner membrane (Kim, 2007). This, in turn, leads to a simultaneous collapse of mitochondrial membrane potential and the elimination of dysfunctional mitochondria. Consistent with this idea, autophagy-dependent clearance of dysfunctional mitochondria by mitophagy is important for organelle quality control and can play a pivotal role in mitochondria related diseases (Gottlieb et al, 2010).

2.2 Mitochondrial dysfunction and disease

Since the first mitochondrial dysfunction was described in 1962 (Luft et al, 1962), biomedicine research has advanced in the understanding of the role that mitochondria play in health, disease, and aging. Besides the inherited mitochondrial diseases, a wide range of seemingly unrelated disorders, such as schizophrenia, bipolar disease, dementia, Alzheimer's disease, epilepsy, migraine headaches, strokes, neuropathic pain, Parkinson's disease, ataxia, transient ischemic attack, cardiomyopathy, coronary artery disease, chronic fatigue syndrome, retinitis pigmentosa, diabetes, hepatitis C, primary biliary cirrhosis, and fibromyalgia (FM), have the common pathophysiological mechanisms of production of mitochondrial ROS resulting in mitochondrial dysfunction (Pieczenik & Neustadt, 2007). As a consequence of these findings, antioxidant therapies hold promise of improving mitochondrial performance in these diseases. Although the underlying characteristic of all of them is lack of adequate energy to meet cellular needs, they vary considerably from disease to disease and from case to case in their effects on different organ systems, age at onset, and rate of progression, even within families whose members have identical genetic mutations. No symptom is pathognomonic, and no single organ system is universally affected. Although a few syndromes are well-described, any combination of organ dysfunctions may occur. However, these diseases most often affect the central and peripheral nervous systems, but can affect any organs or tissues, including the muscles, liver, kidneys, heart, ears, eyes, and endocrine system (Cohen & Gold, 2001) (Table 1).
Damage to mitochondria is caused primarily by ROS generated by the mitochondria themselves. Within the mitochondria, components that are particularly vulnerable to free radicals include lipids, proteins, oxidative phosphorylation enzymes, and mtDNA (Shigenaga et al., 1994; Tanaka et al., 1996). Direct damage to mitochondrial proteins decreases their affinity for substrates or coenzymes and, thereby, decreases their function (Liu et al., 2002). Compounding the problem, once a mitochondrion is damaged, mitochondrial function can be further compromised by increasing the cellular requirements for energy repair processes (Aw & Jones, 1989). Mitochondrial dysfunction can also result in a feed forward process, whereby mitochondrial damage causes additional damage. Generated ROS can be released into cytosol and trigger "ROS-induced ROS-release" (RIRR) in neighboring mitochondria. This mitochondrion-to-mitochondrion ROS-signaling constitutes a positive feedback mechanism for enhanced ROS production potentially leading to significant mitochondrial injury (Zorov et al, 2006).

Organs	Sign and Symptoms
Brain	Developmental delay, mental retardation, autism, dementia, seizures, neuropsychiatric disturbances, atypical cerebral palsy, atypical migraines, stroke, and stroke-like events
Ears	Sensorineural hearing loss, aminoglycoside sensitivity
Eyes	Optic neuropathy and retinitis pigmentosa
Heart	Cardiac conduction defects (heart blocks), cardiomyopathy
Kidneys	Proximal renal tubular dysfunction (Fanconi syndrome); possible loss of protein (amino acids), magnesium, phosphorus, calcium, and other electrolytes
Liver	Hypoglycemia, gluconeogenic defects, nonalcoholic liver failure
Muscles	Hypotonia, weakness, cramping, muscle pain, ptosis, opthalmoplegia
Nerves	Neuropathic pain and weakness (which may be intermittent), acute and chronic inflammatory demyelinating polyneuropathy, absent deep tendon reflexes, neuropathic gastrointestinal problems (gastroesophageal reflux, constipation, bowel pseudoobstruction), fainting, absent or excessive sweating, aberrant temperature regulation
Pancreas	Diabetes and exocrine pancreatic failure
Systemic	Failure to gain weight, short stature, fatigue, and respiratory problems including intermittent air hunger

Table 1. Signs and symptoms associated with mitochondrial dysfunction (Cohen & Gold, 2001).

Furthermore, ROS have an established role in inflammation. Increased levels of Inflammatory mediators such as tumor necrosis factor alpha (TNF-α) and interleukins, have been associated with mitochondrial dysfunction and increased ROS generation (Naik & Dixit, 2011), and it has been hypothesized that abnormal production of cytokines may play a role in the pathogenesis of FM (Wallace et al, 2001). IL-1, IL-6 and IL-8 are dysregulated in the syndrome and therapies directed against these cytokines may be of potential importance in the management of fibromyalgia (Wallace, 2006). However, different studies with conflicting results (Uçeyler et al, 2006; Bazzichi et al, 2007) make necessary more studies to better understand the role of cytokines in FM.

2.3 Mitochondrial dysfunction in FM

Because the main symptoms in FM (pain, stiffness and fatigue) are located in the muscles, muscle biopsies, mostly from the trapezious, have been studied. In most cases, mitochondrial morphologic alterations have been found in muscle biopsies from FM patients. Histochemical analysis demonstrated type II fiber atrophy and the "moth-eaten" appearance of type I fibers. Electron microscopic findings were most impressive, and included, subsarcolemmal mitochondrial accumulation (Kalyan Raman et al., 1984), myofibrillarlysis with deposition of glycogen and abnormal mitochondria (Yunus et al., 1986), low number of mitochondria (Sprott et al., 2004), electrons-dense inclusions and lack of inner membrane (Hénriksson et al., 1982), ragged red fibres (Bengtsson et al., 1988), and single fiber defects of cytochrome-c-oxidase, the complex IV of oxidative phosphorilation (Drewes et al., 1993, Pöngratz & Späth, 1998). The presence of moth-eaten and ragged-red fibres indicates uneven distribution and proliferation of mitochondria. Accumulation of mitochondria is seen in Gomori trichrome staining, and this gives the ragged appearance.

Mitochondrial proliferation may be a compensatory phenomenon in disorders or pathophysiological states affecting oxidative metabolism (Bengtsson, 2002). It is interesting to mention that ragged red fibres, subsarcolemmal mitochondrial accumulation and alteration in ultrastructure, number and size of mitochondrial are typical defects and markers found in genuine mitochondrial diseases (MELAS, MERRF, Kearns-Sayre syndrome, Pearson syndrome, Leigh syndrome, etc) (Haas et al., 2008).

However, red ragged fibers appear to be also related to insufficient blood supply (Heffner & Barron, 1978), and abnormal capillary microcirculation in tender points was found in FM patients (Lund et al, 1986). Microcirculation in the muscle is controlled by the sympathetic nervous system and others local and humoral factors. Therefore, the contribution of the vasoconstrictor activity of the sympathetic nervous system that produces local hypoxia in muscle and fiber damage should be considered in the pathogenesis of red ragged fibers in FM.

^{31}P Magnetic Resonance Spectroscopy (MRS) analysis in muscle has provided objective evidence for metabolic abnormalities consistent with clinical symptoms of weakness and fatigue in patients with FM (Park et al., 1998). The MRS examinations showed phosphocreatine (PCr) and ATP concentrations in muscles of FM patients to be 15% below normal values during rest and exercise. The reduced levels of PCr and ATP in the patients' muscles correlate with clinical observations regarding weakness and pain during exercise. In this study, pain was inversely correlated with ATP and PCr levels. Reduction in ATP levels also has been observed in the erythrocytes of FM patients, suggesting that this may be a more systemic phenomenon than was previously assumed (Russell et al, 1993).

Recently, it has also been noted a decrease of ATP levels in platelets from FM patients (Bazzichi et al., 2008). Blood platelets represent an easily available and simple peripheral model to study bioenergetics alterations in FM. Platelets possess mitochondria, the entire pool of enzymes or proteins involved in oxidative energy production and, consequently, they are active in ATP turnover (Niu et al, 1996). Moreover, platelets present on their plasma membrane either pain/inflammation or neurochemical sites, such as adenosine/monoamine receptors and transporters (Marazziti et al, 1999; Martini et al, 2004), enabling the study of either neurochemistry or ATP energy metabolism in FM. Some recent have shown a significant increase of the platelet peripheral benzodiazepine receptor (PBR) (now named traslocator protein, TSPO, (Papadopoulos et al, 2006) together with high plasma levels of the pro-inflammatory chemokine interleukine 8 (IL-8), low serum cortisol (Bazzichi et al, 2006a) as well as a reduced density and functionality of the platelet serotonin transporter (SERT) (Bazzichi et al., 2006b) in FM patients.

3. Mitochondrial dysfunction, oxidative stress in FM

In general, oxidative stress could be defined as an imbalance between the presence of high levels of ROS and reactive nitrogen species (RNS), and the antioxidative defense mechanisms. These toxic molecules are formed via oxidation-reduction reactions and are highly reactive since they have an odd number of electrons. ROS generated under physiological conditions are essential for life, as they are involved in bactericidal activity of phagocytes, and in signal transduction pathways, regulating cell growth and reduction-oxidation (redox) status (Davies, 1995). ROS includes free radicals, such as hydroxyl and superoxide radicals, and non-radicals, including hydrogen peroxide and singlet oxygen. Oxidative stress and generation of free radicals, as primary or secondary event, have been related in a great number of diseases. It has been suggested that oxidative stress is linked to both the initiation and the progression of Parkinson´s disease (Zhou et al., 2008), and strong evidence exists for early oxidative stress in Huntington´s disease (Stack et al., 2008). Moreover, numerous studies demonstrate that different biomarkers of oxidative-stress-mediated events are elevated in the Alzheimer disease (Praticò, 2008) and renal disease (even in early chronic kidney disease) (Cachofeiro et al., 2008). Moreover, oxidative stress is believed to aggravate the symptoms of many diseases, including hemolytic anemias (Fibach & Rachmilewitz, 2008), amyotrophic lateral sclerosis (Orrell et al., 2008), and metabolic syndrome (Whaley-Connell et al., 2011).

It is known that ROS overproduction induces lipid peroxidation (LP) leading to oxidative destruction of polyunsaturated fatty acids, the main structural component of cellular membranes, and the production of toxic and reactive aldehyde metabolites such as malondialdehyde (MDA) and 4-hydroxynonenal (HNE) (Draper et al., 2000; Esterbauer et al., 1991). These highly cytotoxic metabolites, produced in relatively large amounts, can diffuse from their site of origin to attack distant targets and form covalent bonds with various molecules. Therefore, recognition of LP is of interest, as the deleterious effects of this process might be prevented by administration of scavenging systems or antioxidants.

In recent years, several studies have shown increased level of oxidative stress markers in FM suggesting that this process may have a role in the pathophysiology of this disease (Table 2). High levels of LP and protein carbonyls are two of the most documented oxidative damage markers to be associated with FM.

Author(s)	N(FM/Ctl)	Sample(s)	Parameter
Akkus et al., 2009	30/30	Plasma	Lipid peroxidation (LP)/Vitamins A,C,E/Beta-Carotene
Altindag et al., 2006	20/20	Plasma	Total Antioxidant Status (TAS)
Altindag et al., 2007	42/53	Serum	lipid hydroperoxide (LOOH)/TAS/ free sulfhydryl groups
Bagis et al., 2005	85/80	Serum	LP/Superoxide Dismutase (SOD)
Cordero et al., 2009	40/25	Plasma/BMCs	LP/ROS/Protein carbonyls/CoQ_{10}
Cordero et al., 2010	20/10	Plasma/BMCs	LP/Superoxide Anion/CoQ_{10}
Cordero et al., 2010	2/2	Skin biosies	LP/CoQ_{10}
Chung et al., 2009	48/96	Urine	F2 isoprostanes
Hein et al., 2002	41/46	Serum	Pentosidine
Kaufmann et al., 2008	22/22	Neutrophils	Hydrogen peroxide
Nazıroğlu et al., 2010	32/30	Plasma/ Erythrocytes	LP/Glutathione peroxidase/Vitamins A and E
Ozgocmen et al., 2006	30/16	Serum	LP/ SOD/Xanthine oxidase
Sendur et al., 2009	37/37	Serum	Catalase/Glutathione

Table 2. References about oxidative stress in Fibromyalgia.

Thus, high levels of MDA, a final product of LP, and increased level of protein carbonyls, as result of protein oxidation have been reported in plasma from FM patients. Furthermore, it has been observed that total antioxidant capacity and superoxide dismutase (SOD), catalase and glutathione levels are reduced in FM patients.

Furthermore, increased LP has been described in patients suffering from depression and fatigue, two typical symptoms found in FM patients (Bilici et al., 2001; Vecchiet et al., 2003). Studies on depression have signaled a possible link between depression and LP (Evans, 2003), and the peroxidation-reducing effect of different selective serotonin reuptake inhibitors in major depression has been demonstrated (Bilici et al., 2001). It has been suggested that alterations in phospholipids which are structural components of cell membrane in the brain, may induce changes in membrane microviscosity and, consequently, in various neurotransmitter systems, which are thought be related to the pathology of depression, e.g., serotonin (5-HT), and noradrenaline (Maes et al., 1996; Tsutsumi et al., 1988). LP of cell membranes can modify receptor accessibility, dynamics, ligand binding and action, and therefore altering neurotransmitter functions (Lenaz, 1987). Oxidative stress may also affect the expression of membrane functional proteins and receptors, by interfering with intracellular signalling and receptors turnover, including serotonergic receptors (Maes et al., 2007).

The role of oxidative stress in peripheral neuropathic pain, one of the most prominent symptoms in FM, was recently also tested by assessing the effects of antioxidants (acetyl-L-carnitine, alpha-lipoic acid, and vitamin C) on pain behaviour in a rat model of neuropathic pain induced by the antineoplastic agent oxaliplatin (Joseph et al., 2008). Each agent, administered locally at the site of mechanical nociceptive testing in the skin, markedly inhibited the oxaliplatin-induced hyperalgesia. Finally, ROS also appear to contribute to hyperalgesia induced by PKCε (Joseph et al., 2010), potentially linking this mitochondrial pathway to the hyperalgesic priming model of chronic peripheral pain. Furthermore, ROS are known to be implicated in the etiology of pain by inducing peripheral and central hyperalgesia (Wang et al., 2004). Superoxide plays a major role in the development of pain through direct peripheral sensitization, the release of various cytokines (for example, TNF-α, IL-1β, and IL-6), the formation of peroxynitrite (ONOO-), and PARP activation (Wang et al., 2004). Although, the mechanisms by which increased oxidative stress can affect specifically muscle sensitivity remain to be established, it may be that oxidative damage in muscles results in lowering the threshold of nociceptors locally, thus producing and altered nociception (Fulle et al., 2000).

In recent works, we have examined mitochondrial bioenergetics and antioxidant defenses in Blood mononuclears cells (BMCs) from FM patients. As CoQ deficiency had been suggested to be useful as a mitochondrial dysfunction marker, we addressed mitochondrial dysfunction in BMCs from FM patients and examined whether mitochondrial disturbance could be involved in the pathophysiology of oxidative stress in FM. We analyzed CoQ levels in BMCs and plasma from FM patients. We observed an altered distribution of CoQ levels between plasma and cells (high levels in plasma and low levels in cells) (Cordero et al., 2009). We found that CoQ-deficient BMCs in FM patients showed high level of mitochondrial ROS production and increased levels of LP. To confirm oxidative stress in FM, BMCs of one representative patient were treated with three antioxidants (CoQ, vitamin E, and N-acetilcysteine), and mitochondrial ROS production was examined. Both CoQ and vitamin E, two well known lipophilic antioxidants, induced a significantly reduction of ROS

(Figure 2). These results suggest that ROS were produced in the lipophilic environment of mitochondrial membranes, and that CoQ deficiency may be involved in the oxidative stress observed in FM. In addition, biochemical analysis of citrate synthase indicated a depletion of mitochondrial mass, suggesting selective mitochondrial degradation in BMCs from FM patients. These results were confirmed by electron microscopy that clearly showed autophagosomes where mitochondria are being degraded (Figure 1). Mitophagy can be beneficial for the cells by eliminating dysfunctional mitochondria, but excessive mitophagy can promote cell injury and may contribute to the pathophysiology of FM (Cordero et al., 2010a).

Control Patient

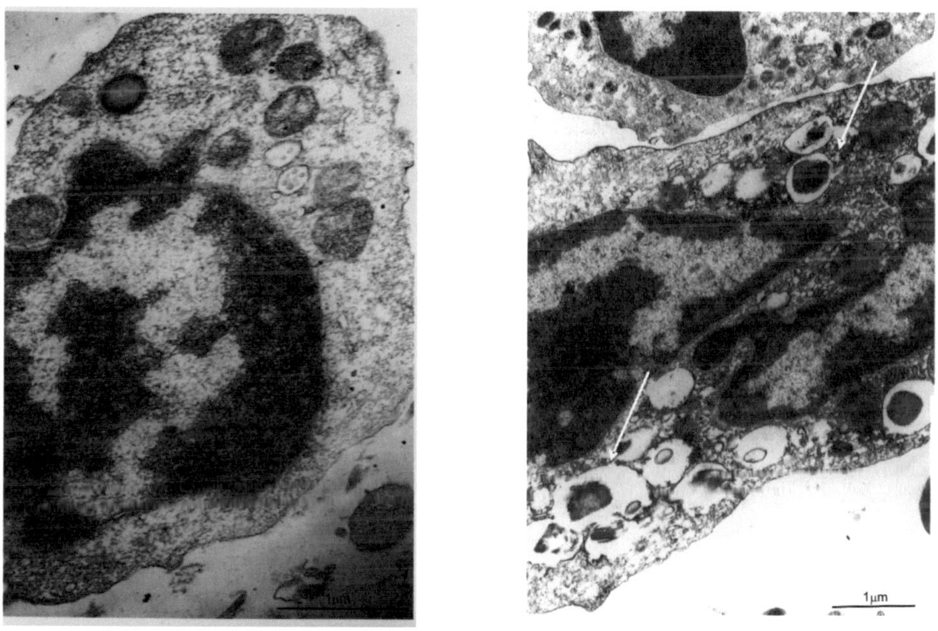

Fig. 1. Ultrastructure of BMCs from FM patients. Control BMCs is showing mitochondria with a typical ultrastructure. Autophagosome with mitochondria (arrows) were present in BMCs from a representative FM patient; Bar =1 μm. (Cordero et al., 2010a).

In general, there is poor correlation between plasma and tissues levels of CoQ, and even patients with genetically proven CoQ deficiency may have plasma CoQ levels at a normal range. However, there is a positive correlation between the content of CoQ in skeletal muscle, dermal fibroblasts and BMCs (Land et al., 2007; Duncan et al., 2005). Furthermore, mitochondrial dysfunction and oxidative stress has also been observed in skin biopsies from FM patients. The biopsies showed CoQ deficiency, increased level of LP, and a decrease in complex II + III and complex IV (Cordero et al., 2010b). Interestingly, it is known that CoQ deficiency induces decreased activities of complex II + III, complex III and complex IV (Quinzii et al., 2008). Furthermore, fibroblasts from skin of some patients with CoQ deficiency syndrome show a higher production of ROS in mitochondria (Quinzii et al., 2008). Therefore, skin fibroblasts and BMCs can be helpful for biochemical diagnosis of mitochondrial defects in FM patients. As is shown in Table 2, most of the studies about oxidative stress markers (MDA levels) in FM have used plasma or serum as experimental sample. However, it has to be taken into account that plasma or serum MDA levels depend on the balance between MDA formation and its detoxification and can be affected for many factors, such as the dilutional effect of plasma, and the renal and/or tissue clearance. Therefore, it would be desirable to measure oxidative stress markers in blood cells rather than in plasma o serum. On this point, it has recently been reported increased levels of hydrogen peroxide in neutrophils from FM patients (Kaufmann et al., 2008). Interesting, there are some discrepancies about the correlation between symptoms and LP and oxidative stress in FM. Significant correlation has been observed between antioxidants levels in plasma and serum, visual analogue scale (VAS) of pain, and morning stiffness (Altindag & Celik, 2007; Sendur et al., 2009). However, Bagis et al. found no correlation between VAS of pain and LP or SOD in serum (Bagis et al., 2005). On the other hand, Ozgocmen et al. found a significant correlation between depression and LP in serum but not between the biochemical parameters and clinical measures of pain and fatigue (Ozgocmen et al., 2006). This controversy could be ascribed to a methodological problem because LP levels may show higher levels and reflect better the degree of oxidative stress if LP measurement is performed in cells rather than in plasma or serum. In fact, we have selected 65 FM patients and evaluated the association between LP and clinical symptoms (Table 3).

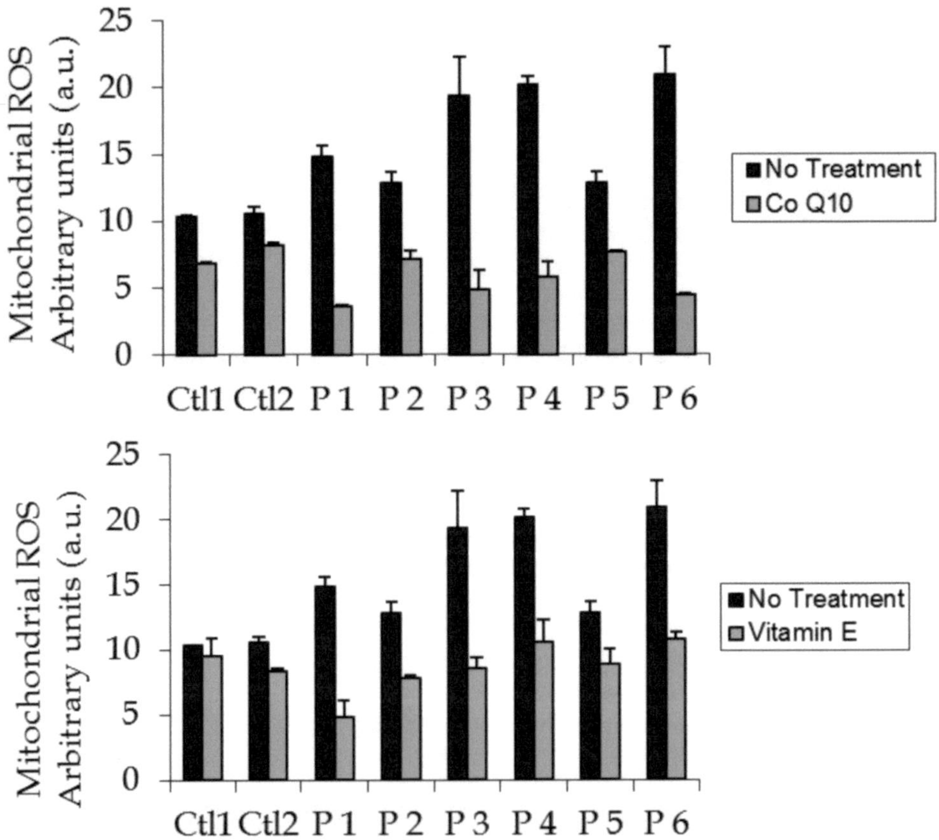

Fig. 2. Effect of CoQ_{10} (up) and Vitamin E (down) about mitochondrial ROS production in BMCs from FM patients compared to healthy patients.

	Patients	Control
Age (years)	45.8±11	44.3±12
Tender points	15.4±2.8	?
Duration of disease (years)	10.4±6.4	?
Sex (males/female)	5/60	5/40
BMI kg/m²	27.4±4.2	23.31±09
VAS Total score	5.9±1.8*	0.5±0.8
FIQ Total score, range 0 - 80	54.5±16*	3±1.6
Pain	7.3±2.2*	0.7±0.3
Fatigue	7.6±1.9*	1.2±0.9
Morning tiredness	6.7±2.2*	1.1±1.0
Stiffnes	5.9±2.3*	0.6±0.1
Anxiety	5.8±2.7*	1±0.9
Depression	5.2±2.7*	1.2±0.8
Beck Depression Inventory	18.5±8.6*	4±1.9

Table 3. Characteristic findings of the FM patients and control group. Values are means ±SD. *P<0.001. (Unpublished data).

We have observed significant correlation between LP levels BMCs or plasma and clinical parameters. However, LP levels in BMCs are better associated than LP levels in plasma to clinical symptoms in FM (Table 4), (**Unpublished data**).

	LP in cells r	**LP in plasma** r
VAS	0.584**	0.452**
FIQ total score	0.823**	0.578**
Pain	0.564**	0.410**
Fatigue	0.617**	0.311*
Morning tiredness	0.574**	0.397**
Stiffness	0.669**	0.402**
Anxiety	0.591**	0.433**
Depression	0.632**	0.561**
Beck Depression Inventory	0.875**	0.579**

Table 4. Correlation between LP and clinical findings in FM patients. *P<0.05; **P<0.01.r, Pearson's Correlation Coefficient (Unpublished data).

Alternatively, as mentioned above, it has been observed low levels of SOD in plasma of FM patients (Bagis et al., 2005, Ozgocmen et al., 2006). SOD is an enzyme presents in all cells that catalyzes the conversion of superoxide free radicals to oxygen and hydrogen peroxide. It is the most powerful antioxidant produced by the body and has been used as a marker of antioxidant defense (Marklund SL (1990). Therefore, SOD deficiency in FM may account for a low protection of cells against ROS damage and support the hypothesis of FM as an oxidative disorder. Besides mitochondrial dysfunction many others objective and measurable biomarkers has been proposed that may facilitate diagnosis and monitor the activity of FM (Dadabhoy et al, 2008).

4. Therapeutic implication of mitochondrial dysfunction in fibromyalgia

An important problem in FM is the moderate effectiveness of pharmacological therapies. In general, about half of all treated patients seem to experience a 30% reduction of symptoms, suggesting that many patients with fibromyalgia will require additional therapies (Staud, 2010). But, in most cases, high incidence of secondary effects is induced by pharmacological therapy, and many drugs may induce mitochondrial damage (Neustadt & Pieczenik, 2008). Mitochondria can be damaged both directly and indirectly by medications. Medications can directly inhibit mtDNA transcription of MRC complexes, damage through other mechanisms MRC components, and inhibit enzymes required for any of the steps of glycolysis and β-oxidation. Indirectly, medications may damage mitochondrial via the production of free radicals, by decreasing endogenous antioxidants such as glutathione and by depleting the body of nutrients required for the creation or proper function of mitochondrial enzymes or MRC complexes. Damage to mitochondria may explain the side effects of many medications. Therefore, treatment with these drugs in patients with mitochondrial dysfunction could be counterproductive. In this respect, amitriptyline is a tricyclic antidepressant frequently recommended in FM treatment. There are evidences to support the short-term efficacy of amitriptilyne 25mg/day in FM (Nishishinya et al, 2008). However, there is no evidence to support the efficacy of amitriptyline at higher doses or for periods longer than 8 weeks. Furthermore, amitriptyline causes a dose-related cytotoxic effect in neurons beginning at clinically relevant concentrations by a mechanism dependent on mitochondrial depolarization (Lirk et al, 2006). Recently, our group has reported that amitriptyline at high doses induced CoQ deficiency and mitochondrial dysfunction in an *in vitro* cellular assay (Cordero et al., 2009). Moreover, CoQ and alpha-tocopherol supplementation prevented the cellular damage induced by amitriptyline. This data suggests that high dose or long-period of amitriptyline treatment should be done with caution monitoring mitochondrial function and, in case of CoQ deficiency, supplementing with CoQ could prevent its adverse effects.

However, since oxidative stress may arise as a consequence of mitochondrial dysfunction or interfere with mitochondrial function, reducing oxidative stress emerges as a form of mitochondrial medicine that could be beneficial in FM patients. Thus, CoQ, frequently used in mitochondrial disease treatment to boost mitochondria function and prevent ROS damage (Tiano et al., 2011; Stack et al., 2008; Zhou et al., 2008; Quinzii & Hirano, 2010), might be an alternative treatment in FM.

Beneficial effects of CoQ administration in FM patients have been observed in a previous pilot study (Lister, 2002). In this study, Lister et al reported beneficial effects of oral CoQ and Gingko Biloba supplementation in FM patients. They observed an important improvement in quality-of-life scores that justified the need for a larger scale clinical trial and further investigations into the possible mechanism of action of CoQ. In our studies, oral CoQ treatment significantly improved clinical symptoms and decrease oxidative stress in several cases of FM (Cordero et al., 2011a; Cordero e al., 2011b). Nevertheless, more controlled clinical trials are needed to provide data on effectiveness of CoQ in FM.

CoQ is a potential drug candidate in the treatment of FM for at least two main reasons. First, CoQ is a mitochondrial cofactor with the potential to improve mitochondrial function. Second, CoQ is a powerful free radical scavenger that can mitigate LP and DNA damage caused by oxidative stress (Lenaz G et al, 2007). Thus, CoQ supplementation has been proven to be beneficial in patient with muscle pain associated with statin treatment (Caso et al., 2007), and migraine prophylaxis (Sandor et al., 2005). Furthermore, CoQ has shown anti-inflammatory and anti-nociceptive activity (Jung et al., 2009), regulating inflammatory gene expression as proinflammatory cytokine TNF-alpha (Schmelzer et al., 2008) which has been demonstrated to have a role in FM (Menzies et al., 2010).

Other antioxidants treatment has been assayed in FM. Melatonin, the pineal hormone with pleiotropic activity is a known powerful antioxidant and anti-inflammatory and increasing experimental and clinical evidence shows its beneficial effects against oxidative/nitrosative stress status, including that involving mitochondrial dysfunction (Acuña Castroviejo et al., 2011).Treatment of FM patients with 3 mg melatonin daily for 30 days significantly improved the tender point count, severity of pain, global physical assessments, and sleep (Citera et al., 2000). Moreover, in a limited number of cases, administration of 6 mg/day melatonin to patients with FMS resulted in normal sleep/wake cycles, normal diurnal activity, lack of pain, and fatigue and claims significant improvement of the behavioral symptoms including lack of depression (Acuna-Castroviejo et al., 2006). Recently, in a double-blind, placebo-controlled clinical study was demonstrated that administration of melatonin, alone or in a combination with fluoxetine, was effective in the treatment of patients with FM (Hussain et al., 2011). The "Myers' cocktail", an intravenous vitamin-and-mineral formula (IVMT) for the treatment of a wide range of clinical conditions, which has vitamin c as antioxidant, has been assayed also, and most subjects experienced relief as compared to baseline, but no statistically significant differences were seen between IVMT and placebo (Ali et al., 2009).

Vitamin D which controls calcium and phosphorus metabolism (Norman et al, 1992) , is also a membrane antioxidant (Wiseman H, 1993) whose deficiency has been linked to chronic pain, muscle weakness (Straube et al, 2009; Zhang et al, 2010) and fibromyalgia (Armstrong et al 2007). Subsequently, studies evaluating the effects of high-dose vitamin D treatment have been demonstrated to improve clinical symptoms in FM(Badsha et al 2009).

One of the most investigated non-pharmacological therapies in the treatment of FM is moderate aerobic exercise, which has found some improvement in patients (Stephens et al., 2008). A long-term combination of aerobic exercise, strengthening and flexibility improves psychological health status and health-related quality of life in patients with fibromyalgia (Sañudo et al., 2011). Aquatic exercise program also has showed an improvement on symptoms an anti-inflammatory effect in FM (Ortega et al., 2010). Interestingly, exercise therapies have been shown to provide significant benefits in patients with mitochondrial diseases because they induce an increase in mitochondrial biogenesis and an increase of the same size, which is proposed as an alternative therapeutic strategy in these conditions (Adhihetty et al., 2007; Safdar et al., 2011a). PGC-1α, an important regulator of mitochondrial biogenesis via regulating transcription of nuclear encoded mitochondrial genes, is increased by exercise (Safdar et al., 2011b). This may explain to some extent the improvement seen in patients with FM. Antioxidant treatment combined with exercise has also been shown to modulate oxidative stress markers in FM patients (Nazıroğlu et al, 2010)

5. Conclusion

Mitochondria and oxidative stress may play an essential role in the pathophysiology of FM. Since oxidative stress may either arise as a consequence of mitochondrial dysfunction or else interfere with mitochondrial function, reducing oxidative stress emerges as a form of mitochondrial medicine that could be beneficial in FM patients.

Our study supports the hypothesis that CoQ deficiency and mitochondrial dysfunction can contribute to cell bioenergetics imbalance, compromising cell functionality in BMCs of FM patients. Abnormal BMCs performance can promote oxidative stress and may contribute to altered nociception in FM. CoQ deficiency in FM patients, or in a subgroup of them, could also be important to initiate CoQ supplementation. Nevertheless, more research is needed to establish a primary causation between mitochondrial dysfunction and FM.

6. Acknowledgements

This work was supported by FIS PI10/00543 grant, FIS EC08/00076 grant, Ministerio de Sanidad, Spain, SAS 111242 grant, Servicio Andaluz de Salud,and by AEPMI (Asociación de Enfermos de Patología Mitocondrial), FEEL (Fundación Española de Enfermedades Lisosomales) and Federación Andaluza de Fibromialgia y Fatiga Crónica (ALBA Andalucía). This group is founded by the Centro de Investigación Biomédica en Red de Enfermedades Raras (CIBERER), ISCIII. We thank Victor and Heather Rice for their collaboration in this manuscript. The authors wish to dedicate this manuscript to FM patients and AFIBROSE (Asociación de Fibromialgia de Sevilla) by their unconditional help.

7. References

Acuna-Castroviejo, D.; Escames, G. &Reiter RJ. (2006). Melatonin therapy in fibromyalgia. *J Pineal Res,* 40,1, (Jan) (98–99), ISSN 0742-3098.

Acuña-Castroviejo, D.; López, L.C.; Escames, G.; López, A.; García, J.A. & Reiter, R.J. (2011). Melatonin-mitochondria interplay in health and disease. *Curr Top Med Chem.* 11, 2, (Jan) (221-40), ISSN 0742-3098.

Adhihetty, P.J.; Taivassalo, T.; Haller, R.G.; Walkinshaw, D.R. & Hood, D.A. (2007). The effect of training on the expression of mitochondrial biogenesis- and apoptosis-related proteins in skeletal muscle of patients with mtDNA defects. *Am J Physiol Endocrinol Metab.* 93, 3, (Sep) (E672-80), ISSN 0193-1849.

Akkuş, S.; Naziroğlu, M.; Eriş, S.; Yalman, K.; Yilmaz, N. & Yener, M. (2009). Levels of lipid peroxidation, nitric oxide, and antioxidant vitamins in plasma of patients with fibromyalgia.*Cell Biochem Funct.* 27, 4, (Jun) (181-5), ISSN 0263-6484.

Ali, A.; Njike, V.Y.; Northrup, V.; Sabina, A.B.; Williams, A.L.; Liberti, L.S.; et al. (2009). Intravenous micronutrient therapy (Myers' Cocktail) for fibromyalgia: a placebo-controlled pilot study. *J Altern Complement Med.* 15, 3, (Mar) (247-57), ISSN 1075-5535.

Alnigenis, M.N.Y. & Barland, P.(2001). Fibromyalgia syndrome and serotonin. *Clin Exp Rheumatol* 19, 2, (Mar-Apr) (205–210), ISSN 0392-856X.

Altindag, O. & Celik, H. (2006). Total antioxidant capacity and the severity of the pain in patients with fibromyalgia. *Redox Rep.* 11, 3, (Jun) (131-5), ISSN 1351-0002.

Altindag, O.; Gur, A.; Calgan, N.; Soran, N.; Celik, H. & Selek, S. (2007). Paraoxonase and arylesterase activities in fibromyalgia. *Redox Rep.* 12, 3, (Jun) (134-8), ISSN 1351-0002.

Armstrong, D.J.; Meenagh, G.K.; Bickle, I.; Lee, A.S.; Curran, E.S. & Finch, M.B. (2007). Vitamin D deficiency is associated with anxiety and depression in fibromyalgia. *Clin Rheumatol.* 26, 4, (Jul) (551-4), ISSN 0770-3198.

Badsha, H.; Daher, M. & Kong, K.O. (2009). Myalgias or non-specific muscle pain in Arab or Indo-Pakistani patients may indicate vitamin D deficiency. *Clin Rheumatol.* 28, 8, (Aug) (971–973), ISSN 0770-3198.

Bagis, S.; Tamer, L.; Sahin, G.; Bilgin, R.; Guler, H.; Ercan, B. & Erdogan, C. (2005). Free radicals and antioxidants in primary fibromyalgia: an oxidative stress disorder?. *Rheumatol Int.* 25, 3, (Apr) (188-90), ISSN 0172-8172.

Bhatia-Kiššová, I. & Camougrand, N. (2010). Mitophagy in yeast: actors and physiological roles. *FEMS Yeast Res.* 10, 8, (Dec) (1023–1034), ISSN 1567-1356.

Bazzichi, L.; Giannaccini, G.; Betti, L.; Italiani, P.; Fabbrini, L.; Defeo, F.; et al. (2006). Peripheral benzodiazepine receptors on platelets of fibromyalgic patients. *Clinical Biochem.* 39, 9, (Sep) (867–72), ISSN 0009-9120.

Bazzichi, L.; Giannaccini, G.; Betti, L.; Mascia, G.; Fabbrini, L.; Italiani, P.; et al. (2006b). Alteration of serotonin transporter density and activity in fibromyalgia. *Arthritis Res Ther* 8, 4, Jun) (R99), ISSN 1478-6354.

Bazzichi, L.; Rossi, A.; Massimetti, G.; et al. (2007). Cytokine patterns in fibromyalgia and their correlation with clinical manifestations. Clin. Exp. Rheumatol. 25, 2, (Mar-Apr) (225–230), ISSN 0392-856X.

Bazzichi, L.; Giannaccini, G.; Betti, L.; Fabbrini, L.; Schmid, L.; Palego, L.; et al. (2008). ATP, calcium and magnesium levels in platelets of patients with primaryfibromyalgia. *Clin Biochem.* 41, 13, (Sep), (1084-90), ISSN 0009-9120.

Bengtsson, A. (2002). The muscle in fibromyalgia. *Rheumatology (Oxford).* 41, 7, (Jul) (721-4), ISSN 1462-0324.

Bengtsson, A.; Henriksson, K.G. & Larsson, J. (1986). Muscle biopsy in primary fibromyalgia. Light-microscopical and histochemical findings. *Scand J Rheumatol.* 15, 1, (Jan) (1-6), ISSN 0300-9742.

Bilici, M.; Efe, H.; Koroglu, M.A.; Uydu, H.A., Bekaroglu, M. & Deger, O. (2001). Antioxidative enzyme activities and lipid peroxidation in major depression: alterations by antidepressant treatments. *Journal of affective disorders.* 64, 1, (Apr) (43-51), ISSN 0165-0327.

Buskila, D.; Cohen, H.; Neumann, L. & Ebstein, R.P. (2004). An association between fibromyalgia and the dopamine D4 receptor exon III repeat polymorphism and relationship to novelty seeking personality traits. *Mol Psychiatry.* 9, 8, (Aug) (730–731), ISSN 1359-4184.

Cachofeiro, V.; Goicochea, M.; de Vinuesa, S.G.; Oubiña, P.; Lahera, V. & Luño, J. (2008). Oxidative stress and inflammation, a link between chronic kidney disease and cardiovascular disease. *Kidney Int.* 111, Suppl, (Dec) (S4-9), ISSN 0098-6577.

Caso, G.; Kelly, P.; McNurlan, M.A. & Lawson, W.E. (2007). Effect of coenzyme q10 on myopathic symptoms in patients treated with statins. *Am J Cardiol.* 99, 10, (May) (1409-12), ISSN 0002-9149.

Citera, G.; Arias, A.; Maldonado-Cocco, J.A.; Lázaro, M.A.; Rosemffet, M.G.; Brusco, L.I.; et al. (2000). The effect of melatonin in patients with fibromyalgia: a pilot study. *Clin Rheumatol.* 19, 1, (Jan) (9–13), ISSN 0770-3198.

Clark, I.A.; Alleva, L.M. & Vissel, B. (2010). The roles of TNF in brain dysfunction and disease. *Pharmacol Ther.* 128, 3, (Dec) (519-48), ISSN 0163-7258.

Cohen, B.H. & Gold, D.R.. (2001). Mitochondrial cytopathy in adults: what we know so far. *Cleve Clin J Med.* 68, 7, (Jul) (625–626, 629–642) ISSN 0891-1150.

Cordero, M.D.; Moreno-Fernández, A.M.; deMiguel, M.; Bonal, P.; Campa, F.; Jiménez-Jiménez, L.M.; et al. (2009). Coenzyme Q10 distribution in blood is altered in patients with fibromyalgia. *Clin Biochem.* 42, 7-8, (May) (732-5), ISSN 0009-9120.

Cordero, M.D.; Moreno-Fernández, A.M.; Gomez-Skarmeta, J.L.; de Miguel, M.; Garrido-Maraver, J.; Oropesa-Avila, M.; et al. (2009). Coenzyme Q10 and alpha-tocopherol protect against amitriptyline toxicity. *Toxicol Appl Pharmacol.* 235, 3, (Jan) (329-37) ISSN 0041-008X.

Cordero, M.D.; De Miguel, M.; Moreno Fernández, A.M.; Carmona López, I.M.; Garrido Maraver, J.; Cotán, D.; et al. (2010a). Mitochondrial dysfunction and mitophagy activation in blood mononuclear cells of fibromyalgia patients: implications in the pathogenesis of the disease. *Arthritis Res Ther.* 12, 1, (Jan) (R17), ISSN 1478-6354.

Cordero, M.D.; Moreno-Fernández, A.M.; Carmona-López, M.I.; Sánchez-Alcázar, J.A.; Rodríguez, A.F.; Navas, P. & de Miguel, M. (2010b). Mitochondrial dysfunction in skin biopsies and blood mononuclear cells from two cases of fibromyalgia patients. *Clin Biochem.* 43, 13-14, (Sep) (1174-6), ISSN 0009-9120.

Cordero, M.D.; Alcocer-Gómez, E.; Cano-García, F.J.; de Miguel, M.; Campa, F.; Bona, P. & Moreno Fernández, A.M. (2011a). The effect of Coenzyme Q10 on symptoms of mother and son with Fibromyalgia Syndrome. *J Muscoskel Pain.* 19, 2, (Apr) (118-9), ISSN 1058-2452.

Cordero, M.D.; Alcocer-Gómez, E.; de Miguel, M.; Cano-García, F.J.; Luque, C.M.; Fernández-Riejo, P.; et al. (2011b). Coenzyme Q(10): A novel therapeutic approach for Fibromyalgia? Case series with 5 patients. *Mitochondrion.* 11, 4, (Jul) (623-5), ISSN 1567-7249.

Cotán, D.; Cordero, M.D.; Garrido-Maraver, J.; Oropesa-Ávila, M.; Rodríguez-Hernández, A.; Gómez Izquierdo, L.; et al.(2011). Secondary coenzyme Q10 deficiency triggers mitochondria degradation by mitophagy in MELAS fibroblasts. *FASEB J.* (May), pp, ISSN 0892-6638.

Crofford, L.J.; Engleberg, N.C. & Demitrack, M.A. (1996). Neurohormonal perturbations in fibromyalgia. *Baillieres Clin Rheumatol.* 10, 2, (May) (365–78), ISSN 0950-3579.

Dadabhoy, D.; Crofford, L.J.; Spaeth, M.; Russell, I.J. & Clauw, D.J. (2008). Biology and therapy of fibromyalgia. Evidence-based biomarkers for fibromyalgia syndrome. *Arthritis Res Ther.* 10, 4, (211), ISSN 1478-6354.

Davies, K.J. (1995). Oxidative stress: the paradox of aerobic life. *Biochem Soc Symp.*61, (1–31), ISSN 0067-8694.

Draper, H.H.; Csallany, A.S. & Hadley, M. (2000). Urinary aldehydes as indicators of lipid peroxidation in vivo. *Free Radic Biol Med.* 29, 11, (Dec) (1071-7), ISSN 0891-5849.

Drewes, A.M.; Andreasen, A.; Schrøder, H.D.; Høgsaa, B. & Jennum, P. (1993). Pathology of skeletal muscle in fibromyalgia: a histo-immuno-chemical and ultrastructural study. *Br J Rheumatol.* 32, 6, (Jun) (479-83) ISSN 0263-7103.

Duncan, A. J.; Heales, S. J.; Mills, K.; Eaton, S.; Land, J. M. & Hargreaves, I. P.(2005). Determination of coenzyme Q10 status in blood mononuclear cells, skeletal muscle, and plasma by HPLC with di-propoxy-coenzyme Q10 as an internal standard. *Clin Chem.* 51, 12, (Dec) (2380-2) ISSN 0009-9147.

Esterbauer, H.; Schaur, R.J. & Zollner, H. (1991). Chemistry and biochemistry of 4-hydroxynonenal, malonaldehyde and related aldehydes. *Free Radic Biol Med.* 11, 1, (Dec) (81-128), ISSN 0891-5849.

Evans, P.H. (1993). Free radicals in brain metabolism and pathology. *Br Med Bull.* 49, 3, (Jul) (577-87), ISSN 0007-1420.

Fibach, E. & Rachmilewitz, E. (2008). The role of oxidative stress in hemolytic anemia. *Curr Mol Med.* 8, 7, (Nov) (609-19), ISSN 1566-5240.

Fulle, S.; Mecocci, P.; Fano, G.; Vecchiet, I.; Vecchini, A.; Racciotti, D.; et al. (2000). Specific oxidative alterations in vastus lateralis muscle of patients with the diagnosis of chronic fatigue syndrome. *Free Radic Biol Med.* 29, 12, (Dec) (1252-9) ISSN 0891-5849.

Gottlieb, R.A. & Carreira, R.S. (2010). Autophagy in health and disease. 5. Mitophagy as a way of life. *Am J Physiol Cell Physiol.* 299, 2, (Aug) (C203-10) ISSN 0363-6143.

Gursoy, S.; Erdal, E.; Herken, H.; Madenci, E.; Alasehirli, B. & Erdal, N. (2003). Significance of catechol-O-methyltransferase gene polymorphism in fibromyalgia syndrome. *Rheumatol Int.* 23, 3, (May) (104–107), ISSN 0172-8172.

Haas, R.H.; Parikh, S.; Falk, M.J.; Saneto, R.P.; Wolf, N.I.; Darin, N.; et al. (2008). The in-depth evaluation of suspected mitochondrial disease. *Mol Genet Metab.* 94, 1, (May) (16-37), ISSN 1096-7193.

Heffner, R.R. & Barron, S.A. (1978). The early effects of ischemia upon skeletal muscle mitochondria. J Neurol Sci. 38, 3, (Oct) (295-315), ISSN 0022-510X.

Hein, G. & Franke, S. (2002). Are advanced glycation end-product-modified proteins of pathogenetic importance in fibromyalgia? *Rheumatology (Oxford).* 41, 10, (Oct) (1163-7), ISSN 1462-0324.

Hénriksson, K.G.; Bengtsson, A.; Larsson, J.; Lindström, F. & Thornell, L.E. (1982). Muscle biopsy findings of possible diagnostic importance in primary fibromyalgia (fibrositis,myofascial syndrome). *Lancet.* 2, 8312, (Dec) (1395), ISSN 0140-6736.

Hussain, S.A.; Al-Khalifa, I.I.; Jasim, N.A. & Gorial, F.I. (2011). Adjuvant use of melatonin for treatment of fibromyalgia. *J Pineal Res.* 50, 3, (Apr) (267-71) ISSN 0742-3098.

Joseph, E.K.; Chen, X.; Bogen, O. & Levine, J.D. (2008). Oxaliplatin acts on IB4-positive nociceptors to induce an oxidative stress-dependent acute painful peripheral neuropathy. *J Pain.* 9, 5, (May) (463–472) ISSN 1526-5900.

Joseph, E.K. & Levine, J.D. (2010). Multiple PKCepsilon-dependent mechanisms mediating mechanical hyperalgesia. *Pain.* 150, 1, (Jul) (17–21) ISSN 0304-3959.

Jung, H.J.; Park, E.H. & Lim, C.J. (2009). Evaluation of anti-angiogenic, anti-inflammatory and antinociceptive activity of coenzyme Q(10) in experimental animals. *J Pharm Pharmacol.* 61, 10, (Oct) (1391-5), ISSN 0022-3573.

Kalyan-Raman, U.P.; Kalyan-Raman, K.; Yunus, M.B. & Masi, A.T. (1984). Muscle pathology in primary fibromyalgia syndrome: a light microscopic, histochemical and ultrastructural study. *J Rheumatol.* 11, 6, (Dec) (808-13) ISSN 0315-162X.

Kaufmann, I.; Schelling, G.; Eisner, C.; Richter, H.P.; Krauseneck, T.; Vogeser, M.;et al. (2008). Anandamide and neutrophil function in patients with fibromyalgia. *Psychoneuroendocrinology.* 33, 5, (Jun) (676-85), ISSN 0306-4530.

Klein, R.; Beansch, M. & Berg, P.A. (1992). Clinical relevance of antibodies against serotonin and gangliosides in patients with primary fibromyalgia syndrome. *Psychoneuroendocrinology.* 17, 5, (Jun) (593–598) ISSN 0306-4530.

Klionsky, D.J. (2005). Autophagy.*Curr Biol.* 15. 8, (Apr) (R282-3) ISSN 0960-9822.

Kim, I.; Rodriguez-Enriquez, S. & Lemasters, J.J. (2007). Selective degradation of mitochondria by mitophagy. *Arch Biochem Biophys.* 462, 2, (Jun) (245-53), ISSN 0003-9861.

Land, J.M., Heales, S.J., Duncan, A.J., Hargreaves, I.P., 2007. Some observations upon biochemical causes of ataxia and a new disease entity ubiquinone, CoQ10 deficiency. *Neurochem Res.* 32, 4-5, (Apr-May) (837-43), ISSN 0364-3190.

Lawrence, R.C.; Felson, D.T.; Helmick, C.G.; Arnold, L.M.; Choi, H.; Deyo, R.A.; et al. (2008). National Arthritis Data Workgroup. Estimates of the prevalence of arthritis and other rheumatic conditions in the United States. Part II. *Arthritis Rheum.* 58, 1, (Jun) (26-35), ISSN 0004-3591.

Lenaz, G. (1987).Lipid fluidity and membrane protein dynamics. *Bioscience reports.* 7, 11, (Nov) (823-37) ISSN 0144-8463.

Lenaz, G.; Fato, R.; Formiggini, G. & Genova, M.L. (2007). The role of Coenzyme Q in mitochondrial electron transport. *Mitochondrion.* 7, Suppl, (Jun) (S8–33) ISSN 1567-7249.

Lirk, P.; Haller, I.; Hausott, B.; Ingorokva, S.; Deibl, M.; Gerner, P. & Klimaschewski, L. (2006). The neurotoxic effects of amitriptyline are mediated by apoptosis and are effectively blocked by inhibition of caspase activity. *Anesth Analg.* 102, 6, (Jun) (1728-33), ISSN 0003-2999.

Lister, R.E. (2002). An open, pilot study to evaluate the potential benefits of coenzyme Q10 combined with Ginkgo biloba extract in fibromyalgia syndrome. *J Int Med Res.* 30, 2, (Mar-Apr) (195-9), ISSN 0300-0605.

Littarru, G.P. & Tiano, L. (2010). Clinical aspects of coenzyme Q10: an update. *Nutrition.* 26, 3, (Mar) (250-4), ISSN 0899-9007.

Luft, R.; Ikkos, D.; Palmieri, G.; Ernster, L. & Afzelius, B. (1962). A case of severe hypermetabolism of nonthyroid origin with a defect in the maintenance of mitochondrial respiratory control: a correlated clinical, biochemical, and morphological study. *J Clin Invest.*41, (Sep) (1776-804), ISSN 0021-9738.

Lund, N.; Bengtsson, A. & Thorborg, P. (1986). Muscle tissue oxygen pressure in primary fibromyalgia. *Scand J Rheumatol.* 15, 2, (165-73), ISSN 0300-9742.

Maes, M.; Smith, R.; Christophe, A.; Cosyns, P.; Desnyder, R. & Meltzer, H. (1996). Fatty acid composition in major depression: decreased omega 3 fractions in cholesteryl esters and increased C20: 4 omega 6/C20:5 omega 3 ratio in cholesteryl esters and phospholipids. *Journal of affective disorders.* 38, 1, (Apr) (35-46), ISSN 0165-0327.

Maes, M.; Mihaylova, I. & Leunis, J.C. (2007). Increased serum IgM antibodies directed against phosphatidyl inositol (Pi) in chronic fatigue syndrome (CFS) and major depression: evidence that an IgM-mediated immune response against Pi is one factor underpinning the comorbidity between both CFS and depression. *Neuro endocrinology letters.* 28, 6, (Dec) (861-7), ISSN 0172-780X.

Menzies, V. & Lyon, D.E.. (2010). Integrated review of the association of cytokines with fibromyalgia and fibromyalgia core symptoms. *Biol Res Nurs.* 11, 4, (Apr) (387-94), ISSN 1099-8004.

Marazziti, D.; Dell'Osso, L.; Presta, S.; Pfanner, C.; Rossi, A.; Masala, I.; et al. (1999). Platelet [3H]paroxetine binding in patients with OCD-related disorders. *Psychiatry Res.* 89, 3, (Dec) (223–8), ISSN 0165-1781.

Martini, C.; Trincavelli, M.L.; Tuscano, D.; Carmassi, C.; Ciapparellim, A.; Lucacchini, A.; et al (2004). Serotonin-mediated phosphorylation of extracellular regulated kinases in platelets of patients with panic disorder versus controls. *Neurochem Int.* 44, 8, (Jun) (627–39), ISSN 0197-0186.

Marklund, S.L. (1990). Expression of extracellular superoxide dismutase by human cell lines. *Biochem J*. 266, 1, (Feb) (213-9), ISSN 0264-6021.

Naik, E. & Dixit, V.M. (2011). Mitochondrial reactive oxygen species drive proinflammatory cytokine production. *J Exp Med*. 208, 3, (Mar) (417-20), ISSN 0022-1007.

Nazıroğlu, M.; Akkuş, S.; Soyupek, F.; Yalman, K.; Çelik, Ö.; Eriş, S. & Uslusoy, G.A. (2010). Vitamins C and E treatment combined with exercise modulates oxidative stress markers in blood of patients with fibromyalgia: a controlled clinical pilot study. *Stress*. 13, 6, (Nov) (498-505), ISSN 1025-3890.

Neeck, G. (2002). Pathogenic mechanisms of fibromyalgia. *Ageing Res Rev*. 1, 2, (Apr) (243–55), ISSN 1568-1637.

Neustadt, J. & Pieczenik, S.R. (2008). Medication-induced mitochondrial damage and disease. *Mol Nutr Food Res*. 52, 7, (Jul) (780-8) ISSN 1613-4125.

Nishishinya, B.; Urrútia, G.; Walitt, B.; Rodriguez, A.; Bonfill, X.; Alegre, C. & Darko, G. (2008). Amitriptyline in the treatment of fibromyalgia: a systematic review of its efficacy. *Rheumatology (Oxford)*. 47, 12, (Dec) (1741–1746), ISSN 1462-0324.

Niu, X.; Arthur, P.; Abas, L.; Whisson, M. & Guppy, M. (1996). Carbohydrate metabolism in human platelets in a low glucose medium under aerobic conditions. *Biochim Biophys Acta*. 1291, 2, (Oct) (97–106), ISSN 0006-3002.

Norman, A.W.; Nemere, I.; Zhou, L.X.; Bishop, J.E.; Lowe, K.E.; Maiyar, A.C.; et al. (1992). 1,25(OH)2-vitamin D3, a steroid hormone that produces biologic effects via both genomic and nongenomic pathways. *J Steroid Biochem Mol Biol*. 41,3-8, (Mar) (231-40), ISSN 0960-0760.

Orrell, R.W.; Lane, R.J. & Ross, M. (2008). A systematic review of antioxidant treatment for amyotrophic lateral sclerosis/motor neuron disease. *Amyotroph Lateral Scler*. 9, 4, (Aug) (195-211), ISSN 1748-2968.

Ortega, E.; Bote, M.E.; Giraldo, E. & García, J.J. (2010). Aquatic exercise improves the monocyte pro- and anti-inflammatory cytokine production balance in fibromyalgia patients. *Scand J Med Sci Sports*. doi: 10.1111/j.1600-0838.2010.01132.x, (Jun), ISSN 0905-7188.

Ozgocmen, S.; Ozyurt, H.; Sogut, S.; Akyol, O.; Ardicoglu, O. & Yildizhan, H. (2006). Antioxidant status, lipid peroxidation and nitric oxide in fibromyalgia: etiologic and therapeutic concerns. *Rheumatol Int*. 26, 7, (May) (598-603), ISSN 0172-8172.

Park, J.H.; Phothimat, P.; Oates, C.T.; Hernanz-Schulman, M. & Olsen, N.J. (1998). Use of P-31 magnetic resonance spectroscopy to detect metabolic abnormalities in muscles of patients with fibromyalgia. *Arthritis Rheum*. 41, 3, (Mar) (406-13), ISSN 0004-3591.

Papadopoulos, V.; Baraldi, M.; Guilarte, T.R.; Knudsen, T.B.; Lacapère, J.J.; Lindemann, P.; et al. (2006). Translocator protein (18 kDa): new nomenclature for the peripheral-type benzodiazepine receptor based on its structure and molecular function. *Trends Pharmacol Sci*. 27, 8, (Aug) (402–9), ISSN 0165-6147.

Pieczenik, S.R. & Neustadt, J. (2007). Mitochondrial dysfunction and molecular pathways of disease. *Exp Mol Pathol*. 83, 1, (Aug) (84-92), ISSN 0014-4800.

Pongratz, D.E. & Späth, M. (1998). Morphologic aspects of fibromyalgia. *Z Rheumatol*. 57, 2, (47-51), ISSN 0340-1855.

Praticò, D. (2008). Oxidative stress hypothesis in Alzheimer's disease: a reappraisal. *Trends Pharmacol Sci*. 29, 12, (Dec) (609-15), ISSN 0165-6147.

Quinzii, C.M.; Lopez, L.C.; Von-Moltke, J.; Naini, A.; Krishna, S.; Schuelke, M.; et al (2008). Respiratory chain dysfunction and oxidative stress correlate with severity of primary CoQ10 deficiency. *Faseb J.* 22, 6, (Jun) (1874-85), ISSN 0892-6638.

Quinzii, C.M. & Hirano, M. (2010). Coenzyme Q and mitochondrial disease. *Dev Disabil Res Rev.* 16, 2, (Jun) (183-8), ISSN 1940-5510.

Rizza, T.; Vazquez-Memije, M.E.; Meschini, M.C.; Bianchi, M.; Tozzi, G.; Nesti, C.;et al. (2009). Assaying ATP synthesis in cultured cells: a valuable tool for the diagnosis of patients with mitochondrial disorders. *Biochem Biophys Res Commun.* 383, 1, (May) (58-62), ISSN 0006-291X.

Rodriguez-Hernandez, A.; Cordero, M.D.; Salviati, L.; Artuch, R.; Pineda, M.; Briones, P.; et al. (2009). Coenzyme Q deficiency triggers mitochondria degradation by mitophagy. *Autophagy.* 5, 1, (Jan) (19-32) ISSN 1554-8627.

Russell, I.J.; Vipraio, G.A. & Abraham, G.E. (1993). Red cell nucleotide abnormalities in fibromyalgia syndrome (abstr). *Arthritis Rheum.* 36,(S223), ISSN 0004-3591.

Russell, I.J.; Orr, M.D.; Littman, B.;Vipraio, G.A.; Alboukrek, D.; Michalek, J.E.; et al. (1994). Elevated cerebrospinal fluid levels of substance P in patients with thefibromyalgia syndrome. *Arthritis Rheum.* 37, 11, (Nov) (1593–601), ISSN 0004-3591.

Safdar, A.; Bourgeois, J.M.; Ogborn, D.I.; Little, J.P.; Hettinga, B.P.; Akhtar, M.; et al. (2011a). Endurance exercise rescues progeroid aging and induces systemic mitochondrial rejuvenation in mtDNA mutator mice. *Proc Natl Acad Sci U S A.* 108, 10, (Mar) (4135-40), ISSN 0027-8424.

Safdar, A.; Little, J.P.; Stokl, A.J.; Hettinga, B.P.; Akhtar, M.; Tarnopolsky, M.A. (2011b). Exercise increases mitochondrial PGC-1alpha content and promotes nuclear-mitochondrial cross-talk to coordinate mitochondrial biogenesis. *J Biol Chem.* 286, 12, (Mar) (10605-17), ISSN 0021-9258.

Sandor, P.S.; Di Clemente, L.; Coppola, G.; Saenger, U.; Fumal, A.; Magis, D.; et al. (2005). Efficacy of coenzyme Q10 in migraine prophylaxis: a randomized controlled trial. *Neurology.* 64, 4, (Feb) (713–5), ISSN 0028-3878.

Sañudo, B.; Galiano, D.; Carrasco, L.; de Hoyo, M.; McVeigh, J.G. (2011). Effects of a prolonged exercise program on key health outcomes in women with fibromyalgia: a randomized controlled trial. *J Rehabil Med.* 43, 6, (May) (521-6), ISSN 1650-1977.

Schmelzer, C.; Lindner, I.; Rimbach, G.; Niklowitz, P.; Menke, T. & Döring, F. (2008). Functions of coenzyme Q10 in inflammation and gene expression. *Biofactors.* 32, 1-4, (Dec) (179-83), ISSN 0951-6433.

Schwarz, M.J.; Offenbaecher, M.; Neumeister, A.; Ewert, T.; Willeit, M.; Praschak-Rieder, N.; et al. (2002). Evidence for an altered tryptophan metabolism in fibromyalgia. *Neurobiol Dis.* 11, 3, (Dec) (434-42), ISSN 0969-9961.

Sendur, O.F.; Turan, Y.; Tastaban, E.; Yenisey, C. & Serter, M. (2009). Serum antioxidants and nitric oxide levels in fibromyalgia: a controlled study. *Rheumatol Int.* 29, 6, (Apr) (629-33), ISSN 0172-8172.

Sprott, H.; Salemi, S.; Gay, R.E.; Bradley, L.A.; Alarcón, G.S.; Oh, S.J.; et al. (2004). Increased DNA fragmentation and ultrastructural changes in fibromyalgic muscle fibres. *Ann Rheum Dis.* 63, 3, (Mar) (245-51), ISSN 0003-4967.

Stack, E.C.; Matson, W.R. & Ferrante, R.J. (2008). Evidence of oxidant damage in Huntington's disease: translational strategies using antioxidants. *Ann N Y Acad Sci.* 1147, (Dec) (79-92), ISSN 0077-8923.

Staud, R. & Spaeth, M. (2008). Psychophysical and neurochemical abnormalities of pain processing in fibromyalgia. *CNS Spectr.* 13, 3 Suppl 5, (Mar) (12-7), ISSN 1092-8529.

Staud, R. (2010). Pharmacological treatment of fibromyalgia syndrome: new developments. *Drugs.* 70, 1, (Jan) (1-14), ISSN 0012-6667.

Stephens, S.; Feldman, B.M.; Bradley, N.; Schneiderman, J.; Wright, V.; Singh-Grewal, D.; et al. (2008). Feasibility and effectiveness of an aerobic exercise program in children with fibromyalgia: results of a randomized controlled pilot trial. *Arthritis Rheum.* 59, 10, (Oct) (1399-406), ISSN 0004-3591.

Straube, S.; Andrew Moore, R.;Derry, S. & McQuay, H.J. (2009). Vitamin D and chronic pain. *PAIN.* 141, 1-2, (Jan) (10–13), ISSN 0304-3959.

Tander, B.; Gunes, S.; Boke, O.; Alayli, G.; Kara, N.; Bagci, H. & Canturk, F. (2008). Polymorphisms of the serotonin-2A receptor and catechol-O-methyltransferase genes: a study on fibromyalgia susceptibility. *Rheumatol Int.* 28, 7, (May) (685-91); ISSN 0172-8172.

Tiano, L.; Padella, L.; Santoro, L.; Carnevali, P.; Principi, F.; Brugè, F.; et al. (2011). Prolonged coenzyme Q(10) treatment in Down syndrome patients, effect on DNA oxidation. *Neurobiol Aging.* Pp., ISSN 0197-4580.

Turrens, J.F. (2003). Mitochondrial formation of reactive oxygen species. *J Physiol.* 552, Pt 2, (Oct) (335-44), ISSN 0022-3751.

Turunen, M.; Olsson, J. & Dallner, G. (2004). Metabolism and function of coenzyme Q. *Biochim Biophys Acta.* 1660, 1-2, (Jan) (171-99), ISSN 0006-3002.

Tsutsumi, S.; Tsuji, K.; Ogawa, K.; Ito, T. & Satake, T. (1988). Effect of dietary salt and cholesterol loading on vascular adrenergic receptors. *Blood vessels.* 25, 5, (209-16), ISSN 0303-6847.

Uçeyler, N.; Valenza, R.; Stock, M.; Schedel, R.; Sprotte, G. & Sommer, C. (2006). Reduced levels of ntiinflammatory cytokines in patients with chronic widespread pain. *Arthritis Rheum.* 54, 8, (Aug) (2656–2664), ISSN 0004-3591.

Vecchiet, J.; Cipollone, F.; Falasca, K.; Mezzetti, A.; Pizzigallo, E.; Bucciarelli, T.; et al.(2003). Relationship between musculoskeletal symptoms and blood markers of oxidative stress in patients with chronic fatigue syndrome. *Neuroscience letters.* 335, 3, (Jan) (151-4), ISSN 0304-3940.

Wallace, D.J.; Linker-Israeli, M.; Hallegua, D.; Silverman, S.; Silver, D. & Weisman, M.H. (2001). Cytokines play an aetiopathogenetic role in fibromyalgia: a hypothesis and pilot study. *Rheumatology (Oxford).* 40, 7, (Jul) (743-9), ISSN 1462-0324.

Wallace, D.J. (2006). Is there a role for cytokine based therapies in fibromyalgia. *Curr Pharm Des.* 12, 1, (17-22), ISSN 1381-6128.

Wang, Z.Q.; Porreca, F.; Cuzzocrea, S.; Galen, K.; Lightfoot, R.; Masini, E.; et al. (2004). A newly identified role for superoxide in inflammatory pain. *J Pharmacol Exp Ther.* 309, 3, (Jun) (869-78), ISSN 0022-3565.

Werle, E.; Fisher, H.; Muller, A.; et al. (2001). Antibodies against serotonin have no diagnostic relevance in patients with fibromyalgia syndrome. *J Rheumatol.* 28, 3, (Mar) (595–600, ISSN 0315-162X.

Whaley-Connell, A.; McCullough, P.A. & Sowers JR. (2011). The role of oxidative stress in the metabolic syndrome. *Rev Cardiovasc Med.* 12, 1, (21-9), ISSN 1530-6550.

Wiseman, H. (1993). Vitamin D is a membrane antioxidant. Ability to inhibit iron-dependent lipid peroxidation in liposomes compared to cholesterol, ergosterol and tamoxifen and relevance to anticancer action. *FEBS Lett.* 326, 1-3, (Jul) (285-8), ISSN 0014-5793.

Wolfe, F.; Smythe, H.A.; Yunus, M.B.; Bennett, R.M.; Bombardier, C.; et al. (1990) The American College of Rheumatology 1990 Criteria for the Classification of Fibromyalgia. Report of the Multicenter Criteria Committee. *Arthritis Rheum.* 33, 2, (Feb) (160-72), ISSN 0004-3591.

Yunus, M.; Masi, A.T.; Calabro, J.J.; et al. (1981). Primary fibromyalgia (fibrositis): clinical study of 50 patients with matched normal controls. *Semin Arthritis Rheum.* 11, 1, (Aug) (151-71), ISSN 0049-0172.

Yunus, M.B.; Kalyan-Raman, U.P.; Kalyan-Raman, K. & Masi, A.T. (1986). Pathologic changes in muscle in primary fibromyalgia syndrome. *Am J Med.* 81, 3A, (Sep) (38-42), ISSN 0002-9343.

Yunus, M.B.; Kalyan-Raman, U.P. & Kalyan-Raman, K. (1988). Primary fibromyalgia syndrome and myofascial pain syndrome: clinical features and muscle pathology. *Arch Phys Med Rehabil.* 69, 6, (Jun) (451-4), ISSN 0003-9993.

Zhang, R. & Naughton, D.P. (2010). Vitamin D in health and disease: current perspectives. *Nutr J.* 8, 9, (Dec) (65), ISSN 1475-2891.

Zhou, C.; Huang, Y. & Przedborski, S. (2008). Oxidative stress in Parkinson's disease: a mechanism of pathogenic and therapeutic significance. *Ann N Y Acad Sci.* 1147, (Dec) (93-104), ISSN 0077-8923.

Zorov, D.B.; Juhaszova, M. & Sollott, S.J. (2006). Mitochondrial ROS-induced ROS release: an update and review. *Biochim Biophys Acta.* 1757, 5-6, (May-Jun) (509-17), ISSN 0006-3002.

Part 2

Definition and Diagnosis of Fibromyalgia

6

The Affective-Motivational Domain of the McGill Pain Questionnaire Discriminates Between Two Distinct Fibromyalgia Patient Subgroups – A Preliminary Study Based on Self-Organizing Maps

Monika Salgueiro and Jon Jatsu Azkue
University of the Basque Country, Spain

1. Introduction

For two decades, FMS was diagnosed based on the criteria established on 1990 by the American College of Rheumatology, i.e. by the identification of widespread pain of at least 3 months duration, and pain and tenderness evoked by palpation in at least 11 of 18 specific bodily sites (Wolfe et al., 1990). Recently, the ACR preliminary diagnostic recommendations for fibromyalgia abandoned the tender point examination criterion, and were established as the following 3 conditions: (i) Widespread Pain Index (WPI) equal to or greater than 7 and Symptom Severity Score (SS) equal to or greater than 5, or WPI ranging between 3 and 6 and the SS equal to or greater than 9; (ii) steady presence of symptoms for at least 3 months; and (iii) the absence of a disorder that would otherwise explain the pain.

Despite being highly prevalent, the etiology and pathophysiology of FMS remains obscure. Diverse factors contributing to this condition have so far been identified, however none has been proven unequivocally responsible. Due to limited understanding of the etiology and pathophysiology of FMS, a diversity of pharmacological and nonpharmacological interventions alone and in combination have been attempted to treat the symptoms, largely on a trial-and-error basis. However, this condition remains largely refractory to treatment.

1.1 Overcoming clinical heterogeneity by subgrouping patients

Clinical variability within individuals whose disease manifestations meet the ACR criteria is enormous. While chronic pain represents a clinical manifestation that all FMS patients share in common, pain may be largely the only symptom experienced by some individuals with FMS while others present with a myriad of psychological manifestations, and even other disease entities such as chronic fatigue syndrome, irritable bowel syndrome or temporomandibular dysfunction also are frequently found in patients with FMS (Aaron & Buchwald, 2001).

Because of this heterogeneity of symptoms, authors have suggested that there is a need for empirically derived subgroups. These subgroups should not only be based on easily measurable variables but also clinically relevant in assisting to target more specific therapeutic interventions to individuals (Bennett et al., 1991; Masi & Yunus, 1991; Turk et al., 1996). Indeed, there is a growing awareness regarding the need to identify subgroups of chronic pain patients on the basis of biological and psychosocial characteristics, as a strategy to overcome some of the problems derived from patient heterogeneity. On the one hand, if one assumes patient homogeneity and relies on measuring group means to assess treatment outcome, false negative outcomes may be magnified. Specificity should thus be optimized by interventions in patient subgroups characterized by operating pathogenic vectors, and intervention outcome may be more specifically assessed if targeted at homogeneous patient subpopulations even when underlying mechanisms are not well understood. On the other hand, assuming patient homogeneity when prescribing treatment will not provide relief or amelioration of symptoms for a significant proportion of patients, and subsets of patients may be receiving treatment for symptoms without necessity.

In chronic pain patients, numerous studies have attempted to respond to this need for patient subgrouping by empirically identifying patient subsets based on psychological characteristics and psychopathology. Turk and collaborators (1996) were among the first to identify chronic pain patients subgroups by using cluster analysis techniques, based on the Multidimensional Pain Inventory. Importantly, FMS patients classified into different subgroups according to those criteria may respond differently to treatment (Turk et al., 1998). FMS patients have also been classified empirically on the basis of psychopathological profiles by using the 90 Symptoms Checklist (SCL-90) (Williams et al., 1995). Indeed, a growing literature has suggested that psychological mechanisms may be important correlates in the establishment and maintenance of FMS (Boissevain & McCain 1991), and previous observations have shown that certain psychological variables, in combination with pain sensitivity indices, may best distinguish FMS patient subsets (Giesecke et al., 2003).

There have been several other attempts or proposals. Müller and collaborators (2007) proposed that FMS may be categorized as primary or secondary. According to the authors, primary fibromyalgia may be much more common than the secondary type and is characterized by the absence of any definitive organic factor triggering the syndrome, whereas in secondary fibromyalgia the underlying disease, such as inflammatory rheumatic processes or collagenosis can be easily diagnosed. Based on clinical manifestations, they defined four subtypes of patients with primary fibromyalgia that may reportedly benefit from different treatment strategies. Patients with fibromyalgia with extreme sensitivity to pain but with no associated psychiatric conditions would be best treated by administration of the 5-HT3 receptor-antagonist tropisetron, whereas patients with fibromyalgia and comorbid, pain-related depression, or patients with depression with concomitant fibromyalgia syndrome would both ideally benefit from treatment with antidepressants. Finally, the focus of treatment should be on psychotherapy in the fourth group, which was composed of patients with fibromyalgia due to somatization (Müller et al., 2007). In secondary fibromyalgia, the initial focus should be on influencing the underlying disease, and for example a regimen of daily administration of 5 mg tropisetron for 5 days was found by these authors to be as effective in fibromyalgia patients whose pain was not relieved by non-steroidal analgesics, as in patients suffering from associated progressive scleroderma or other collagen-related diseases (Müller et al., 2007).

In one-hundred fifteen patients with FMS, Thieme and co-workers (2004) assessed mental status by using the DSM-IV, as well as pain, impact of pain, anxiety and depression, posttraumatic stress disorder-like symptoms, and sexual and physical abuse. Based on the Structured Clinical Interview for DSM-IV, axis I disorders were grouped into (i) anxiety and (ii) mood disorders, the latter including major depressive episode, major depressive disorder, and dysthymic disorder. The authors conducted a multivariate analysis of variance (MANOVA) to assess differences among the groups with no axis I disorders, anxiety disorders, and mood disorders, and one-way ANOVAs were conducted following a significant MANOVA. Their analysis yielded three psychosocial subgroups that were termed Dysfunctional, Interpersonally Distressed, and Adaptive Copers. The Dysfunctional and Interpersonally Distressed groups mainly reported anxiety and mood disorders, respectively, while the Adaptive Copers group exhibited little comorbidity. These findings were further suggestive that, rather than being a homogeneous diagnosis, varying proportions of FMS patients present with comorbid anxiety and depression dependent on psychosocial characteristics, emphasizing the importance of assessing patients with FMS for the presence of affective distress and not treating them as a homogeneous group. A recent, additional attempt to identify FMS patient subgroups was that advanced by Hasset and co-workers (2008). These authors proposed a classification based on affect balance styles, which considered four categories, including (i) healthy subjects with high levels of positive affects (PA) and low levels of negative affect (NA), (ii) low (low PA/low NA), reactive (high PA/high NA), and depressive (low PA/high NA). The authors found high prevalence of depressive and reactive affect balance styles in patients with FMS, and reported an association between affect balance style and psychiatric comorbidity.

Giesecke and co-workers (2003) proposed a classification scheme based on variables from three clinical domains, namely pain sensitivity and hyperalgesia, cognitive variables including coping strategies and catastrophizing, as well as mood and negative affective states such as depression and anxiety. They assessed pressure pain sensitivity by the use of an algometer, whereas cognitive evaluation was largely conducted by means of the Coping Strategies Questionnaire and psychological distress was assessed by using the Center for Epidemiologic Studies Depression Scale and the State-Trait Personality Inventory. By using agglomerative hierarchic cluster analysis and multiple analyses of variance to confirm that each variable used in the analysis was differentiated by the cluster solution, these authors claimed the existence of three clinical profiles that were distinctly different in terms of anxiety, depression, catastrophizing, control over pain and pressure-pain sensitivity. One subgroup of patients was distinguished by moderate mood scores, moderate levels of catastrophizing and perceived control over pain, and low levels of pain sensitivity. A second subgroup exhibited poorer mood scores, highest scores on the catastrophizing subscale, the lowest values for perceived control over pain, and high levels of tenderness. The third group displayed normal mood, very low levels of catastrophizing, high perceived control over pain, but increased pain sensitivity.

More recently, de Souza and collaborators (de Souza et al., 2009) relayed on hierarchical clustering analysis techniques of selected items of the Fibromyalgia Impact Questionnaire (FIQ), a self-administered report that comprehensively reflects the functional capacity and the occurrence and severity of typical symptoms of the disease (Burckhardt et al., 1991). These authors proposed a two-cluster group solution based on differences in scores on specific items of the FIQ, where so termed type II subjects reported higher scores on anxiety, depression and morning tiredness items over type I subjects (de Souza et al., 2009). A major strength of this classification scheme is that it assesses the most prevalent complaints reported by persons with FMS while being easy to administer in the clinical setting. These two subgroups scored differently on Pain Catastrophyzing Scale scores, Interference and Life Control subscales of the Multidimensional Pain Inventory, or the SF-36 mental component summary. This classification scheme highlighted the contribution of emotional status to clinical variability in FMS patients and provided an avenue for refining treatment by addressing psychological comorbidities in discrete subpopulations. Observations by de Souza's group have been replicated in a larger sample, and differences in depressive and anxiety symptoms have been corroborated by using the Beck Depression Inventory and the Hospital Anxiety and Depression Scale, respectively (Calandre et al., 2010).

Although depression and anxiety have been extensively studied in persons with FMS, also general psychopathology is prevalent in fibromyalgia relative to other chronic conditions (Epstein et al., 1999; Gormsen et al., 2010; Shaver et al., 1997), acute disorders (Gatchel, 1996) or asymptomatic subjects (Arnold et al., 2008; Dersh et al., 2002; Hellström & Jansson, 2001; Winfield, 2000) and, for example, post-traumatic stress disorder may affect 50% of subjects suffering from FMS (Fietta et al., 2007), whereas obsession, somatization disorders and alexithymia are reportedly present in up to 55%, 71%, and 47%, respectively (Centonze et al., 2004). High prevalence of positive testing for bipolar disorder among patients with FMS has also recently been reported (Wilke et al., 2010).

1.2 Does the McGill Pain Questionnaire capture clinically relevant information?

Verbal rating scales are among the most commonly used methods for assessing pain intensity. Due to the ease with which they can be administered and because they are generally easy to comprehend, compliance rates for verbal rating scales are comparable to those obtained for other measures of pain intensity (Jensen et al., 1986, 1989). In addition, verbal rating scales are positively co-related to other measures of pain intensity and are sensitive tools to assess treatment outcome (Fox & Melzack, 1976; Ohnhaus & Adler, 1975; Rybstein-Blinchik, 1979), and thus can be considered as valid indicants of pain intensity (Ahles et al., 1984; Downie et al., 1978; Jensen et al., 1986, 1989; Kremer et al., 1981; Littman et al., 1985; Ohnhaus & Adler, 1975; Woodforde & Merskey, 1972).

The McGill Pain Questionnaire (MPQ) (Chen & Treede, 1985; Dubuisson & Melzack, 1976; Melzack, 1975, 1985) is a widely employed verbal pain assessment tool. The MPQ comprises 20 subscales or sets of verbal descriptors designed to evaluate the sensory, affective, evaluative and miscellaneous dimensions of pain.

A number of studies have shown that the MPQ can capture clinical characteristics of pain that predict the occurrence of distinct underlying pathogenic mechanisms. For example, Gruschka and Sessle (1984) administered the MPQ to 102 patients presenting with toothache, in order to test the sensitivity of this tool to distinguish between dental pain originating from reversible or irreversible tooth pulp inflammation. In that study, patients suffering from pain derived from necrotic or irreversibly inflamed tooth pulp scored significantly higher on the total Pain Rank Index, as well as on the sensory, evaluative and miscellaneous Pain Rank Indices, although no differences were found as to the affective-motivational dimension, in such a manner that the MPQ alone was capable of predicting diagnosis in 73% of patients. Melzack and co-workers (1986) used a similar approach to distinguish between trigeminal neuralgia and atypical facial pain patients, and they were able to correctly classify 91% of the patients based on 7 descriptors of the MPQ.

In the present study, we hypothesized that the MPQ might indeed be capable of capturing distinctive pain characteristics in homogeneous subgroups of patients with FMS. However, understanding complex relationships between subjects across multiple variables may be difficult, and in particular the MPQ comprises a considerable number of items to address several dimensions of the complex experience of pain. Data visualization techniques attempt to solve these difficulties by transforming high-dimensional data into lower dimensional topologies that may be better understood visually. A relatively novel approach to discover similarity relations between data objects in high-dimensional signal space is the Self-Organizing Maps (SOMs), a relatively recent methodology for unsupervised machine learning. A SOM is an artificial intelligence technique, first introduced by Teuvo Kohonen (1995), that is commonly used to non-linearly project multidimensional input data into a bi-dimensional output space, thus providing the investigator with a visual image of the data that better depicts their structure and relationships. A simple SOM layout typically consists of an input neuron layer, a competition layer and an output layer. The SOM builds a two-dimensional regular lattice where the location of a map unit, or neuron, is indicative of relative distances between data points in the high-dimensional space, in such a way that mapping or data structure from the high-dimensional space onto map units are preserved. Based on this property, the SOM can serve as a very efficient, powerful and highly visual tool for cluster analysis of high-dimensional data (Kohonen, 1995). This has been the selected use for the SOM in the present work. The SOM algorithm, also known as the Kohonen feature map algorithm, is a well known artificial neural network algorithm that is based on unsupervised learning, in contrast to many other neural networks using supervised learning.

2. Methods

2.1 Setting and data collection

A total of 30 consecutive FMS subjects (28 females, 2 males) aged 29-82 years (58±10.38) participated in the study. Subjects were recruited through advertisements, FMS associations and informative presentations. All participating subjects had been diagnosed with FMS on according to the 1990 ACR classification criteria at least one year prior to the initiation of this study. The study was approved by the Ethical Review Board of the University of the Basque Country, and all participants provided their written, informed consent.

Subjects were interviewed individually by a trained psychologist. Sociodemographic data were collected first, and self-applied questionnaires were then administered for clinical assessment in the same session, including the MPQ, FIQ, Medical Outcomes Study - Short Form 36 (SF-36) and SCL-90-R (c.f. below). Participants were not asked to discontinue medication prior to study.

We used the validated Spanish translation of the MPQ (Lázaro et al., 1994), which includes a total of 78 descriptors of pain. In this version, five out of the 78 descriptors are computed into a Present Pain Index (PPI) measuring overall pain intensity. A zero score was assigned here if no descriptor was chosen. PRIs on sensory, affective-motivational, evaluative and miscellaneous dimensions of pain were obtained from weighted MPQ item scores. A total, cumulative PRI was also calculated. Subjects were also asked to assess the level of pain using the Visual Analog Scale (VAS), that is, by placing a mark on a line 10 cm long at an appropriate distance between two endpoints signaling no pain at all and the most intense pain imaginable (Huskisson, 1974; Scott & Huskisson, 1976).

The Fibromyalgia Impact Questionnaire (FIQ) is a 20-term, self-administered questionnaire that assesses both psychological and physical symptoms of FMS and allows for a broad quantification of interference of the condition on daily living tasks and quality of life (Burckhardt et al., 1991). It is quickly administered and easily evaluates a broad range of clinical characteristics associated to FMS. The total FIQ score can adopt values ranging from 0 to 10, where higher scores denote greater severity or impact. The FIQ is held to be a sensitive tool to monitor progression of the FMS and response to treatment in clinical trials (Dunkl et al., 2000). We used the Spanish language version validated by Monterde and co-workers (2004).

The Spanish version of the SF-36 (Alonso et al., 1995) was administered to assess overall health status. The physical component summary (PCS) and the mental component summary (MCS), whose values can range from 0 to 100 where higher values relate to better quality of life, were calculated.

The 90 Symptom Checklist Revised (SCL-90-R) (Derogatis, 1994) is a self-reported inventory that inquiries patients on the level of distress generated by 90 hypothetical situations, and provides scores on 9 psychopathological dimensions, namely somatization, obsessive-compulsive, interpersonal sensitivity, depression, anxiety, hostility, phobic anxiety, paranoid ideation and psychoticism, plus additional scores assessing the severity of the above dimensions. SCL-90-R has proved useful for assessing emotional status in chronic pain patients, since it is an inexpensive way to access a significant amount of data on a range of psychopathological dimensions (Torres et al., 2010). Here, we used the validated version for Spanish language (González de Rivera et al., 2002). For each subscale, composite normalized T-scores with a mean of 50 and standard deviation of 10 were obtained on the basis of standardized norms for nonpsychiatric patients provided in the test manual (González de Rivera et al., 2002). The global severity index (GSI), positive symptom total (PST) and positive symptom distress index (PSDI) were also obtained.

2.2 Analysis with Self-Organizing Maps

Raw scores from all 20 MPQ items were normalized prior to SOM analysis, and scores from items contributing to the four studied subscales were fed to separate SOMs. Geometry of the two-dimensional SOMs was set to 10 rows x 20 columns, and the grid of the SOM units was selected as hexagonal from the beginning. Competition layer neurons were initialized randomly as a pre-processing step, and the SOM was then trained by iteratively calculating similarity measures as Euclidean distances.

The unified distance matrix (U matrix) method is the most commonly used method to render the SOM. The U matrix consists of a regular neuron grid where each cell or node represents a neuron, and distances between neighboring neurons provide a bi-dimensional representation of the distance of the corresponding reference vectors. The basic idea of this method is based on the principle of representing the distance matrix between the all the reference vectors associated to the SOM units by using colorings or shades of gray, where a dark shade of gray between neurons denotes a large distance and thus a gap between the vectors in the input space, whereas light shades represent close relationships between neighboring units. Analogously, light areas in a U matrix can be seen as a representation of clusters whereas dark areas denote cluster separators. Thus a U matrix representation of a SOM can be a practical way to easily visualize clusters in the input space when no a priori information exists on the input data. Here, U matrices yielded by the four maps were used to decide on the presence of patient clusters on each analyzed MPQ dimension and the convenience of further statistical analyses.

2.3 Characterization of FMS patient subpopulations

Subjects located on each side of a cluster separator were considered as belonging to separate clusters. Subsequently, the resulting patient clusters were compared as to a number of clinical and psychological determinants, in order to verify the capability of the SOM analysis to discriminate between clinically distinct patient subpopulations. Firstly, we wished to know whether FMS patient subgroups discriminated by the SOM analysis on the basis of their scoring profiles on a given pain dimension also exhibited differential profiles on other dimensions of pain or differences in intensity of experienced pain. Secondly, we compared identified patient clusters in terms of severity of the disease and impact on health-related quality of life. Finally, we assessed whether or not patients with FMS belonging to different clusters also differed in associated psychological comorbidity.

3. Results

We used self-organizing maps to test whether verbal pain descriptors on the MPQ that address sensory, affective-motivational and evaluative dimensions of pain would be able to discriminate between clinically distinct FMS patients.

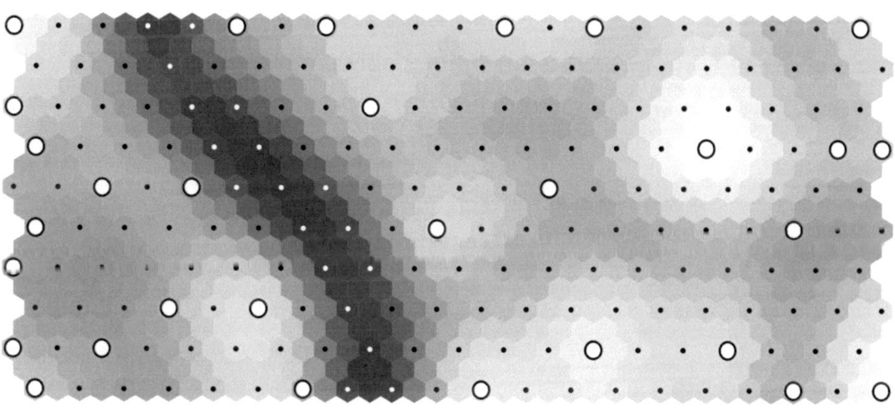

Fig. 1. U-matrix representation of responses from patients with FMS on MPQ items addressing the affective-motivational dimension of pain. Neurons of the network are marked as dots, whereas open circles indicate neurons representing patients with FMS. The cluster separating zone, represented here by the dark gap, segregates two patient subgroups which are termed here as cluster 1 (right side of the map) and cluster 2 (left side).

SOM analysis of the MPQ items corresponding to the sensory, evaluative and miscellaneous domains of the test failed to suggest the existence of clinically distinct patient subsets. Therefore, no further analysis was conducted on these dimensions. However, SOM analysis of the responses to items 15 through 17, i.e. those addressing the affective-motivational domain of pain, consistently revealed the existence of two distinct subject clusters (figure 1), which will be referred to as *cluster 1* and *cluster 2* hereafter.

Cluster 1 comprised 17 out of the 30 subjects of the sample, whereas 13 subjects were found to fall within cluster 2. General characteristics of both clusters are provided in table 1. No statistically significant differences were found between clusters with respect to age, gender, years since diagnosis, or active status. The two FMS patient clusters were indistinguishable in terms of MPQ scores on sensory-discriminative, evaluative and miscellaneous dimensional scores (table 2). As well, no significant between-clusters differences were found as to Present Pain Index scores or intensity of pain as measured by the VAS.

	cluster 1 n=13	cluster 2 n=17
Mean age (s.d.)	51.85 (7.76)	50.35 (12.21) (1)
Women (%)	100	88.2 (2)
Mean years from diagnosis	4.76 (1.32)	3.94 (0.95) (1)
Active status	23.2%	41,2% (2)

Table 1. Sociodemographics of patients with FMS classed as cluster 1 and cluster 2. Differences between groups were not significant ((1) Student's *t* test; (2) Mann-Whitney *U* test).

	cluster 1	cluster 2
McGill Pain Questionnaire		
Sensory PRI	21.00 ±4.912	22.15 ±4.298
Affective-motivational PRI	4.41 ±1.906	6.08 ±1.935 *
Evaluative PRI	7.71 ±2.054	7.85 ±2.267
Miscellaneous PRI	3.24 ±0.903	3.46 ±0.660
Total PRI	36.35 ±8.299	39.54 ±5.348
Number of words	14.00 ±1.852	16.54 ±1.450 **
PPI	2.82 ±0.809	3.08 ±1.038
Pain intensity (VAS)	6.76 ±1.15	7.05 ±0.92
Total FIQ score	6.83 ±1.296	7,90 ±1.277 *
SF-36		
Physical component Summary	39.81±9.206	29.77±7.224 **
Mental component Summary	50.60±25.436	22.22±9.149 **

Table 2. Summary of measures of pain and assessment of clinical status in FMS patient clusters. * $p<0.05$ and ** $p<0.01$ as compared to cluster 1.

Not surprisingly, the two subpopulations segregated by SOM analysis on the basis of affective dimensional scoring profiles differed significantly ($p<0.05$ at the Student's *t* test) in their affective Pain Rating Index (table 2). Specifically, cluster 1 patients exhibited lower scores (4.41 ±1.906) that cluster 2 patients (6.08 ±1.935). In addition, the total number of words chosen by cluster 1 patients was significantly lower (14.00 ±1.852) than cluster 2 patients (16.54 ±1.450). Clinical status in both patient clusters was assessed here by using the FIQ and the SF-36. Cluster 2 patients exhibited significantly more severe FMS in terms of total FIQ scores

(7,90±1.277) as compared to cluster 1 patients (6.83±1.296, $p<0.05$ at the Student's *t* test). Health-related quality of life was also significantly poorer in cluster 2 patients as assessed by the physical component summary (29.77±7.224) and the mental component summary (22.22±9.149) derived from the SF-36 items, relative to cluster 1 patients (39.81±9.206 and 50.60±25.436, respectively.

Psychological status was assessed by means of the SCL-90-R. The profile exhibited by FMS patients classed as cluster 1 was fully within normal levels for nonpsychiatric population (figure 2). In contrast, subjects in cluster 2 exhibited generally higher levels of psychopathology. Composite normalized T-scores in these patients were significantly higher and well above the diagnostic cutoff value of 60 on dimensions somatization, obsession-compulsion, depression, anxiety, paranoia, and psychoticism. On the global severity index (GSI), positive symptom total (PST) and positive symptom distress index (PSDI), cluster 2 patients also scored above the diagnostic level, although differences with respect to cluster 1 scores were not statistically significant.

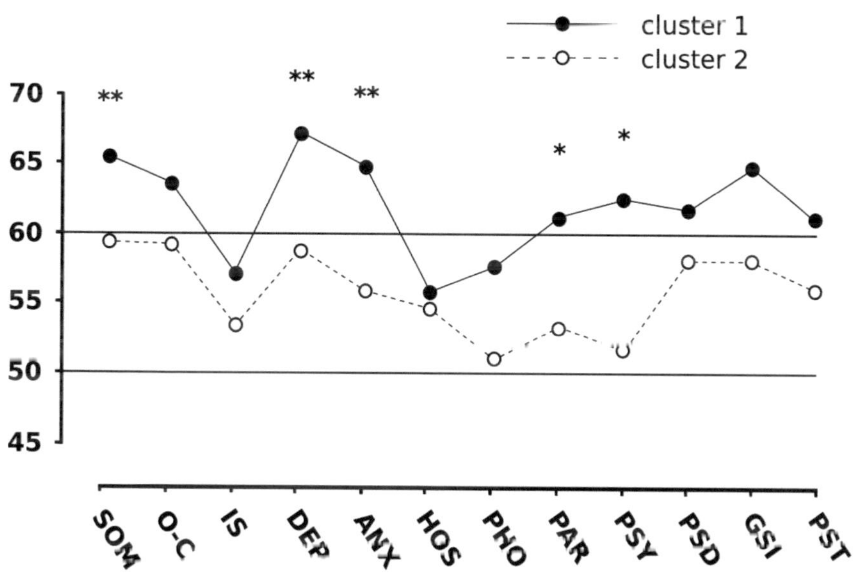

Fig. 2. Psychological profiles based on composite, normalized T-scores of the SCL-90-R in the two FMS patient clusters identified by SOM analysis. Scores of cluster 2 patients are generally higher and above diagnostic level on subscales somatization, obsession-compulsion, depression, anxiety, paranoia, and psychoticism.

4. Discussion

4.1 What is in the affective-motivational dimension of pain?

Pain descriptors have long been analyzed to differentiate medical from psychiatric diagnoses or to differentiate among medical diagnostic categories. Back in the 70's, Veilleux and Melzack (1976) proposed that the pattern of sensory and affective pain descriptors used by patients may be clinically useful, since patients with persisting pain complaint within a psychiatric population used more sensory than affective descriptors to describe their pain, whereas patients with rather transient pain used more affective than sensory descriptors.

These investigators suggested that a pattern dominated by sensory descriptors may indicate traditional methods of pain control, whereas psychiatric approaches may be more appropriate when affective descriptors prevail (Veilleux & Melzack, 1976).
Pain affect can be defined as the emotional arousal and disruption engendered by the pain experience. Arousal produced by pain, which may be experienced as frightening or distressing, can interfere in daily activities and habitual modes of response. Measures of pain affect can be complex and heterogeneous, and can be different from measures of pain intensity (Gracely, 1992; Jensen et al., 1989). In a sense, pain may not be different from other sensory-perceptual experiences in that, in the same way that identifying a pitch or a melody can be distinguished from what that sound means to us or how it makes us feel, pain intensity or location may be considered as different from its affective or motivational interpretation. The emotional response to pain is closely related to the meaning the patient gives to the pain experience. Hence, pain is likely to engender more suffering when it is perceived as indicative of threat to the subject's physical, psychological, or social integrity than when it is not considered to be a threat. It is thus not hard to understand that among cancer patients, the emotional aspects of pain can eventually come to dominate the whole clinical picture. According to their affective dimensional scores, cluster 2 FMS patients as characterized here are thus likely to associate negative affect to their perception of pain and disease.
The Affective subscale of the McGill Pain Questionnaire (MPQ) is by far the most widely used measure of pain affect. The results of the present study demonstrate that segregation of patients with FMS on the basis of the affective-motivational dimension of pain results in reliable segregation on other indicators of physical and psychological status. Patients classed as cluster 2, which exhibited higher scores on the pain affect scale, also reported poorer clinical status as revealed by the FIQ and the MCS and PCS, and showed psychological comorbidities in a number of subscales addressed by the SCL-90-R. In contrast, patients of cluster 1 showed no psychological distress and lower levels of severity of disease. A reasonable interpretation of this finding is that the affective dimension of the MPQ provides a good indication of psychological distress and mood disturbances associated to FMS. In connection with this, it may be argued that, since mood is known to modulate perceived pain intensity (Von Graffenried et al., 1998), higher scores on the affective dimension could be considered to indicate the contribution of pain affect to perceived pain intensity. However, we failed to find significant differences between clusters as to sensory-perceptual or evaluative dimensions of pain and intensity of pain as measured by the VAS. This appears to suggest that psychological factors may be influencing the clinical status of patients with FMS irrespective of sensory aspects and intensity of pain. As a consequence, interventions addressing associated psychopathology might be expected to improve FMS and health-related quality of life without necessarily alleviating the perceived levels of pain.

4.2 Clues to therapeutic intervention

It is reasonable to assume that treatment can be enhanced if it is appropriately tailored to the patient characteristics and that therapeutic interventions targeted at specific needs identified in patient subgroups should be substantially more effective than therapies prescribed just by matching to medical diagnosis. This is why research into the heterogeneity and individual variability in patients with FMS is of such great importance, as is the identification of distinctive features of relevant patient subgroups. The clinical practice should also respect this heterogeneity and should aim at developing interventions that best match the needs of the individual patient. There has been a growing awareness of the importance of psychological factors in reports of pain (Gatchel & Turk, 1999). For example, physical factors fail to predict pain severity or life interference in patients with low back pain or rheumatoid arthritis, whereas assessment of helplessness and hopelessness can help to predict reported pain and behavior in response to pain (Flor & Turk, 1988). In addition, variables other than medical diagnoses and identified physical pathology may affect treatment outcome (Jensen et al., 1994, 2001). In particular, perceptions of control over pain is inversely correlated with pain, depression, and physical disability, whereas catastrophizing and beliefs about being disabled may be directly associated.

It has long been suggested that high affective scores indicate the necessity of a psychotherapeutic approach (Veilleux & Melzack, 1976). Psychotherapy appears to be mandatory in subgroups of patients displaying measurable distortions of pain affect. An implication of the present results is that only one subset of patients with FMS, namely the one termed here as cluster 2, is likely to benefit from a psychotherapeutic intervention. Thus, in these patients, criteria used to tailor psychotherapeutic interventions may allow for different factors in a biopsychosocial model (Keefe et al., 2004). Treatment tailored to the shared cognitive behavioral factors may enhance treatment effects (Prins et al., 2001; Bazelmans et al., 2006). Indeed, in patients with chronic pain, treatment outcomes tend to be affected by specific cognitive behavioral factors such as passive pain coping and helplessness (Nicassio et al., 1995; Rollman & Lautenbacher, 1993). Cognitive-behavioral therapies can successfully modify key elements of the fear-avoidance model in patients with chronic low-back pain (de Jong et al., 2005; Vlaeyen et al., 2001, 2002). Likewise, factors shared in common by specific FMS patients subsets should be taken into account. For example, approaches aimed at pain acceptance that proved relevant in chronic pain should be considered for FMS (Evers et al., 2001; McCracken et al., 2005). Meta-analyses have concluded that cognitive-behavioral therapy in combination with physical exercise training is the most effective non-pharmacological treatment for FMS (Goldenberg et al., 2004; Hadhazy et al., 2000; Rossy et al., 1999; Sim & Adams, 2000). Behavioral reactions such as avoidance, cognitive reactions including increased attention to bodily sensations and catastrophizing, and physiological reactions such as increased neurovegetative arousal and muscle tension, all are appropriate adaptive short-term reactions. However, they may become detrimental in the long term in response to chronic pain (Evers et al., 2001; Flor & Birbaumer, 1990; Turk & Flor, 1999). For example, muscle tension and neurovegetative hyperactivity, which are physiological reactions to pain, may lead to higher levels of pain and functional disability in the long term if misinterpreted as evidence of physical harm (Bortz, 1984; Flor & Birbaumer, 1990). Catastrophizing is also an important cognitive factor in chronic pain (McCracken & Gross, 1993; Vlaeyen et al., 1995). Exaggeration of negative interpretations of pain may lead to increased levels of pain-related fear, and this may in turn result in increased attention to bodily sensations and hypervigilance to pain (McCracken, 1997). Finally, sustained avoidance of physical activities can worsen musculoskeletal condition, thereby contributing to exacerbate pain (Bortz, 1984). On this basis, a number of treatments have been aimed at modifying the patient's pain experience and disability. Although studies show that the cognitive-behavioral approach alone does not generally provide significant benefits to FMS patients over group educational or physical exercise programs, it may indeed be beneficial for patients in great psychological distress (Van Koulil et al., 2007).

5. Conclusions

The results of this pilot study based on SOM analysis suggest that the affective-motivational domain of the MPQ discriminates between two FMS patient subpopulations with distinctly different levels of illness severity and associated psychopathology. According to these findings, psychotherapeutic intervention on one subset of patients with FMS with associated psychopathology may result in better clinical status and improved quality of life.

6. Acknowledgement

Supported by Ayudas a Grupos de Investigación del Sistema Universitario Vasco (Gobierno Vasco/ Eusko Jaurlaritza). Monika Salgueiro was supported by Gobierno Vasco/ Eusko Jaurlaritza. Awarded *VI Premio Fibromialgia 2010* by *Fundación FF y Ciencia SER.*

7. References

Aaron L.A. & Buchwald, D. (2001) A review of the evidence for overlap among unexplained clinical conditions. *Ann Intern Med,* Vol. 134, pp. 868–881.

Ahles, T.A.; Ruckdeschel, J.C. & Blanchard E.B. (1984) Cancer-related pain-II. Assessment with visual analogue scales. *J Psychosom Res,* Vol. 28, pp. 121-124.

Alonso, J.; Prieto, L. & Antó, J.M. (1995) La versión española del SF-36 Health Survey (Cuestionario de Salud SF-36): un instrumento para la medida de los resultados clínicos. *Medicina Clínica (Barcelona),* Vol. 104 pp. 771-776.

Bazelmans, E.; Prins, J.B. & Bleijenberg, G. (2006) Cognitive behaviour therapy for relatively active and for passive chronic fatigue syndrome patients. *Cogn Behav Pract,* Vol. 13, pp. 157–166.

Bennett, R.M.; Campbell, S.; Burckhardt, C.; Clark, S.; O'Reilly, C. & Wiens, A. (1991) A multidisciplinary approach to fibromyalgia management: balanced approach provides small but significant gains. *J Musculoskeletal Medicine,* Vol. 8, pp. 21–32.

Bortz, W.M. (1984) The disuse syndrome. *West J Med,* Vol. 141, pp. 691-694.

Burckhardt, C.S.; Clark, S.R. & Bennett, R.M. (1991) The fibromyalgia impact questionnaire: development and validation. *Journal of Rheumatology,* Vol. 18, pp. 728-733.

de Jong, J.; Vlaeyen, J.W.; Onghena, P;, Goossens, M.E.; Geilen, M. & Mulder, H. (2005) Fear of movement/(re)injury in chronic low back pain: education or exposure in vivo as mediator to fear reduction? *Clin J Pain,* Vol. 21, pp. 9–17.

Derogatis, L.R. (1994) *SCL-90-R, Symptom Checklist 90 Revised.* Minneapolis: NCS Pearson.

Downie, W.W.; Leatham, P.A.; Rhind, V.M.; Wright, V.; Branco, J.A. & Anderson, J.A. (1978) Studies with pain rating scales. *Ann Rheum Dis,* Vol. 37, pp. 378-381.

Dunkl, P.R.; Taylor, A.G.; McConnell, G.G.; Alfano, A.P. & Conaway, M.R. (2000) Responsiveness of fibromyalgia clinical trial outcome measures. *Journal of Rheumatology,* Vol. 27, pp. 2683-2691.

Epstein, S.A.; Kay, G.; Clauw, D.J.; Heaton, R.; Klein, D. & Krup, L. (1999) Psychiatric disorders in patients with fibromyalgia. A multicenter investigation. *Psychosomatics*, Vol. 40, pp. 57-63.

Evers, A.W.; Kraaimaat, F.W.; van Lankveld, W.; Jongen, P.J.; Jacobs, J.W. & Bijlsma, J.W. (2001) Beyond unfavorable thinking: the illness cognition questionnaire for chronic diseases. *J Consult Clin Psychol*, Vol. 69, pp. 1026-1036.

Evers, A.W.; Kraaimaat, F.W.; van Riel, P.L. & Bijlsma, J.W. (2001) Cognitive, behavioral and physiological reactivity to pain as a predictor of long-term pain in rheumatoid arthritis patients. *Pain*, Vol. 93, pp. 139-146.

Flor, H. & Turk, D.C. (1988) Chronic back pain and rheumatoid arthritis: predicting pain and disability from cognitive variables. *J Behav Med*. Vol. 11, pp. 251- 265.

Flor, H.; Birbaumer, N. & Turk, D.C. (1990) The psychobiology of chronic pain. *Adv Behav Res and Therapy*, Vol. 12, pp. 47–84.

Fox, E.J. & Melzack, R. (1976) Trancutaneous electrical stimulation and acupuncture: comparison of treatment for low-back pain. *Pain*, Vol. 2, pp. 141-148.

Gatchel, R.J. (1996) Psychological disorders and chronic pain: cause and effect relationships. In:. *Psychological approaches to pain management: a practicioner´s handbook*. Gatchel, R.J. & Turk, D.C. (Eds.), pp. 33-54. New York: Guildfor Publications.

Gatchel, R.J. & Turk, D.C, eds. (1999) *Psychosocial Factors in Pain: Critical Perspectives*. New York, NY: Guilford Press.

Goldenberg, D.L.; Burckhardt, C. & Crofford, L. (2004) Management of fibromyalgia syndrome. *JAMA*, Vol. 292, pp. 2388-2395.

González de Rivera, J.L.; Cuevas, C.; Rodríguez, M. & Rodríguez, F. (2002) *SCL-90-R Cuestionario de 90 Síntomas*. Madrid: TEA Ediciones.

Gormsen, L.; Rosenberg, R.; Bach, F.W. & Jensen, T.S. (2010) Depression, anxiety, health-related quality of life and pain in patients with chronic fibromyalgia and neuropathic pain. *European Journal of Pain*, Vol. 14, pp. 127.e1-127.e8.

Gracely, R.H. (1992) Evaluation of multi-dimensional pain scales. *Pain*, Vol. 48, pp. 297-300.

Grushka, M. & Sessle, B.J. (1984) Applicability of the Mc-Gill Pain Questionnaire to the differentiation of `toothache' pain. *Pain*, Vol. 19, pp. 49-57.

Hadhazy, V.A.; Ezzo,J.; Creamer, P.; Berman, B.M. & McCain, G.A. (2000) Mind-body therapies for the treatment of fibromyalgia. A systematic review. *J Rheumatol*, Vol. 27, pp. 2911-2918.

Huskisson, E.C. (1974) Measurement of pain. *Lancet*, Vol. 2, pp. 1127-1131

Jensen, M.P.; Karoly, P. & Braver, S. (1986) The measurement of clinical pain intensity: a comparison of six methods. *Pain*, Vol. 27, pp. 117-126.

Jensen, M.P.; Karoly, P.; O'Riordan, E.F.; Bland, F.Jr. & Burns, R.S. (1989) The subjective experience of acute pain: an assessment of the utility of 10 indices. *Clin J Pain*, Vol. 5, pp. 153-159

Jensen, M.P.; Turner, J.A. & Romano, J.M. (1994) Correlates of improvements in multidisciplinary treatment of chronic pain. *J Consult Clin Psychol*, Vol. 62, pp. 172–179.

Jensen, M.P.; Turner, J.A. & Romano, J.M. (2001) Changes in beliefs, catastrophizing, and coping are associated with improvements in multidisciplinary pain treatment. *J Consult Clin Psychol*, Vol. 69, pp. 655–662.

Keefe, F.J.; Rumble, M.E.; Scipio, C.D.; Giordano, L.A. & Perri, L.M. (2004) Psychological aspects of persistent pain: current state of the science. *J Pain*, Vol. 5, pp. 195–211.

Kohonen, T. (1995). *Self-Organizing Maps. Series in Information Sciences*. Vol. 30. Springer, Heidelberg.

Kremer, E.; Atkinson, J.H. & Ignelzi, R.J. (1981) Measurement of pain: patient preference does not confound pain measurement. *Pain*, Vol. 10, pp. 241-248.

Littman, G.S.; Walker, B.R. & Schneider, B.E. (1985) Reassessment of verbal and visual analog ratings in analgesic studies. *Clin Pharmacol Ther*, Vol. 38, pp. 16-23.

Masi, A.T. & Yunus, M.B. (1991) Fibromyalgia: which is the best treatment? A personalized, comprehensive, ambulatory, patient-involved management programme. *Baillieres Clin Rheumatol*, Vol. 4, pp. 333-370.

McCracken, L.M. & Gross, R.T. (1993) Does anxiety affect coping with chronic pain? *Clin J Pain*, Vol. 9, pp. 253-259.

McCracken, L.M. (1997) "Attention" to pain in persons with chronic pain: A behavioral approach. *Behav Ther*, Vol. 28, pp. 271-284.

McCracken, L.M.; Vowles, K.E. & Eccleston, C. (2005) Acceptance-based treatment for persons with complex, long standing chronic pain: a preliminary analysis of treatment outcome in comparison to a waiting phase. *Behav Res Ther*, Vol. 43, pp. 1335-1346.

Melzack, R.; Terrence, C.; Fromm, G. & Amsel, R. (1986) Trigeminal neuralgia and atypical facial pain: use of the McGill Pain Questionnaire for discrimination and diagnosis. *Pain*, Vol. 27, pp. 297-302.

Monterde, S.; Salvat, I.; Montull, S. & Fernández-Ballart, J. (2004) Validación de la versión española del Fibromyalgia Impact Questionnaire. *Revista Española de Reumatología*, Vol. 3, pp. 507-513.

Müller, W.; Schneider, E.M. & Stratz, T. (2007) The classification of fibromyalgia syndrome. *Rheumatol Int*, Vol. 27, pp. 1005-1010.

Nicassio, P.M.; Schoenfeld, S.K.; Radojevic, V. & Schuman, C. (1995) Pain coping mechanisms in fibromyalgia: relationship to pain and functional outcomes. *J Rheumatol*, Vol. 22, pp. 1552-1558.

Ohnhaus, E.E. & Adler, R. (1975) Methodological problems in the measurement of pain: a comparison between the verbal rating scale and the visual analogue scale. *Pain*, Vol. 1, pp. 379 381.

Prins, J.B.; Bleijenberg, G.; Bazelmans, E.; Elving, L.D.; de Boo, T.M.; Severens, J.L.; van der Wilt, G.J.; Spinhoven, P. & van der Meer, J.W. (2001) Cognitive behaviour therapy for chronic fatigue syndrome: a multicentre randomised controlled trial. *Lancet*, Vol. 357, pp. 641-647.

Rollman, G.B. & Lautenbacher, S. (1993) Hypervigilance effects in fibromyalgia: pain experience and pain perception. In: *Progress in fibromyalgia and myofascial pain*. Vaeroy, H. & Merksey, H. (Eds.) pp. 149-159. Amsterdam: Elsevier.

Rossy, L.A.; Buckelew, S.P.; Dorr, N.; Hagglund, K.J.; Thayer, J.F.; McIntosh, M.J.; Hewett, J.E. & Johnson, J.C. (1999) A meta-analysis of fibromyalgia treatment interventions. *Ann Behav Med*, Vol. 21, pp. 180-191.

Rybstein-Blinchik, E. (1979) Effects of different cognitive strategies on chronic pain experience. *J Behav Med*, Vol. 2, pp. 93-101.

Scott, J. & Huskisson, E.C. (1976) Graphic representation of pain. *Pain*, Vol. 2, pp. 174-184.

Shaver, J.L.F.; Lentz, M.; Landis, C.A.; Heitkemper, M.M.; Buchwald, D. & Woods NF. (1997) Sleep, psychological distress, and stress arousal in women with fibromyalgia. *Research in Nursing & Health*, Vol. 20, pp. 247-257.

Sim, J. & Adams N. (2002) Systematic review of randomized controlled trials of nonpharmacological interventions for fibromyalgia. *Clin J Pain*, Vol. 18, pp.324-336.

Torres, X.; Bailles, E.; Collado, A.; Taberner, J.; Gutierrez, F. & Peri, J.M. (2010) The Symptom Checklist-Revised (SCL-90-R) is able to discriminate between simulation and fibromyalgia. *Journal of Clinical Psychology*, Vol. 66, pp. 774-790.

Turk, D.C.; Okifuji, A.; Sinclair, J.D. & Starz, T.W. (1996) Pain, disability, and physical functioning in subgroups of patients with fibromyalgia. *J Rheumatol*, Vol. 23, pp. 1255-1262.

Turk, D.C.; Okifuji, A.; Sinclair, J.D. & Starz, T.W. (1998) Differential responses by psychosocial subgroups of fibromyalgia syndrome patients to an interdisciplinary treatment. *Arthritis Care Res*, Vol. 11, pp. 397–404.

Turk, D.C. & Flor, H. (1999) Chronic pain: a biobehavioral perspective. In: *Psychosocial factors in pain: critical perspectives*. Gatchel, R.J. & Turk, D.C. (Eds.) pp. 18-34. New York: Guilford Press.

van Koulil, S.; Effting, M.; Kraaimaat, F.W.; van Lankveld, W.; van Helmond, T.; Cats, H.; van Riel, P.L.; de Jong, A.J.; Haverman, J.F. & Evers, A.W. (2007) Cognitive-behavioural therapies and exercise programmes for patients with fibromyalgia: state of the art and future directions. *Ann Rheum Dis*, Vol. 66, pp. 571-581.

Veilleux, S. & Melzack, R. (1976) Pain in psychotic patients. *Exp. Neural*, Vol. 52, pp. 535-543.

Vlaeyen, J.W.; Kole-Snijders, A.M.; Boeren, R.G. & van Eek, H. (1995) Fear of movement/(re)injury in chronic low back pain and its relation to behavioral performance. *Pain*, Vol. 62, pp. 363-372.

Vlaeyen, J.W.; de Jong, J.; Geilen, M.; Heuts, P.H. & van Breukelen, G. (2001) Graded exposure in vivo in the treatment of pain-related fear: a replicated single-case experimental design in four patients with chronic low back pain. *Behav Res Ther*, Vol. 39 pp. 151–166.

Vlaeyen, J.W.; de Jong, J.; Geilen, M.; Heuts, P.H. & van Breukelen, G. (2002) The treatment of fear of movement/(re)injury in chronic low back pain: further evidence on the effectiveness of exposure in vivo. *Clin J Pain*, Vol. 18, pp. 251–261.

Von Graffenried, B.; Adler, R.; Abt, K.; Nuesch, E. & Spiegel, R. (1978) The influence of anxiety and pain sensitivity on experimental pain in man. *Pain*, Vol. 4, pp. 253-263.

Williams, D.A.; Urban, B.; Keefe, F.J.; Shutty, M.S. & France, R. (1995) Cluster analyses of pain patients' responses to the SCL-90R. *Pain*, Vol. 61, pp. 81–91.

Wilke, W.S.; Gota, C.E. & Muzina, D.J. (2010) Fibromyalgia and bipolar disorder: a potential problem? Bipolar Disord, Vol. 12, pp. 514-522.

Wolfe, F.; Smythe, H.A.; Yunus, M.B.; Bennett, R.M.; Bombardier, C.; Goldenberg, D.L.; Tugwell, P.; Campbell, S.M.; Abeles, M. & Clark, P. (1990) The American College of Rheumatology 1990 Criteria for the Classification of Fibromyalgia. Report of the Multicenter Criteria Committee. *Arthritis Rheum*, Vol. 33, pp. 160-172.

Wolfe, F.; Clauw, D.J.; Fitzcharles, M.A.; Goldenberg, D.L.; Katz, R.S.; Mease, P.; Russell, A.S.; Russell, I.J.; Winfield, J.B. and Yunus, M.B. (2010) The American College of Rheumatology Preliminary Diagnostic Criteria for Fibromyalgia and Measurement of Symptom Severity. *Arthritis Care Res,* Vol. 62, pp. 600-610.

Woodforde, J.M. & Merskey, H. (1972) Some relationships between subjective measures of pain. *J Psychosom Res,* Vol. 16, pp. 173-178.

7

The Difficulties in Developing and Implementing Fibromyalgia Guidelines

M. Reed and M. Herrmann
Otto von Guericke University,
Magdeburg,
Germany

1. Introduction

Fibromyalgia syndrome (FMS) is increasing in prevalence both in Europe and the rest of the developed world. In 1995 the prevalence was estimated in the USA as roughly 1 in 50 with a female to male ratio of 6:1 (Wolfe F. et al, 1995). In 1998 the German health survey identified similar figures. In 2010 a survey of five European countries estimated it to be now as high as 1 in 30 (Branco et al, 2010), with a sex ratio of 2 to 1. This ratio varies widely in different countries being as high as forty to one in Brazil (Senna et al, 2004). The presence of the sex ratio and it's variability is one of many hotly debated areas in FMS. Why prevalence of the syndrome should be increasing is another area under discussion, although it is not as controversial as to how such a subjective condition should be accurately diagnosed and effectively treated; there have been many guidelines constructed by experts to assist in these two domains and this is the focus of the following chapter.

2. Guidelines

It might be useful before beginning to focus on fibromyalgia to simply ask the question, why do we have guidelines? In Britain alone, there are thousands of guidelines and protocols, some of which are formulated within a particular clinic, some for the wider region and some nationally; many give little more than general advice such as antibiotic guidelines in urinary tract infections, whereas others are expected to be adhered to more strongly, such as deep venous thrombosis protocols. Is it possible for a doctor to even be aware of the existence of every one of these guidelines? Indeed, it could be asked whether our obsession with guidelines is encouraging our healthcare practitioners to simply learn to follow stepwise instructions rather than to think for themselves? In the future will doctors be replaced by robots who are quicker, more accurate and likely to follow protocols better? The answer is surely "No!" if only because the doctor themselves exerts a great healing effect simply through their own person, that is through the interaction of the doctor-patient relationship (Balint, 1957).

It is also hard to imagine how any computer algorithm would be able to adapt guidelines to the benefit of the patient in front of them; this is where guidelines and protocols differ of course; a protocol must be followed and if it isn't one is open to legal challenge in the event of future problems. They are well suited to complex drug regimens such as the setting up of chemotherapy; such repetitive tasks benefit from protocols to assist practitioners in not missing important steps. But this is different to the role of guidelines. The Institute of Medicine defines guidelines as "systematically developed statements to assist practitioner and patient decisions about appropriate health care for specific clinical circumstances." (Field and Lohr, 1990). In other words they are devised in order to help clinicians make better decisions. When reading any guideline it is this that the physician should ask him or herself.

2.1 Why are guidelines useful to doctors?

A good guideline helps the physician by addressing only the major decisions and making the evidence clear at each stage (Jackson and Feder, 1998). Ideally it should be as concise as possible for obvious reasons. This is why NICE (the National Institute of Clinical Excellence, in the UK) publishes 'long' guideline versions and 'short' versions. Moreover, good guidelines help doctors in their continuing professional development (CPD) by providing clinical overviews and identifying knowledge gaps. However, despite their apparent usefulness, research has shown that guidelines are not very effective at changing physician practice (Woolf S.W., et al 1999). Other important benefits to guidelines are empowering patients and informing health policy. Even the existence of a guideline will give publicity to a condition and bring it to the attention of politicians.

It has been shown that guidelines can improve medical care (Grimshaw and Russell, 1993) but this naturally pre-supposes they are of high quality. Even with the best intentions however, guideline committees can miss evidence, and they can allow their subjective opinions to override what might be good evidence. This also extends to their recommendations. Attempts to minimise this by having large committees ameliorate this to a degree but are themselves open to "group-think", whereby a subtle peer pressure leads group members to agree with each other and advocate ideas that may not be right. This is the great danger with guidelines that are based on 'group consensus' but it is reduced by having members from as wide and different backgrounds as possible. Finally guidelines often have conflicting interests. Good guidelines have any potential conflicts of interest (normally financial) stated at the end. However, some guidelines are specifically constructed to address matters such as cost. This raises such thorny issues as the value of a life, or the value of an extra healthy year (Sassi, 2006). Finally, doctors also may be influenced even unwittingly by other priorities or special interests.

Guidelines are supposed to encourage more consistent practice and reduced variation (both geographical and between doctors) but where does this leave individualised medicine, which is the bedrock of modern primary care (Schuster et al, 2006)? Even guidelines that are accurate, up-to-date, well-evidenced and patient-centred may conflict with other guidelines and may undermine the years of training physicians have undertaken.

It has been said that guidelines are only useful when scientific evidence can provide an answer; given this, it is arguable that there is not a great deal of point in publishing guidelines in the absence of useful evidence. Such guidelines would only really be able to give recommendations based on consensus opinion, that is, the opinion of people who have not met our patients, but only their own. It is this which is the main problem in the construction of guidelines for fibromyalgia syndrome.

3. Guidelines for diagnosis

In response to the huge amount of published work concerning FMS over the decades there have been a number of attempts to elucidate the best ways to diagnose and treat it. Thus far, there have proven no physical tests that are consistently positive in people who complain of the symptoms characteristic of the disorder. In other words, there is no "gold standard" that enables physicians to rule in or rule out the diagnosis.

3.1 There are no useful diagnostic tests

In actual fact, there has been only one biochemical marker that research has shown to be commonly raised (as much as threefold) in the condition and that is substance P, a neurotransmitter , thought to be involved in pain perception, as measured in cerebrospinal fluid (Russell et al, 1994). However, there are a number of reasons why this is not likely to prove a useful test: firstly there is the sheer practicality of measuring the level of a neurotransmitter, as it would need invasive testing - there is no peripheral blood correlate; secondly it does not always appear to be raised and the levels do not correlate with subjective pain - so whilst there is an apparent relationship any test is not likely to be specific or sensitive enough to be useful. It might be theorised that actually what the raised biochemical marker is telling us is the same as what the patient is saying, that is to say that they are suffering pain. Any subjective feeling should have a neurochemical correlate, if we can but find it.

3.1.2 Altered central nervous system processing

Indeed functional magnetic resonance imaging scans of patients suffering with FMS shows that areas of the brain correlating with pain perception (such as the primary and secondary somatosensory cortices) are more greatly activated than in normal controls (Gracely et al, 2002). In this study similar subjective pain perception resulted in similar cortical activation in FMS patients compared with controls, whereas similar pressure application led to far greater cortical activation in the FMS patients. But what does this actually tell us? It tells us that the patient is truly perceiving pain; it tells us that are not lying to their doctors. But even doctors who question the validity of the entire syndrome would surely never suggest that patients are making up their symptoms; exaggerating, misinterpreting, misunderstanding, perhaps, but our thoughts and feelings need a physical medium to function; if we measure the workings of that medium we ought to see correlative changes.

This is given further weight by work which suggests that the threshold for perceiving pain is lower at a spinal level in certain chronic pain conditions, such as chronic knee pain, whiplash, and fibromyalgia (Lim et al, 2011). In other words, patients with chronic pain conditions are more sensitive, a description which would hopefully not generate offence.

3.2 Diagnostic guidelines

In the absence of any tests there have been other attempts to create useful diagnostic guidelines. The first widely accepted ones were those constructed by the American College of Rheumatology (ACR), originally to enable consistent research, ironically (Wolfe et al, 1990).

3.2.1 ACR criteria 1990

These guidelines were simple to use in that they defined FMS as widespread persistent pain of more than 3 months duration in three out of four body quadrants, as well as the axis (spine or sternum). Moreover, physical examination was included, namely concerning pressure points, of which 11 of 18 needed to be positive for a diagnosis. Physicians were also supposed to exclude other causes; whilst this is obviously wise, it is now recognised that other diseases can trigger FMS, although the links are weak; only 10% of lupus sufferers fulfil FMS criteria for example (Taylor et al, 2000). Despite its simplicity however, it was not widely used by primary care doctors and many patients who were given diagnoses of FMS did not fulfil the 1990 ACR criteria. Even primary care doctors who did perform tender point examinations often did so incorrectly, for reasons which are likely to be manifold, although a scepticism with the 1990 guidelines is probably part of it. Such scepticism was formalised by a study which showed that sham points were nearly as discriminative as 'ACR points' and that pain at three 'ACR points' was equivalent to any greater number for diagnostic accuracy (Harden et al, 2007).

The ACR paper built on previous work, of course, the most foundational being that of Yunus, often known as the father of fibromyalgia. It was he who first applied the term "primary fibromyalgia" in a systematic way to a syndrome of muscular aches, tiredness, anxiety, poor sleep and subjective swelling. Before him the term "fibrositis" had been used for nearly a century (Gowers, 1904). This referred unsystematically to patients with hard to define muscular pains, although it was Hench, in 1976, who suggested the term 'fibromyalgia' was more appropriate as there was no evidence of inflammation (implied by the suffix '-itis'). It was Yunus who then adopted it (Yunus, 1981) and formally applied it the condition under discussion.

As Wolfe elaborates in the 2010 paper (Wolfe et al, 2010) the original 11/18 tender points from the 1990 ACR criteria have not been used consistently as a diagnostic tool. Indeed he was one of the first authors to say they should actually not be (Wolfe, 2003). At least 25% of patients with a diagnosis of FMS have never satisfied the 1990 list, and hence new criteria were proposed which examine a wider range of symptoms, and are quantifiable using two parameters: the wide-spread pain index (WPI) and the symptom severity scale (SS).

3.2.2 Proposed new criteria from 2010

The chosen symptoms are these: fatigue, waking unrefreshed, cognitive symptoms and somatic symptoms. The minimum duration is again three months, and must not be explained by another physical disease process. Experienced General Practitioners (GPs) and rheumatologists will know well the huge variety of potential symptoms the last two categories comprise. It is immediately obvious that such fluid and subjective measures are likely to mean that some patients will slip in and out of a diagnosis of fibromyalgia, but this is, of course, what is actually seen, as patients move from good to bad phases and back again. Usefully, as Wolfe pertinently observes at the end of his 2010 editorial (Wolfe, 2010), these new criteria now enable us to "study fibromyalgia syndrome ... without the requirement for belief in its existence."

This table summarises the two diagnostic strategies:

Guideline	Pain	Regions	Other symptoms	Other conditions
1990 ACR	11 out of 18 tender-points	3 out of 4 quadrants +axis	Not diagnostic	Exclude diagnosis
Proposed 2010 - Criteria 1	WPI ≥ 7 in the last week	Any of 19 areas.	SS scale ≥ 5	Concurrent diagnosis acceptable.
- Criteria 2	WPI ≥ 3 in the last week	Any of 19 areas.	SS scale ≥ 9	Concurrent diagnosis acceptable.

Table 1.

The new criteria obviate the need for a physical examination, which seems sensible, as there is not actually anything physical to find. The WPI roughly correlates to the tender points notion except that patients are *asked* if particular areas have been painful over the last week. A score of 7 (out of 19) or more is diagnostic if accompanied by a SS score of 5 or greater; alternatively an SS greater than or equal to 9 with a WPI of 3-6 also qualifies; it will be noted that the second scale has much overlap with chronic fatigue. Each of the four symptom categories is scored 0-3, giving a potential score of 12. It is worth noting that these new provisional criteria classify nearly 90% of patients as having FMS who fulfilled the original tender point criterion. The new criteria also enable easier longitudinal patient assessment and make it simpler for GPs to assess patients. At this stage the new criteria are still provisional depending on feedback and validation through research, but the improvement they offer is clear and they are likely to be official before too long. There are voices of concern however; Smythe (Smythe 2011) for example, suggests that in discarding the 'objective measure' of tender points, the scope and hence number of those diagnosed is likely to rise even further. Diagnosis is now based purely upon subjective feelings with the huge potential for disease mongering that this creates (see below). Also the diagnostic strategies will include different types of patient. He also questions the validity of allowing overlap with other conditions and symptom constellations; in essence, the new criteria push us to think of FMS no longer as simply a pain condition with associated symptoms. This has implications for many of the guidelines below whose emphasis was actually upon pain rather than these other subjective disturbances.

3.3 Part of the somatoform spectrum

The lack of abnormalities on physical examination, and the absence of objective diagnostic tests lead many physicians to question whether FMS should have its own diagnostic category or whether it should simply be categorised in the spectrum of somatoform illnesses (Ciccone and Natelson, 2003). Even Wolfe, the lead author of the paper defining the 1990 ACR fibromyalgia classification criteria, having first in 2003, questioned the use of the tender points classification he helped to devise, now feels that FMS should not be recognised as a disease at all but as a greater than normal physical response to stress, depression, and anxiety; he has written that it should be recognised simply as part of the human condition (Wolfe 2010), although most authors would argue that it is, at the very least, at the far end of the spectrum. Indeed many of the opponents of the FMS diagnostic category feel that the label may legitimise sickness behaviour and can slow recovery (Payer, 1992).

Whilst some authors question the validity of the diagnosis at all, preferring to class it within the wider "functional somatic syndromes set" (Kanaan et al, 2007) there have also, conversely, been attempts to actually sub-divide FMS. For example Mueller in 2007 (Mueller et al, 2007) advocated the following four sub-divisions:

- extreme sensitivity to pain but no associated psychiatric conditions
- fibromyalgia and comorbid, pain-related depression
- depression with concomitant fibromyalgia syndrome
- fibromyalgia due to somatization

They even suggest that these sub-divisions may ("may" being the operative word) respond to different therapies; this will require a great deal of further research to refute or deny. Perhaps the most logical position is that FMS is an 'illness' ie a constellation of symptoms, rather than a disease, which is normally understood as being based on a pathophysiological disorder (Wilke, 1999).

3.4 The value of receiving a diagnosis is controversial

Whilst anecdotally, many GPs might agree that labels appear to legitimise sickness behaviour, there is little evidence that the presence or absence of the fibromyalgia label has any effect on the course of the disease over the years (White et al, 2002). Indeed there is little which does alter the course of FMS, leading to the observation that most cases of FMS are 'resistant' cases! (Wilke, 1995). Neither inpatient rehabilitation nor early retirement seem to affect it. Indeed most patients are physically able to work full-time in jobs requiring light activities (Raspe et al, 1994). What does appear to improve is patient satisfaction with their health status once they receive the diagnosis; it seems that, whilst the label might not help people recover or function better, most are happy to at least "know" what the problem is, something that is consistently observed across a range of conditions, whether improvement then follows or not (Choy et al, 2010).

Indeed this reflects the fact that prognostication has historically been a very important physician skill, although it is less so these days (perhaps naturally enough, when many diseases can actually be cured rather than simply observed); prognosis is still highly valued by patients, though (Christakis, 1999). However, with FMS patients it appears this satisfaction is short-lived; in the UK it has been found that visits to the doctor, for any reason, are reduced for about two years following diagnosis, but that they rise to even higher levels after that, typically over twice that of controls ie all other patients (Hughes et al, 2006).

3.5 Aetiology is controversial

Another element which contributes to the difficulty in creating diagnostic guidelines is the unknown aetiology, indeed the absence of any plausible mechanism. What appears to be true is that psychosocial factors predominate (both in the causality and also in effective treatment) but this is very hard to investigate. The long-term, refractive nature of the condition makes comparative studies very hard to achieve. Patients also tend to have strong beliefs regarding their diagnosis, treatment and prognosis which creates resistance to many proffered treatments and difficulties with blinding studies. In particular patients are very resistant to psychological explanations or solutions and for the most part reject them. Indeed it is notoriously easy to offend patients with FMS; the notion of number needed to offend (NNO) is useful here: 40% of patients with medically unexplained symptoms find the terms 'psychosomatic' or 'medically unexplained' offensive in the UK, which is an NNO of 2-3; this is less than those offended by the term 'hysterical' but more than the mere 12% (NNO 9) who object to the term 'functional' (Stone et al, 2002).

It is probably true that many patients fear not being taken seriously by the doctor, or adjudged to be abusing their time. Every day, patients apologise to their GPs for wasting their valuable time (to which the conscientious GP replies "but without you I would be unemployed!"). When someone with fibromyalgia, who experiences daily pain is told their problem is " all in the mind" and they can solve it by simply "pulling yourself together" it is unsurprising that they would both disagree and be offended; of course, few GPs would phrase it like this, but it is this that patients with FMS often hear.

The most widely accepted hypothesis of causality is that of increased brain sensitivity to pain stimuli. However, this is not really a explanation at all, as it is just another way of stating what patients tell us. The question is, of course, how could such a sensitivity change arise? Numerous associations have been noted as statistically significant, such as pre-morbid anxiety (Netter and Hennig, 1998), abuse (Haeuser et al, 2011), negative life events (Anderberg et al, 2000), genetics (Rapael et al, 2004), and neck trauma (Buskila et al, 1997). Levels of dopamine and serotonin have also been observed to be deranged in some studies, although inconsistently; once again this raises the question of what is actually being measured: cause or effect (or neurological substrate)?

So, diagnosis and causality are controversial. There are no tests. There are no objective signs. All the symptoms are subjective. However, anyone who feels that great controversy has thus far been demonstrated 'ain't seen nothing yet'.

4. Treatment guidelines

In 1987 Goldenberg prophesied that fibromyalgia was "emerging but controversial condition" (Goldenberg, 1987). He was proved quickly right. Attempts at making guidelines for FMS were made even before Goldenberg's prophetic comments. Indeed, in 1986, Hench created an early incarnation of guidelines (Hench, 1986) and suggested that treatment should include physical, behavioural, psychological, and pharmacological means. Moreover he observed that no drug therapy had proved particularly successful, the tendency being only to provide temporary relief from pain. He suggested that studies aimed at defining the cause have linked it to sleep disorders, neurogenic mediators, immune mechanisms, muscle disease, and psychological disturbances. His final conclusion was that after the establishment of an initial therapy program, patients should assume the major responsibility for management.

Nowadays, the phrase 'responsibility for management' would be put differently but the attitude that patients should be listened to, their preferences taken into account and that they should be fully involved in their management plan is a very vogue attitude that is unlikely to change (Woolf S.H., 1997). In so far as it aids compliance or even advances concordance this stance is admirable in many conditions (Bissell et al, 2004). For example, there are numerous anti-hypertensives for which an important part of the treatment plan is how a particular person feels about side-effects and overall efficacy; hence advocating choice is very valid in this condition. Moreover different patients will respond differently to different drugs; indeed the placebo effect itself is likely to be different with different medications and different patients (Zhang et al, 2008). However, there is sadly no evidence that advocating patient choice or encouraging patients to seek therapies with which they are happier has any effect in improving outcomes in FMS. It is likely that this reflects the absence of truly effective options. This raises the question of whether it is sensible, or even ethical, to offer our patients a choice between cognitive behavioural therapy and reflexology, between an exercise program and homeopathy?

4.1 Guideline for the management of fibromyalgia syndrome pain in adults and children 2005

Goldenberg himself, in late 2004, (Goldenberg et al, 2004) observed that despite there being in existence papers concerning treatment for decades, stepwise guidelines were absent in the medical literature. This was following the otherwise well received American Pain Society (Buckhardt et al, 2005), to which Goldenberg himself had just contributed. A patient version of the guidelines was also produced, aimed at improving patients' knowledge on the aetiology and effective treatment options available, thus promoting self-management. The guidance graded the diagnostic and treatment evidence-base according to the well-recognised Oxford Centre for Evidence-based medicine system (Oxford University, 2009), that is 1-5 for evidence levels and A-D for Grades of Recommendation.

The guideline presented six diagnostic recommendations (four according to panel consensus), nineteen broad treatment recommendations for adults (including ten consistent with Grade A level) and ten for children (all according to panel consensus, but logically extrapolated from the adult evidence). This guideline was first formulated in 2004 and has since been updated in 2007, in response to new pharmacological research concerning antidepressants (Guideline central, 2007). It appears at first glance very impressive but the lack of useful treatments is laid bare by the fact that two of the level A recommendations are 'do not's (steroids and non-steroidal anti-inflammatories) and two are appeals to use many different strategies. This leaves the following A graded advice: gentle exercise, balneotherapy, sedatives, tricyclics, education and cognitive behavioural therapy. It is no wonder that physicians responded to such advice with frustration.

4.1.1 Goldenberg's stepwise guideline 2005

Doctors, particularly general (or family) practitioners often like to think in terms of step-wise management plans. This has the advantage of encouraging recognised 'best practice', simplifying treatment decisions and making the treatment journey easier for patients to understand. A good example of such stepwise treatment plans is the British Thoracic Societies guideline for asthma which is so useful for GPs in the United Kingdom (BTS, 2011). Presumably partly as a consequence of the perceived flaws in the guideline above, Goldenberg and two of the other authors constructed a step-wise treatment algorithms in an effort to guide doctors (particularly family practitioners) and patients more clearly along a treatment path, a journey which most research suggests is inevitable. It delineated a three step process. The first step should involve diagnosis, education and treatment of any co-morbid conditions; this is followed by tricyclic antidepressants (TCAs), exercise and cognitive behavioural therapy (CBT); thirdly, the following drugs were suggested as second-line: tramadol, selective serotonin uptake inhibitors (SSRIs), anti-convulsants or a combination of medications. If all else fails then refer to rheumatology, the pain team or a psychiatrist; their guidance does not extend to suggesting what might then subsequently be done, although most wily primary care doctors will already have a fairly shrewd idea.

4.2 Arnold's stepwise guidelines 2006

The theme of a stepwise approach was taken up by Arnold following the APS guidelines (Arnold L.M., 2006). She was critical that the guidelines took little consideration of post-morbid conditions leading her to feel other, newer antidepressants may often be more appropriate. She was also critical of the emphasis placed on exercise, pointing out that for such patients there are many barriers to this such as pain, stress, depression and support. Given that compliance with exercise advice is very low, she advises that any barriers should be addressed first in order to enable exercise to be better pursued; indeed she suggests CBT as one means of boosting self-efficacy to achieve this.

The down-side of this advice is that whilst psychosocial issues are recognised to contribute to aetiology, most GPs will also recognise them as exacerbating factors and often even excuses; this can contribute to the sad state of affairs where patients suffering from FMS are perceived as 'yes-but's, a type of heart-sink patient more formally known as 'manipulative health rejecters', by their doctors (O' Dowd, 1988). The term 'heart-sink' has now entered the English language and the World English Dictionary defines it thus: “a patient who repeatedly visits his or her doctor's surgery often with multiple or non-specific symptoms, and whose complaints are impossible to treat”. This description will sound very familiar to those caring for patients with FMS; indeed one of the reasons for trying to define the disease, and ease the route to diagnosis, was to shorten the distressing period during which patients lived in the limbo time of non-diagnosis. This is true of any condition of course, but particularly FMS.

Thus Arnold's stepwise approach starts in the same way as the Goldenberg et al's above. In essence, she then switches steps 2 and 3, advising antidepressants and/or anti-convulsants following this, particularly an SSRI +/- pregabalin +/- low dose TCA. Only following this does she recommend using CBT and exercise. To a degree, this change in emphasis reflects the fact that even well-informed doctors, relying on the evidence, will have different opinions as to what is the best treatment, a situation magnified, of course, by orders of magnitude in our patients.

4.3 The Canadian consensus 2003

By way of contrast in the way of guidelines is the Canadian Consensus document for FMS from 2003 (Jain et al, 2003). It is written in a very different style to most guidelines; it does not grade evidence nor does it reject unproven therapies; the style is that of experienced physicians giving sensible, practical advice. Moreover it does not recommend any therapies above others, or suggest any kind of stepwise approach. Although referred to as a guideline it is more of a review article. As a result their suggestions regarding both aetiology and treatments are great and wide-ranging in number. Most of them are rather circumspect, and often preceded thus: “may occur”, “may result”, “may reduce”, “may protect”. The rationale is that all patients are individuals and different causal factors will be present in different people; equally different treatments may be efficacious in different individuals.

This document does, however, they also make the point that studies tend to show FMS persists over decades and most people gradually get worse. Perhaps this is the reason why subsequent guidelines have restricted themselves to describing only causes and cures with a strong evidence base, whilst acknowledging that patient preference must be considered. The authors of the Canadian Consensus are also firm in their belief that CFS, FMS and depression are completely different disease entities, which is another controversial position. In a way, the title itself illustrates one of the problems of this document: it is a "consensus". To be sure, this illustrates the lack of available evidence, and if we can't rely on the opinion of those with decades of experience then on whose can we rely? However, many authors feel that an overemphasis on expert opinion is dangerous, particularly when constructing guidelines that may be followed at a national or even international level (Herrmann and Klement, 2008).

4.4 German guidelines 2008

In Germany guideline development for FMS was initiated by the Deutschen Interdisziplinäre Vereinigung für Schmerztherapie (DIVS - German interdisciplinary Society for pain therapy) and the Arbeitsgemeinschaft der Wissenschaftlichen Medizinischen Fachgesellschaften (AWMF - Association of the Scientific Medical Societies in Germany); thirteen different medical and psychological Scientific Societies contributed, including the Deutschen Gesellschaft für Allgemeinmedizin und Familienmedizin (DEGAM - German Society of General Practice and Family Medicine). Two patient self-help groups also partook (Haeuser et al, 2008). Notably absent however was input from public health and societies of social medicine, which is unfortunate as in the case of FMS, these domains are probably more central than with most conditions.

The rationale for the development of the guidelines was as follows (in common with previous guidelines presumably): the high prevalence in the general population, the high associated disease-related costs, and the conflicting data on treatment effectiveness (Schiltenwolf et al, 2008). The hope, as with all guidelines, was that the advice would provide patients and physicians help in selecting among the alternatives available. This it surely does; however, like most other guidelines, the best it can probably hope for is to "help patients live with fibromyalgia syndrome (FMS) pain" (in the words of the "Guide for Adults with Fibromyalgia Syndrome Pain").

A summary of the DIVS/AWMF advice is now presented. The management suggested is stepwise and based on patient preferences and informed decision making. The first step is concerned with diagnosis, education, and treatment of co-morbid conditions (such as depression): first treatments (basic therapy) should be exercise, CBT and amitryptiline. The authors emphasised that they did not feel the ACR criteria of 11/18 tender points were necessary, which is becoming increasingly accepted. Indeed, pain can also be felt in other areas and is known to fluctuate. They also emphasised the importance of fatigue, sleep disturbance, and a feeling of stiffness and swelling in the limbs. These criteria were arrived at by strong consensus (defined as over 95% agreement). FMS is currently a diagnosis of exclusion so consensus is probably the best that could be hoped for here. Although some authors argue for a diagnosis of inclusion (Khasnis and Wilke, 2010), this does not yet appear possible.

Step 2 involves doing many of these things together (multicomponent treatment). Step 3 comprises either giving no further treatment or encouraging self-management (exercise, stress reduction), or further multicomponent therapy. Included in the latter is further pharmacological therapy (duloxetine or SSRI, pregabalin, tramadol +/- paracetamol), further CBT, psychotherapy, including hypnotherapy, and warm bath treatments. These latter two straddle the border between mainstream and complementary therapies. Further complementary medicine suggestions are homeopathy or vegetarian diet, which take into account the desire patient's generally have for further passive treatments, but which have no evidence for efficacy.

The authors emphasised the difficulty of using or even finding controlled studies on which to base the evidence, for the following reasons. Firstly it is virtually impossible to conduct studies on care coordination of the long term nature needed in FMS. Secondly the healthcare systems of different countries are sufficiently different as to make evidence from international sources of dubious applicability to Germany.

The German guidelines describe the foundations of therapy as being cognitive behavioural therapy, aerobic exercise, antidepressants (such as amitriptyline), and anticonvulsants (such as pregabalin). These are the only treatments with a solid evidence base and it has been shown with controlled-studies that they reduce pain and increase capacity for activities of daily living (Haeuser et al, 2009, Arnold et al, 2011). There is also good evidence that "multimodal therapy" (ie exercise activities with psychotherapeutic input) is effective in the medium term, but this is really just another way of describing the extension of the foundations of treatment. In terms of long term care the guidelines are based purely on consensus opinion as the evidence base is so poor. Other guidelines are constructed in a similar manner ie full of good advice from experts, but little is ever based on good evidence (given that there isn't any). The implication that this advice is factual (even when the process of consensus is transparent) has worrying implications; patients may read such guidelines and not understand the levels of evidence. Harm may follow, not least in potential mistrust that could damage the doctor-patient relationship.

4.4.1 The majority of patients with FMS are cared for by general practitioners

As Herrmann and Klement emphasised in 2008: such a guideline not only will have difficulties being relevant to primary care, given the emphasis of the treatment options, but there is also a great risk of what they call "disease mongering" ie the expansion of the number of people who might spuriously fall into the defined group; this is because FMS has no binding definition, a controversial aetiology and no way of objectively testing for it.

If subjective illness is redefined as disease then it becomes harder to define individual cases according to symptomatology. The DEGAM guidelines above are actually more oriented to this sort of subjectivity anyway (compare their tiredness and back pain guidelines). This principle is known as hermeneutical case-finding. By classifying purely subjective complaints as diseases, the focus switches from a picture of a pathological process to a subjective one, concerned more with disturbances of functionality. In a way the personal needs of an individual are restructured into a disease requiring treatment, the external agency aspect is disease-mongering.

Given this, it is remarkable that most diagnoses are made in primary care and that most rheumatologists normally concur with the diagnosis (Porter, 2009). It would seem wise, given this and the long term nature of the condition that primary care doctors should have overall responsibility for managing patients with the condition. Indeed many rheumatologists refer patients back to GPs after a diagnosis has been made, although whether this is wise is questioned by some who suggest rheumatologists should retain ownership (Shir and Fitzcharles, 2009).

4.5 There are no NHS guidelines in Britain

Particular criticism was levelled at the AWMF guidelines due to the fact they were largely based upon consensus opinion, and not even always strongly concordant opinion either, rather than on scientific evidence. Given the paucity of useful evidence it is difficult to see what the guideline authors could do however, except perhaps take the route the National Institute of Clinical Excellence (NICE) in the UK have chosen, which is not to publish a guideline at all. NICE articulate an oft heard worry concerning FMS, that is, given the paucity of evidence-based treatments, will these guidelines actually help people or will they risk encouraging illness behaviour? This is known as disease mongering in some quarters, which is meant to imply external agencies having an undue influence on the self-perception of disease; this is very different to *malingering* which implies the patient is deliberately fabricating their symptoms for secondary gain.

It should be emphasised here that despite all the criticism of FMS as a real entity, guidelines as being incomplete and the research being flawed, no implication that patients are making up or exaggerating their symptoms is ever intended by researchers. It is always agreed that patients' suffering is real; it is the causes and hence the treatment which cannot be agreed upon. It has been noted that when guidelines are produced for a condition such a fibromyalgia, one whose aetiology is unknown, which has no agreed diagnostic criteria, and where there is a poor evidence base for treatment, then the possibility of over-diagnosing patients is great. It has also been noted that joblessness, sick pay and pension requests are all positively correlated with FMS. The consequent harm of disease mongering to patients' long-term psychological functioning and indeed to society itself is thus potentially very large.

Those in favour of NICE producing guidelines point to many other NICE guidelines that also only suggest cheap, low technology treatments such as Chronic Fatigue Syndrome and Low Back Pain (NICE, 2007). One of the functions of NICE, after all, is to *reject* treatments that may appear attractive but that are actually too expensive and/or ineffective. A NICE guideline would surely help primary care doctors become up-to-date with this difficult condition, even if no new treatments were suggested. Indeed this is one of the chief aims of the German DIVS guidelines, although whether they achieved this is questionable.

4.6 Primary care is best placed to care for patients with FMS

The DEGAM, as Germany's pre-eminent GP organisation, was ideally placed to contribute to primary care guidelines for FMS. DEGAM is an opinion leader in primary care and emphasises the following aspects of primary care, with particular relevance to Germany:

- GPs are experts in first contact health care.
- As such they perform a filtering and steering function for all their patients' medical problems.
- GPs look at patients holistically: social, psychological, spiritual, ecological and physical aspects, and all these aspects may be involved in their treatment.
- GPs are involved in emergency, acute and long-term care as well as prevention and education.
- The basis of good GP care is the long-term relationship with the patient.

From this short but lucid description of the essence of general practice, which applies to anywhere that primary care is pursued, it can be seen that GPs are ideally placed to care for patients with FMS. Indeed it is hard to imagine how a specialist could manage the task, except in an advisory role. As such FMS guidelines ought to be principally aimed at GPs, but a continual criticism is the lack of applicability to primary care.

4.7 EULAR guidelines 2007

Other organisations have published guidelines with similar aims to those above. In 2007, the European League Against Rheumatism (EULAR) published a well-researched report, basing their guidance on 39 pharmacological and 59 non-pharmacological studies, following which they gave ten broad recommendations. These recommendations were specifically not weighted in terms of order and there was no step-wise advice for a treatment plan, which for many physicians limits their usefulness. They were as follows:

1. Recognition of FMS as a complex condition requiring good understanding of psychosocial context and daily functioning as well as pain levels.
2. A multidisciplinary approach is recommended according to symptoms and discussion with the patient.
3. Exercise: aerobic/strength
4. Heated pool treatment
5. Relaxation/Rehabilitation/Physiotherapy
6. Cognitive Behavioural Therapy
7. Tramadol
8. Antidepressants
9. Tropisetron, pramipexole or pregabalin
10. Other analgesics such as paracetamol but not opiates.

The pharmacological recommendations 7-9 were based upon randomised controlled double-blind trials. Recommendations 1,2,6 and 10 were purely on the basis of consensus expert opinion. Recommendations 3,4 and 5 were based on randomised, blinded, crossover trials of varying quality. This list illustrates well the lack of evidence for diagnosis and treatment and how most is purely based on the opinion of 'experts'; this is presumably better informed than that of 'non-experts', but of course expert opinion has a notoriously poor track record. 'Experts' have naturally always listened to other 'experts' opinions and historically speaking this has led to the perpetuation of either useless or even harmful treatments for centuries or more. More pertinently, it often leads to the rejection of new ideas, although hopefully physicians are more open minded these days.

This advice was formulated at about the same time as the German AWMF advice and it can be seen that it is similar, especially in the broad lack of evidence-based advice and the over-reliance on expert consensus.

4.8 Guidelines differ

It has been noted that guidelines tend to differ. Haeuser attributes this (unsurprisingly) to the composition of the panels among other things (Haeuser et al, 2010). The EULAR guidelines were predominantly formulated by rheumatologists and have an emphasis on pharmacological treatments; this is probably also partly because they placed the highest credence on randomised controlled trials. The APS and the AWMF also considered these (especially looking at meta-analyses) but also placed great importance on systematic reviews, since many treatments are less easily assessable using RCTs; as such they emphasise CBT and exercise more.

4.9 Overlap with chronic fatigue and depression

In the UK, as already stated although there is no specific FMS guideline, NICE have, however, published guidance for dealing with chronic fatigue syndrome (NICE, 2007). Whilst superficially different from FMS the epidemiology is very similar. For those with an FMS diagnosis, the presence of extreme fatigue is almost always present, as would be expected with a condition where sleep disorder is a major component. For those with CFS, studies show a co-morbidity of 33-75% with FMS (some studies quote an increased odds ratio of twenty to one - Aaron et al, 2001). In FMS the annual prevalence of depression approaches 25% (Offenbaecher et al, 1998), and with CFS it is even higher. Many people, particularly patients themselves maintain their FMS or CFS caused their depression, rather than vice versa. However, many doctors now believe that FMS, CFS and major depression have much overlapping aetiology and risk factors, as well as symptomatology. This is illustrated by the fact the first degree relatives of those with FMS have as high a risk of developing major depression as if their relatives had depression instead of FMS (Raphael et al, 2004). This is consistent with the hypothesis that FMS is a depression spectrum disorder. Indeed there are some physicians who even feel the term itself is redundant and only serves to legitimise sickness behaviour. This depends on how the label is used however. If it serves to help explain to patients their illness is not progressive and that withdrawal from normal activities will not promote healing, it may be useful (Goldenberg, 1995). In that respect it is no more a label than particular psychosocial issues or personality traits are.

4.10 Alternative and complementary medical options

Another criticism of most guidelines is the notable absence of suggested alternative medical therapeutic options. Whilst there is also little demonstrable evidence for these other treatments, patients themselves are normally keen to pursue them, so at the very least 'harmless' (is any treatment truly harmless?) or popular therapies could be discussed.

There have indeed been huge numbers of studies of complementary and alternative medical therapies in FMS; some guidelines have mentioned these. Indeed, the Canadian consensus was particularly voluble in its listing of such treatments; unfortunately the best they could say was that they "may" help. Given that patients themselves are so keen to pursue these avenues, it does seem wise not to ignore them in guidelines. Indeed some studies put the usage of CAM among FMS patients at over 50% (Sarac and Gur, 2006). A study in Washington in 2007 found that FMS users who used CAM actually had worse symptoms than those who didn't (Lind et al, 2007). Whether this is cause or effect, is, of course, hard to ascertain. The economic study did, however, find that CAM users did not have higher health expenditures than the non-CAM users, because, although they visited healthcare practitioners more often, they saw their actual doctors (who are more expensive) less. This finding, already intuitive to many doctors, could be expected to be one reason why CAM in FMS might be advocated by the medical profession. Any doctors who feel it is unethical can take heart from a study in another psychosomatic condition (IBS) which showed that even telling patients they are receiving a placebo does not negate the actual placebo effect *(Kaptchuk et al, 2010). Indeed this likely reflects the fact that good medical care is more than simply supplying medications (or other interventions); time, concern for the person and a relationship with the practitioner are also essential parts of the healing process.*

Countless claims for alternative therapies have been made over the years, one of the most persistent being for guaifenesin. It has been strongly advocated by St. Amand and has been widely believed by many in the FMS community for decades. At the very least, a strong belief is likely to have a powerful placebo effect and that, given the few treatment options with known benefit, might be useful part of a physician's armoury against FMS. Bennett conducted a year-long RCT investigating the claims of guaifenesin (Bennet and Clark, 1996). The results were negative. Disgracefully, this led to the study being unpublished, which is inexplicable given the importance, although Bennett reported on it himself in an review editorial in "Arthritis and Rheumatism" (Bennett, 1996). (This illustrates well the dangers of positive publication bias). Guaifenesin is actually still very popular among FMS patients and indeed still strongly advocated by St. Amand himself who gives many excuses for why the original trial failed (despite actually being an external advisor throughout). Bennett's concluding comment tells us a lot about how many complementary medical techniques work: "We have shown the placebo is just as effective as the placebo!"

5. Future developments

There are a number of important themes that are absent in all the guidelines mentioned above, representing new ideas, as well as old.

It is worth mentioning the new "wonder drug" milnacipran, an SNRI similar to duloxetine. The growth consulting company "Frost and Sullivan" predicted the market would be worth at least four hundred million dollars in Europe alone in 2013 (Frost and Sullivan 2008). It is now licensed in America. A study in 2010 (Haeuser et al, 2010) showed it to be similar in efficacy to duloxetine and pregabalin over a 6 month period with respect to pain reduction (30%); whether it has any useful long term effect remains to be seen.

There is a new concept developing in primary care which is known as quarternary prevention; in essence it refers to the old adage of "do no harm" and has particular relevance to FMS. It is the notion of preventing our patients from becoming over-medicalised; this is for their own benefit, as well as society's (Kuehlein et al, 2010). It was first mentioned, prophetically, by Jamoulle in 1986 (Jamoulle, 1986) but it is only in recent years that it is becoming more and more apparent as a real risk. Patients with FMS are the paradigmatic example of this: they eternally seek tests and solutions with which it is very tempting for their GPs to collude; future guidelines should emphasise that this is actually harmful, and that GPs should be using their unique relational position with their patients to prevent this over-medicalisation (Jamoulle, 2011).

The concept of disease-mongering, which includes the expansion of the diagnostic definitions (see above) as well as the increasing of 'public awareness' is particularly relevant with FMS. The field is ripe for drug companies to exploit by promoting knowledge of the disease as well as new curative drugs. An analagous example is 'metabolic syndrome' a pre-diabetic state which it might be imagined could be a prompt to weight loss and exercise, but of course, offers potentially billions of pounds to drug companies who have thus vigorously promoted in and encouraged 'awareness' amongst doctors and patients.

Finally, the age-old concept of the placebo effect probably needs more emphasis in future guidelines; since most treatments either have no effect or a weak effect, it is tempting to suggest that more emphasis should be placed on the utilization of the placebo. However, it could also be argued that this goes against the concept of quarternary prevention mentioned above, and that it may undermine the only known effective treatments of CBT and exercise.

6. Conclusion

Fibromyalgia Syndrome is a controversial condition whose aetiology is unknown, which has no agreed diagnostic criteria, and for which there is a poor evidence base for treatment. Indeed, even the very existence of it as a condition is questioned by many senior researchers and practitioners in the field. Given this, the production of meaningful and useful guidelines is much more difficult than for most medical conditions, and credit should be given to those who have tried as there is no doubt that, whether we agree with all the recommendations or not, their use can help doctors in their mission to help patients cope with their pain, which is ultimately what our medical care is all about.

Without clear aetiology and with the known strong placebo effect of various therapies, it is difficult to create evidence-based guidelines; it is inevitable that the composition of the guideline group, each with their own unique backgrounds will have a strong influence. For the German guidelines to be valid. Following points had a strong influence on the recommendations of the S3 guideline quality:

- the composition of the interdisciplinary consensus conference mainly by specialists, the methodological structure and the frame of the consensus procedure (structure quality).
- the group consensus process structured by a majority vote concerning by the treatment and evaluation of the meagre evidence

The "interdisciplinarity" of the guideline group masks area-specific interpretations and particular interests. The utilitarian concept effectively ignores the concept of authentication and legitimation. It is essential, not only for the scientific credibility of guidelines but also for the therapeutic process. Each form of mistrust interrupts that process, not only for the development of guidelines but also concerning patient therapy. The outworking of this is, of course, different when considering the treatment process as compared to the guideline debate. It is fragile because it is essentially unconscious and so is easily lost. Conquering this mistrust is something that needs to be repeatedly managed by doctors in their dealings with their patients. Interference from the outside provokes mistrust if recommendations are not evaluated and scientifically legitimized in the best way possible.

There are no guidelines where the interests of the pharamaceutical industry are accurately considered. The relevance of the economic side is very important. Traditionally illnesses of the central nervous system are a great challenge for medicine and research. Fibromyalgia is no exception: so far no effective long-term therapy is available for this chronic pain illness. The interest and the commitment of the pharmaceutical sector is correspondingly great. It is possible that a break-through is be approaching, since some active substances in clinical studies of later phases have proved promising. In view of these developments the management consultation company Frost and Sullivan (http://www.pharma.frost.com) projected that the European market for Fibromyalgia treatment is possible to be worth 406.3 million dollars in the year 2013, as compared to 73.4 million dollars in the year 2007.[1]

The establishment of the milnacipran in the drug market is a good example. Since its introduction in Austria in 1998 milnacipran struggled to achieve an important market share. Consequently, manufacturers and distributors relinquished their market entrance into additional European Union member states, although in principle permission from one

[1] http://www.ishoof.com/?p=25: Access 02.12.2011

member country of the EU allows marketing in the entire European Union area. In a second attempt the finance and marketing partners prepared a broad introduction into the market in the USA in 2004. According to stock exchange reports investors made available several hundred million US Dollars to promote milnacipran for the indication of fibromyalgia syndrome. According to reports at the time Forest Cypress supported further development to the region of two hundred and fifty million dollars as advance and milestone payments (finanzen.net)[2].

7. References

Aaron L.A., Herrell R., Ashton S., Belcourt M., et al. (2001). Comorbid clinical conditions in chronic fatigue: a co-twin control study. *J Gen Intern Med.* Vol 16(1), (Jan 2001), 24-31

Anderberg U.M., Marteinsdottir I., Theorell T., von Knorring L. (2000). The impact of life events in female patients with fibromyalgia and in female healthy controls. *Eur Psychiatry*, 15 (5), (Aug 2000), 33–41

Arnold L. (2006). Biology and therapy of fibromyalgia. New therapies in fibromyalgia. *Arthritis Res Ther* (June 2006) Vol 8(4), 212

Arnold L., Gergana Z., Sadosky A. et al. (2011). Correlations between fibromyalgia symptom and function domains and patient global impression of change: a pooled analysis of three randomised, plaecbo-controlled trials of pregabalin. *Pain Medicine* Vol 12(2), (Feb 2011), 260-267

Balint M (1957). *The doctor, his patient and the illness.* International Universities Press , Michigan

Bennett R.M. (1996) Review: Fibromyalgia and the Disability Dilemma: A New Era in understanding a Complex, Multidimensional Pain Syndrome. *Arthritis and Rheumatism* Vol39 (10), (Oct 1996), 1627-34.

Bennett R. and Clark S. (1996). A Randomized, Prospective, 12 Month Study To Compare The Efficacy Of Guaifenesin Versus Placebo In The Management Of Fibromyalgia. *Orlando American College of Rheumatology meeting (unpublished)*

Bissell P., May C.R., Noyce P.R. (2004). From compliance to concordance: barriers to accomplishing a re-framed model of health care interactions. *Soc Sci Med.* Vol 58(4), (Feb 2004), 851-62

Branco J.C., Bannwarth B., Failde I., et al (2010). Prevalence of fibromyalgia: a survey in five European countries. *Semin Arthritis Rheum.* Vol 39 (6), (Jun 2010), 448-53

British Thoracic Society. (May 2011). Asthma Guidelines, In: *British Thoracic Society,* Available from: <http://www.brit-thoracic.org.uk/Portals/0/Clinical%20Information/Asthma/Guidelines/sign101%20June%202011.pdf>

Buckhardt C.S., Goldenberg D., Crofford L. et al. (2005). *Guideline for the management of fibromyalgia syndrome pain in adults and children.* American Pain Society, Glenview

[2] http://www.wallstreet-online.de/diskussion/809332-1-10/cypress-bio-erhaelt-millionenvorschuss-fuer-medikament; Access 02.12.2011

Buskila D., Neumann L., Vaisberg G., Alkalay D., Wolfe F. (1997) Increased rates of fibromyalgia following cervical spine injury. A controlled study of 161 cases of traumatic injury. *Arthritis Rheum* Vol 40 (3), (Mar 1997), 446–52

Carville S.F., Arendt-Nielsen S., Bliddal H. (2008). EULAR evidence based recommendations for the management of fibromyalgia syndrome. *Ann Rheum Dis* Vol 67 (4), (april 2008), 536-4

Choy E., Perrot S., Leon T,. (2010). A patient survey of the impact of fibromyalgia and the journey to diagnosis. *BMC Health Services Research,* Vol 10 (April 2010), 102

Christakis N., (1999). *Death Foretold: Prophecy and Prognosis in Medical Care,* University of Chicago Press, Chicago

Ciccone D.S. and Natelson B.H. (2003). Comorbid Illness in Women With Chronic Fatigue Syndrome: A Test of the Single Syndrome Hypothesis. *Psychosomatic Medicine* Vol 65 (2), (April 2003), 268-275

Field M.J. and Lohr K.N. (Eds.). (1990). *Clinical Practice Guidelines: Directions for a New Program.* National Academy Press, Washington

Frost and Sullivan (2007). In *Frost.com,* available from http://www.frost.com/prod/servlet/press-release.pag?docid=148807077

Goldenberg D.L. (1987). Fibromyalgia syndrome. An emerging but controversial condition. *JAMA* Vol 257 (20) (May 1987), 2782–7

Goldenberg D.L. (1995). Fibromyalgia: why such controversy? *Ann Rheum Dis.* Vol 54 (1), (Jan 1995), 3–5

Goldenberg D.L., Burckhardt C., Crofford L. (2004). Management of Fibromyalgia Syndrome. JAMA Vol 292 (19), (Nov 2004), 2388-2395

Gowers W.R, (1904). Lumbago: it's lessons and analogues. *BMJ* (Jan 1904), 117-121

Gracely R.H., Petzke F., Wolf J.M., Clauw D.J. (2002). Functional magnetic resonance imaging of pain in patients with primary fibromyalgia. *Arthritis and Rheumatism* Vol 46 (5), (Feb 2002),1333-43

Grimshaw J.M. and Russell I.T. (1993). Effect of clinical guidelines on medical practice: a systematic review of rigorous evaluations. *Lancet* Vol 342 (8883), (Oct 1993), 1317-1322

Haeuser W., Arnold B., Eich W., Felde E. et al. (2008). Management of fibromyalgia syndrome--an interdisciplinary evidence-based guideline. [German]. *German medical science GMS ejournal* Vol 6, (Dec 2008)

Haeuser W., Bernardy K., Uceyler N., Sommer C. (2009) Treatment of Fibromyalgia Symptoms with antidepressants. *JAMA* Vol 301 (2), (Jan, 2009), 198-209

Haeuser W., Kosseva M., Uceyler N., (2011). Emotional, physical, and sexual abuse in fibromyalgia syndrome: a systematic review with meta-analysis. *Arthritis Care Res* Vol 63 (6), (June 2011), 808-820

Haeuser W., Petzke F., Sommer C., (2010). Comparative efficacy and harms of duloxetine, milnacipran, and pregabalin in fibromyalgia syndrome. *J Pain* Vol 11 (6), (June 2010), 505-521

Haeuser W., Thieme K.,Turk D.C. (2010). Guidelines on the management of fibromyalgia syndrome - a systematic review. *European journal of pain,* Vol 14 (1)(January 2010), 5-10

Harden R.N., Revivo G., Song S. et al (2007). A critical analysis of the tender points in fibromyalgia. *Pain Med.* Vol 8 (2), March 2007), 147-56

Hench P.K. (1976). *Nonarticular rheumatism. Arthritis Rheum Vol 19 (6 suppl), (1976), 1081-1088*

Hench P.K. and Mitler M.M. (1986). Fibromyalgia. Management guidelines and research findings. *Postgraduate medicine* Vol 80 (7), (Nov 1986), 57-64

Herrmann M. and Klement A. (2008). Limits in Transforming Evidence into Guidelines - Analysis on the Example of the Interdisciplinary Guideline About Fibromyalgia Syndrome [German]. *Z Allg Med Vol 84 (10), (Oct 2008), 436- 443*

Hughes G., Martinez C., Myon E., et al (2006). The impact of a diagnosis of fibromyalgia on health care resource use by primary care patients in the UK: an observational study based on clinical practice. *Arthritis Rheum*Vol 54 (1), (Jan 2006), 177-183

Jackson R. and Feder G. (1998). Editorial: Guidelines for clinical guidelines. A simple, pragmatic strategy for guideline development. *British Medical Journal* Vol 317 (7256), (August 1998), 427

Jain A.K., Carruthers B.M., van de Sande M.I., et al (2003) Fibromyalgia Syndrome:Canadian Clinical Working Case Definition, Diagnostic and Treatment Protocols - A Consensus Document. *Journal of Musculoskeletal Pain* Vol 11 (4), (April 2003), 3 - 107

Jamoulle M., (1986). *Les informa-g-iciens*. Namur University Press , Namur

Jamoulle M., (2011). Quarternary Prevention: first, do no harm. *11th Congress of the Sociedade Brasileira de Medicina de Familia e Comunidade.* (June 2011)

Janis, I. L. (1982). *Groupthink: Psychological Studies of Policy Decisions and Fiascoes.* (2nd Edition). Houghton Mifflin, New York

Kanaan R.A., Lepine J.P., Wessely S.C. (2007). The association or otherwise of the functional somatic syndromes. *Psychosom Med* 69 (9), (Dec 2007), 855–9

Kaptchuk T.J., Friedlander E., Kelley J.M (2010). Placebos without deception: a randomized controlled trial in irritable bowel syndrome. *PLoSone.* Vol 5 (12), (Dec2010), e15591.

Khasnis A. and Wilke W.S. (2010). Diagnosing fibromyalgia: Moving away from tender points. *J Musculoskel Med* Vol 27, (April 2010), *155-162*

Kuehlein T., Sghedoni D., Visentin G., et al (2010). Quarternary prevention: a task of the general practitiner. *Primary Care* Vol 10 (18), (2010), 350-

Lim E.C., Sterling M., Stone A., Vicenzino B. (2011). Central hyperexcitability as measured with nociceptive flexor reflex threshold in chronic musculoskeletal pain: A systematic review. *Pain.* Vol 152 (8), (Aug, 2011), 1811-20

Lind B., Lafferty W., Tyree P. et al. (2007). Use of complementary and alternative medicine providers by fibromyalgia patients under insurance coverage. *Arthritis Rheum* Vol 57 (1), (Feb 2007), 71-76

Mueller W., Schneider E.M., Stratz T. (2007). The classification of fibromyalgia syndrome. *Rheumatol Int.* Vol 27 (11), (Sept 2007), 1005–10

National Guideline Clearinghouse. (Nov 2007). Guideline for the management of fibromyalgia syndrome pain in adults and children, In: *Guideline Central,* available from: <http://www.guidelinecentral.com/_webapp_1825321/Guideline_for_the_management_of_fibromyalgia_syndrome_pain_in_adults_and_children>

National Institute for Health and Clinical Excellence. (Aug 2007). Chronic Fatigue Guidance, In: *NICE*, available from: http://www.nice.org.uk/nicemedia/live/11824/36193/36193.pdf

Netter P. and Hennig J. (1998) The fibromyalgia syndrome as a manifestation of neuroticism? [German]. *Zeitung fuer Rheumatologie* Vol 57 (2), (1998), 105-8

O' Dowd T. C. (1988). Five years of heartsink patients in general practice. *BMJ* Vol 297(6647), (Aug 1988), 528-30

Offenbaecher M., Glatzeder K., Ackenheil M. (1998). Self-reported depression, familial history of depression and fibromyalgia (FM) and psychological distress in patients with FM. *Zeitschrift für Rheumatologie* Vol 57(2), (June 1998), 94-6

Payer L (1992). *Disease-Mongers: How Doctors, Drug Companies and Insurers Are Making You Feel Sick.* John Wiley, New York

Porter D., Nelson Hospital (May 2009). Fibromyalgia in Adults. *Diagnosis and Management* Advice for GPs, In: New Zealand Govt Health Files, available from: <http://www.nmdhb.govt.nz/filesGallery/New%20Website/03Health%20Concerns/FibromyalgiaMay09.pdf>

Raphael K.G., Janal M.N., Nayak S. et al (2004). Familial aggregation of depression in fibromyalgia: a community-based test of alternate hypotheses. *Pain* 110(1), (Jul 2004), 449-60

Raspe H., Cellarius J., Mau W., Wasmus A., von Gierke S. (1994) Gesundheitswesen. 1994 Nov;56(11):596-8. Guidelines for social medicine assessment and evaluation of primary fibromyalgia. [German]

Russell I.J., Orr M., Littman B., et al (1994). Elevated cerebrospinal fluid levels of Substance p in patients with the fibromyalgia syndrome. *Arthritis and Rheumatism* 37 (Dec 1994) Issue 11, 1593-1601

St. Amand R. P. (2009). The Guaifenesin Protocol, In: Fibromyalgia Treatment Centre. Available from: <http://www.fibromyalgiatreatment.com/GuaiProtocol.htm>

Sarac A.J. and Gur A. (2006). Complementary and alternative medical therapies in fibromyalgia. *Curr Pharm Des* Vol 12(1), (2006), 47-57

Sassi F (2006). Calculating QALYs, comparing QALY and DALY calculations. *Health Policy Plan* Vol 21 (5), (Sept 2006), 402-408

Schiltenwolf M., Eich W., Schmale-Grete R., Haeuser W. (2008) Aims of the guidelines for diagnostic and treatment of fibromyalgia syndrome [German]. *Schmerz,* Vol 22 (3), (June 2008), 241-243

Schuster R.J., Terwoord N.A., Tasosa J. (2006). Changing physician practice behavior to measure and improve clinical outcomes. *Am J Med Qual* Vol 21 (6), (Nov 2006), 394-400

Senna E.R., De Barros A.L.P., Silva E.O., et al (2004). Prevalence of rheumatic diseases in Brazil: a study using the COPCORD approach. *J Rheumatol* Vol 31 (Nov 2004), 594-7

Shir Y. and Fitzcharles, M.J. (2009). Should Rheumatologists Retain Ownership of Fibromyalgia? *Rheumatology* Vol 36 (4), (April 2009), 667-670

Smythe H. (2011). Unhelpful Criteria Sets for "Diagnosis" and "Assessment of Severity" of Fibromyalgia. *J Rheumatol* Vol 38 (June 2011), 975-978

Stone J., Wojcik W., Durrance D., et al. (2002). What should we say to patients with symptoms unexplained by disease? The "number needed to offend". *BMJ* Vol 325 (7378), 1449-1450

Taylor J., Skan J., Erb N., et al (2000). Lupus patients with fatigue - is there a link with fibromyalgia syndrome. *Rheumatology* Vol 39 (6), (June 2000), 620-623

University of Oxford. (March 2009). Levels of Evidence, In: *Centre for Evidence-based Medicine*, Available from: http://www.cebm.net/index.aspx?o=1025

White K.P., Nielson W. R., Harth, M et al (2002). Does the label "fibromyalgia" alter health status, function, and health service utilization? A prospective, within-group comparison in a community cohort of adults with chronic widespread pain. *Arthritis Rheum*. 47 (3), (June 2002), 260–5

Wilke W.S. (1995). Treatment of "resistant" fibromyalgia. *Rheum Dis Clin North Am*. Vol 21 (1), (Feb 1995), 247-60

Wilke W.S. (1999). The clinical utility of fibromyalgia. *J Clin Rheumatol* Vol 5 (2), (April 1999), 97–103

Wolfe F. Editorial (2010). New Amercian College of Rheumatology Criteria for Fibromyalgia: A 20 Year Journey. *Arthritis Care and Research* Vol 62 (5), (May 2010), 583-584

Wolfe F. (2003). Stop using the American College of Rheumatology criteria in the clinic. *J. Rheumatol* Vol 30 (8), (August 2003), 1671–2

Wolfe F., Clauw D., Fitzcharles M. et al (2010). The American College of Rheumatology preliminary diagnostic criteria for fibromyalgia and measurement of symptom severity. *Arthritis Care Res* , Vol 62 (5), (May 2010), 600-610

Wolfe F., Ross K., Anderson J. et al (1995). The prevalence and characteristics of fibromyalgia in the general population. *Arthritis Rheum* (Sept 1995), Vol 38 (1), 19-28

Wolfe F., Smythe H.A., Yunus M. et al (1990). The American college of rheumatology 1990 criteria for the classification of fibromyalgia. *Arthritis and Rheumatism* Vol 33 (2), (Feb 1990), 160–172

Woolf S.H. (1997). Shared decision-making: the case for letting patients decide which choice is best. *J Fam Pract* Vol 45 (Dec 1997), 205-208

Woolf S.W., Grol R., Hutchinson A. et al (1999) Education and debate: Clinical guidelines Potential benefits, limitations, and harms of clinical guidelines. *British Medical Journal* Vol 318 (7182), (February 1999), 527-530

Yunus M. (1981). Primary Fibromyalgia (Fibrositis): Clinical Study of 50 Patients with Matched Normal Controls. *Sem. Arth and Rheum* Vol 11, (Nov 1981), 151-171.

Zhang W., Robertson J.,Jones A.C. (2008). The placebo effect and its determinants in osteoarthritis: meta-analysis of randomised controlled trials. *Ann Rheum Dis Vo*l 67 (12), (Dec 2008), 1716-1723

8

Alexithymia in Fibromyalgia Syndrome

Ercan Madenci and Ozlem Altindag
Gaziantep University Research Hospital,
Department of Physical Medicine and Rehabilitation,
Gaziantep
Turkey

1. Introduction

Fibromyalgia Syndrome (FMS) is one of the most frequent rheumatologic conditions, its main characteristic being chronic diffuse musculoskeletal pain and hyperalgesia, likely due to altered central processing of pain (1). Fibromyalgia symptoms are not restricted to pain. Besides the painful setting, these patients often complain of fatigue, sleep disturbances, morning stiffness, and paresthesia on the extremities, subjective edema sensations and cognitive disturbances. Many patients experience cognitive dysfunction known as "fibrofog" which may be characterized by impaired concentration. Cognitive symptoms in these patients may be exacerbated by the presence of depression, anxiety, sleep problems, endocrine disturbances, and pain.

Association with other comorbidities is often found, contributing to the suffering and decline in quality of life of these patients. Among the most frequent comorbidities found, it can be mentioned depression, anxiety, chronic fatigue syndrome, myofascial syndrome, irritable bowel syndrome and nonspecific urethral syndrome (2).

Various researches show that there were relation between chronic pain and cognitive dysfunction in FM patients. Managing chronic pain may take some cognitive effort and this may interfere with performance on cognitive tasks. This is important because even though FMS patients report cognitive symptoms, physicians and scientists must consider the possibility that because FM patients experience many symptoms, there may be a tendency to mistake normal, everyday lapses on cognition as something more serious (3).

2. Relational issues and alexithyma

It has been argued that alexithymia is one of the major disturbances about cognitive difunction. The core disturbance in alexithymia is a deficit in the cognitive processing and regulation of emotions. Thus, some of the emotional and behavioural disturbances associated with fibromyalgia may reflect the presence of alexithymia. (4)

The high rate of alexithymia as well as anxiety and depression may be seen in FMS patients. Several reports reveal the existence of a relationship between alexithymia, depression and anxiety. Hendryx et al. (5) suggested that alexithymia is multidimensional feature. They also reported that alexithymia was related to generalized anxiety and depression. Anxiety and depression are often precedes or manifest after fibromyalgia.

3. The etiology of alexithymia

Although environmental, neurological, and genetic factors are each involved, the role of genetic and environmental factors for developing alexithymia is still unclear.

Alexithymia is considered to be a personality trait. It seems to have a connection with several chronic conditions including FMS. Muftuoglu et al. (6) studied alexithymic features and affective states in migraine patients and found that migraine patients were significantly more depressive, anxious, and alexithymic than the control group. Also Yucel et al. (7) studied depression and alexithymia in patients with tensiontype headache. They found that compared to healthy controls, the subjects with headaches had significantly higher scores on measures of depression and alexithymia. However, to date, a clear correlation between alexithymia and pain has not been established (8).

The term "alexithymia" derived from the Greek words "a" (lack), "lexis" (word), and "tymos" (emotion). It was first introduced by Sifneos in 1972 (9). The original definition of alexithymia is difficulties to identifying and describing feelings (10). Alexithymia is a multifaceted personality construct characterized by a reduced ability the emotions and a tendency to utilize an externally-focused.

Some researchers hypothesized that alexithymia is associated with brain abnormalities (11, 12). Neuroimaging studies found that alexithymia may be associated with a higher level cognitive deficit in estimating emotional inputs rather than a lack of neuronal response in structures representing lower level processing of emotional stimuli (13). Biopsychosocial model including psychological factors as well as factors related to perturbation of the autonomic nervous system and the HPA-axis explained a substantial part of variance of pain in the fibromyalgia patients. Psychological distress was strongly associated with perceived pain, and only affective pain was found to be associated with autonomic reactivity.

An early theory of pain localization in the CNS was the gate control theory (14). A goal of this theory was to provide mechanisms whereby one differentiates between innocuous and noxious stimulation. Inherent to this undertaking was the effort to explain two cognitive domains of pain processing; sensory-discriminative and affective-motivational. The concept of a duality in CNS pain processing was amplified by Albe-Fessard et al. (15) who suggested these domains are differentially localized in the thalamus. The lateral nuclei were thought to mediate sensory-discrimination and the medial nuclei the affective-motivational component of processing. Although its likely these functions are subserved in part by the two thalamic divisions as proposed, accounting for multiple aspects of the conscious experience of pain with thalamic mechanisms alone is difficult. Particularly in light of the prominent projections of the medial and lateral thalamic nuclei to different parts of the cerebral cortex and the role of the cortex in anticipation and memory of many events including those associated with painful experiences. Indeed, one of the main functions of cortical pain processing is integrating the pain experience with other cortical information processing functions. A complete model of the medial and lateral systems, therefore, requires consideration of many cortical areas that are involved in such functions in addition to thalamic sites. (16)

It has long been known that ablations of cingulate cortex (cingulotomy lesions) or its underlying white matter (cingulumotomy lesions) alleviate pain in chronic conditions. In Neurobiology of Cingulate Cortex and Limbic Thalamus (17), it was proposed that projections of the midline and intralaminar thalamic nuclei to Anteror cingulate cortex provided a cortical substrate for the affective-motivational aspects of pain processing. Nociceptive stimuli may evoke brain responses longer than the stimulus duration often partially detected by conventional neuroimaging. Fibromyalgia patients typically complain of severe pain from gentle stimuli.

Anterior cingulate cortex, for example, has been shown to play an essential role in pain perception and pain control, brain imaging studies in FMS patients point to alterations in regional cerebral blood flow (18, 19).

Early studies showed evidence that there may be an interhemispheric transfer deficit among alexithymics; that is, the emotional information from the right hemisphere is not being properly transferred to the language regions in the left hemisphere, as can be caused by a decreased corpus callosum, often present in psychiatric patients who have suffered severe childhood abuse.

A neuropsychological study indicated that alexithymia may be due to a disturbance to the right hemisphere of the brain, which is largely responsible for processing emotions (20). In addition, another neuropsychological model suggests that alexithymia may be related to a dysfunction of the anterior cingulate cortex (21). These studies have some shortcomings however, and the empirical evidence about the causes of alexithymia remains inconclusive (22). It has been reported that there were strong focus by clinicians on neurophysiological at the expense of psychological explanations for the genesis and operation of alexithymia, and introduced the alternative term "disaffectation" to stand for psychogenic alexithymia. The disaffected individual had at some point "experienced overwhelming emotion that threatened to attack their sense of integrity and identity", to which they applied psychological defenses to pulverize and eject all emotional representations from consciousness (23).

Alexithymia has also been found to be related to dysfunction in the posterior cingulate cortex during various mental imagery conditions (24). Lane et al. (25) stressed the core feature of alexithymia as a deficit in conscious awareness of emotions (e.g., differentiating, symbolizing emotions and appreciating complexity in the experience of self and other). Thus, alexithymia refers to impairment in not only affective but also cognitive emotional processing.

All attempts at elucidating the etiology of alexithymia must be considered theoretical and speculative. Since the clinical criteria and measurement of alexithymia have not been standardized, it seems premature to be focusing exclusively on etiological models. However, theoretical speculations may lead to testable hypotheses, which could then serve to validate the concept itself.

4. Prevalence of alexithymia

In studies of the general population the degree of alexithymia was found to be influenced by age, but not by gender; the rates of alexithymia in healthy controls have been found at. 4.7%-8.3%. Thus, several studies have reported that the prevalence rate of alexithymia is less than 10% (26). A less common finding suggests that there may be a higher prevalence of alexithymia amongst males than females, which may be accounted for by difficulties some males have with "describing feelings", but not by difficulties in "identifying feelings" in which males and females show similar abilities.

Studies reported a distinct gender difference in the prevalence of alexithymia: men were alexithymic almost twice as often as women. A similar gender difference was found in the previous study using the Beth Israel Questionnaire (BIQ). The study by Parker et al. (27) using the Toronto Alexithymia Scale (TAS), and covering three samples of young adults in Canada, the United States, and Germany, showed similar correlations between alexithymia and gender. On the other hand, the Italian study by Pasini and coworkers (28) using the TAS showed no gender difference in alexithymia, but it must be pointed out that their study group consisted of healthy volunteers. It seems likely that alexithymia is a personality trait that is more typical of men than women.

5. Clinical approach to alexithymia

Individuals with alexithymia are typically unable to identify, understand, or describe their own emotions. Psychiatric and psychosomatic patients with alexithymia are unable to talk about feelings due to a lack of emotional selfawareness (29). Self-awareness is a fundamental aspect of empathy because the individual's recognition of their own feelings is the basis for identification with the feelings of others.

Alexithymia is defined by Parker

1. difficulty identifying feelings and distinguishing between feelings and the bodily sensations of emotional arousal
2. difficulty describing feelings to other people
3. constricted imaginal processes, as evidenced by a scarcity of fantasies
4. a stimulus-bound, externally oriented cognitive style.

It supposed to be there were two kinds of alexithymia, "primary alexithymia" which is an enduring psychological trait that does not alter over time, and "secondary alexithymia" which is state-dependent and disappears after the evoking stressful situation has changed. These two manifestations of alexithymia are otherwise called "trait" or "state" alexithymia. The study regarding with alexithymia and depression by Steinweg et al. (30) reported that "Aleithymia may be more of a *state* than a *trait*".

Several authors have suggested that alexithymia was elevated in numerous psychosomatic and medical conditions such as rheumatoid arthritis, low-back pain, and hypertension. Furthermore, alexithymia has been shown to prospectively predict the maintenance of summarization over a period of two years. Alexithymia is especially related with medical disorders thought to be primarily caused or maintained by psychological factors.

There is some evidence suggesting that alexithymia is related to chronic pain. In addition to chronic, other aspects of pain such as severity may be related to alexithymia. It has proposed that chronic pain consist two components sensory and affective. Alexithymia is related to the affective-rather than the sensory-dimension of chronic pain, and this association could be mediated by increased depression. Several studies have been reported that affective dimension of the pain was predominant in fibromyalgia. Depressive mood is the best predictor of the affective dimension of the pain, whereas depression is a significant clinical component of fibromyalgia.

Depression has been widely studied in both the chronic pain and alexithymia fields. Chronic pain is often comorbid with depression, and alexithymia is also substantially related to depression and may predispose to it. These observations suggest that depression may mediate the correlation between alexithymia and chronic pain. In a various study relation was found between alexithymia anxiety and depression, which may indicate that alexithymia scores may be associated with psychological distress. They argued that treatment-seeking behavior may be more commonly. In the light of the several literatures it thought to be the psychiatric evaluation of fibromyalgia patients is as important as the locomotor system evaluation.

Fibromyalgia is considered one of the more difficult chronic pain syndromes to deal with. The chronic pain in fibromyalgia had been severe enough to disrupt all aspects of the patient's life. It has been proposed that abnormal pain sensitivity and pain inhibition in fibromyalgia. Abnormal pain perception might be related to abnormal levels of serotonin and norepinephrine, which are key neurotransmitters in endogenous pain inhibitory pathways.

According to the model of neuropathic pain, sensitivity of pain is reduced when the source is eliminated. In contrast, the source of sensory input among patients with fibromyalgia remains unknown. There is increasing evidence that fibromyalgia is characterized by an augmentation of sensory input that is mediated by central nervous system (CNS) events similar to those associated with neuropathic pain conditions. For this reason, most investigators involved in fibromyalgia research refer to central augmentation of sensory input rather than central sensitization when they discuss the pathophysiology of fibromyalgia.

Patients with fibromyalgia have difficulty to identify self expression. It has been suggested that difficulty to description of complaints may associated with alexithymic personality characterized by inability to differentiate affective from somatic states (31). Patients with alexithymia have also manifest cognitive characteristics such as paucity of fantasy, imagery, or daydreaming. Consequently, due to the difficulty to experience and express emotions, alexithymia has been linked with somatosensory amplification, which is tendency to focus on somatic sensations.

Fibromyalgia is thought to arise from miscommunication among nerve impulses in the central nervous system, in other words, the brain and spinal cord. It has been postulated that a disconnection between neocortex and the limbic system occurs and causes difficulty verbalizing feelings and regulating affect. Alexithymia is thought to impede the regulation of negative emotions, resulting in increased negative affect, sympathetic hyperarousal, and impaired immune status, which may contribute to development or exacerbation of somatic disease and pain (32).

Alexithymia does not emerge as a product of pain, but rather is intrinsic to this illness. Alexithymia has been reported to be associated with enhanced sensitivity not only to somatic sensations, but also to externally induced pain. Increased alexithymia may effects the nature of assessment and treatment the chronic pain.

6. Measurement of alexithymia

Along with the clinical descriptions of alexithymia, there have been attempts to objectify and operationalize the concept in some measurable way.

Alexithymia may be measured as a valid and reliable clinical phenomenon (12). However, it is a difficult concept to operationally and only few instruments are sufficiently reliable and valid. Several scales are used to measure alexithymia but only the Beth Israel Questionnaire

(BIQ) and the Toronto Alexithymia Scale (TAS) can be regarded as having sufficient psychometric properties.

Sifneos (29) devised the Beth Israel Psychosomatic Questionnaire (BIQ) to measure alexithymic characteristics. It is a 17-item forced-choice questionnaire completed by the therapist. The items are derived from the clinical criteria described previously.

The TAS is a 20-item self-report scale with strong psychometric properties (33). Subjects were asked to indicate the extent to which they agreed or disagreed with each statement on a five-point Likert scale. The results are expressed as TAS-20 global scores, as well as three subscales measuring difficulty in identifying feelings and distinguishing them from bodily sensations of emotion, difficulty expressing feelings, and externally oriented thinking.

The TAS-20 has 3 subscales:

1. Difficulty Describing Feelings
2. Difficulty Identifying Feeling
3. Externally-Oriented Thinking

7. Treatment in alexithymia

In treatment situations, therapists find such patients dull and boring (34). Although the concept was derived from experience with patients having classic psychosomatic illness, this clinical presentation has also been described in patients who somatize (35), who have substance abuse disorders (36), and who have experienced severe Psychosomatic condition.

To treatment the alexithymia physicians should focus on developing skills and knowledge (vocabulary for emotions, practicing emotional self-awareness). The suffering associated with fibromyalgia is intense, and physicians continue to struggle with the management even to the extent of prescribing narcotic pain medications where not ideally indicated (37). It proposed that much of this suffering will remain unspeakable and untreated until management strategies actively consider the prevalence of alexithymia and associated mood disorders.

The therapeutic approach to patients with elevated alexithymia in fibromyalgia includes treatment modalities that have the potential for increasing emotional awareness and the capacity to regulate and modulate instinctual tensions and emotional arousal through cognitive process (38). Such as fantasizing, dreaming, as well as reflective thinking and the verbal communication and sharing off feelings.

The basic idea is that during the process of therapy mental representations of difficult situations in patients' lives need to be constructed by putting into words the chain of events that makes up the difficult situation making the patient's appraisal of the difficult situation explicit; and addressing affective responses and discussing the patient's way of dealing with the difficult situation. These three steps reflect principles that should be integrated in therapy. In practice, these three steps necessarily get intermingled, and we advise against considering them in terms of a protocol or in terms of therapeutic phases (33).

The therapist should first start by asking the patient about his actual life and by exploring what goes wrong, with the idea of creating a conversation. Practically speaking, such a conversation focuses on the concrete situations the patient presents as important and difficult (trouble at home, an incident with a colleague...). In addressing these events, the therapist primarily strives to elicit a factual account that is coherent with the patient's story. The aim is to construct the chain of events that makes up specific difficulties, and to create an elementary narrative about these difficulties. To construct such narrative, the therapist converses with the patient: many open questions are asked, summaries of what has been said are frequently given, and a relaxed atmosphere is established. It is important that the therapist strives to articulate the difficult situation in the patient's own terms; explicitly recognizes the patient's trouble in dealing with the situation; and invites him/her to further investigate this problem in therapy. Therapy itself is defined as a place for the investigation of difficulties in life.

As a result, pain and other physical symptoms may influence by different factors including alexithymia in patients with fibromyalgia. It thought to be treatment of alexithymia can be beneficial in patients with fibromyalgia. Bu it is not possible to fully determine the interaction between fibromyalgia and alexithymia.

Fibromyalgia is a chronic pain syndrome with steadily fluctuating musculo-skeletal pain as the main symptom. However, despite intensive research, the primary mechanisms underlying the etiopathogenesis of fibromyalgia remain elusive.

8. Conclusion and future directions

Alexithymia is a clinical concept that refers to individuals who have difficulty describing feelings, have an impoverished fantasy life, and demonstrate a particular style of interpersonal relationships. Whether these individuals recognize their emotions but cannot express them verbally, or simply cannot recognize these signals from their bodies, is not clearly understood.

The concept itself needs to be more precisely defined, and reliable methods of measurement must be developed and validated. If agreement can be reached upon the clinical definition and measurement, further studies can be undertaken to evaluate which factors are important in the etiology and maintenance of alexithymic behavior.

In patients with FMS who show the alexithymic trait, attention should be given to the establishment of a secure therapeutic alliance. Capacities for regulating emotions on a verbal level can only be acquired within an environment based on trust and perceived by the patient as safe. The high comorbidity of psychiatric disorders underscores the importance of psychiatric interventions in patients with FMS.

9. References

[1] Pedrosa Gil F, Weigl M, Wessels T, Irnich D, Baumüller E, Winkelmann A. Parental bonding and alexithymia in adults with fibromyalgia. Psychosomatics 2008;49(2):115-22.

[2] Yunus MB. Fibromyalgia and overlapping disorders: a unifying concept of central sensitivity syndromes. Semin Arthritis Rheum 200;36:339–56.

[3] Glass J, Park D. Cognitive Function & Fibromyalgia. Monday, April 28, 2003.

[4] Julie D, Henry A, Louise H, Phillips B, John R, Crawford B, et all. Summersc Cognitive and psychosocial correlates of alexithymia following traumatic brain injury. Neuropsychologia 2006;4462–72.
[5] Hendryx M.S, Haviland M.G, Shaw D.G. Dimensions of alexithymia and their relationships to anxiety and depression. Journal of Personality Assessment, 1991;56(2): 227-37.
[6] Neyal M.M, Herken H, Demirci H, Virit O, Neyal A. Alexithymic features in migraine patients. European Archives of Psychiatry & Clinical Neuroscience 2004;254(3):182.
[7] Yucel B, Kora K, Ozyalcin S, Alcalar N, Ozdemir O, Yucel A. Depression, automatic thoughts alexithymia and assertiveness in patients with tension-type headache. Headache 2002;42:194–9.
[8] Yalug I, Selekler M, Erdogan A, Kutlu A, Dundar G, Ankaralı H, Aker T. Correlations between alexithymia and pain severity, depression, and anxiety among patients with chronic and episodic migraine. Psychiatry and Clinical Neurosciences 2010; 64(3):231–8.
[9] Stotsky BA. Sodium butabarbital for emotional disorders and insomnia. Dis Nerv Syst 1972;33(12):798-803.
[10] Lesser IM. A review of the alexithymia concept. Psychosom Med 1981;43(6):531-43.
[11] Hoppe KD, Bogen JE. Alexithymia in twelve commissurotomized patients. Psychotherapy and psychosomatics 1977;28(1–4):148–55.
[12] Nemiah, CJ. Alexithymia and Psychosomatic Illness. Journal of Continuing Education 1978;39:25–37.
[13] Berthoz S, Artiges E, Van de Moortele P, et al. Effect of Impaired Recognition and Expression of Emotions on Frontocingulate Cortices: An fMRI Study of Men With Alexithymia. American Journal of Psychiatry 2002;159(6): 961-7.
[14] Wall PDl, Melzack R. The challenge of pain. 2nd ed. New York: Penguin Books; 1996;pp:61–9.
[15] Albe-Fessard D, Berkley KJ, Kruger L, Ralston HJ, Willis WD. Diencephalic mechanisms of pain sensation. Brain Res Brain Res Rev 1985;9:217–96.
[16] Lenz FA, Rios M, Zirh A, Chau D, Krauss, Lesser RP. Painful stimuli evoke potentials recorded over the human anterior cingulate gyrus. J Neurophysiol 1998;79(4):2231-4.
[17] Vogt BA, Nimchinsky EA, Hof PR. Human cingulate cortex: Surface features, flat maps, and cytoarchitecture. J Comp Neurol 1997;359:490-506.
[18] Mountz JM, Bradley LA, Modell JG, Alexander RW, Triana-Alexander M, Aaron LA et al. Fibromyalgia in women. Abnormalities of regional cerebral blood flow in the thalamus and the caudate nucleus are associated with low pain threshold levels. Arthritis Rheum 1995;38(7):926-38.
[19] Gracely RH, Petzke F, Wolf JM, Clauw DJ. Functional magnetic resonance imaging evidence of augmented pain processing in fibromyalgia. Arthritis Rheum 2002;46: 1333–43.

[20] Jessimer M, Markham R. Alexithymia: a right hemisphere dysfunction specific to recognition of certain facial expressions? Brain and Cognition 1997;34(2):246–58.
[21] Lane RD, Ahern GL, Schwartz GE, Kaszniak AW. Is alexithymia the emotional equivalent of blindsight? Biol Psychiatry 1997;42(9):834–44.
[22] Tabibnia G, Zaidel E. Alexithymia, interhemispheric transfer, and right hemispheric specialization: a critical review. Psychotherapy and Psychosomatics 2005;74(2):81–92.
[23] Maclaren K. Emotional Disorder and the Mind-Body Problem: A Case Study of Alexithymia. Chiasmi International 2006;8:139–55.
[24] Fukunishi I, Berger D, Wogan J, Kuboki T. Alexithymic traits as predictors of difficulties with adjustment in an outpatient cohort of expatriates in Tokyo. Psychological Reports 1999;85(1):67–77.
[25] Lane RD, Ahern GL, Schwartz GE, Kaszniak AW. Is alexithymia the emotional equivalent of blindsight? Biol Psychiatry 1997;42:834-44.
[26] Salminen JK, Saarijärvi S, Äärelä E, Toikka T, Kauhanen J. Prevalence of alexithymia and its association with sociodemographic variables in the general population of Finland. Journal of Psychosomatic Research 1999;46(1):75–82.
[27] Parker, JDA; Taylor, GJ; Bagby, RM. The Relationship Between Emotional Intelligence and Alexithymia. Personality and Individual Differences 2001;30:107–15.
[28] Pasini, R.D. Chiaie, S. Seripa and N. Ciani, Alexithymia as related to sex, age, and educational level: results of the Toronto Alexithymia Scale in 417 normal subjects. Compr Psychiatry 1992;33:42–6.
[29] Sifneos PE, Cambridge MA. Short-term psychotherapy and emotional crisis. Harvard University Press, 1972.
[30] Donald L. Steinweg, F.A.C.P., Apostolos P., Dallas, William S. Rea. Fibromyalgia: Unspeakable Suffering, A Prevalence Study of Alexithymia. Psychosomatics 2011:52:255–62.
[31] Haviland MG, Warren WL, Riggs ML. An observer scale to measure alexithymia. Psychosomatics 2000;41(5):385–92.
[32] Beales DL, Dolton R. Eating disordered patients: Personality, alexithymia, and implications for primary care. British Journal of General Practice 2000;50:21–6.
[33] Ira M. A review of the alexithymia concept. Lesser Psychosom Med 1981;43(6):531-43.
[34] Hosoi M, Molton IR, Jensen MP, Ehde DM, Amtmann S, O'Brien S, Arimura T, Kubo C. Relationships among alexithymia and pain intensity, pain interference, and vitality in persons with neuromuscular disease: Considering the effect of negative affectivity. Pain 2010;149(2).273-7.
[35] Huber A, Suman AL, Biasi G, Carli G. Alexithymia in fibromyalgia syndrome: associations with ongoing pain, experimental pain sensitivity and illness behavior. J Psychosom Res 2009;66(5):425-33.
[36] Gur A. Physical therapy modalities in management of fibromyalgia. Curr Pharm Des 2006;12(1):29-35.
[37] Bradley LA. Pathophysiology of fibromyalgia. Am J Med 2009;122(12):22-30.

[38] Madenci E, Herken H, Keven S, Yagiz E, Gursoy S. Alexıthymıa in patıents wıth fıbromyalgıa syndrome. Turkiye Klinikleri J Med Sci 2007;27:32-5.

9

Diagnosis of Fibromyalgia Syndrome: Potential Biomarkers and Proteomic Approach

Federica Ciregia[1], Camillo Giacomelli[2], Laura Giusti[1], Antonio Lucacchini[1] and Laura Bazzichi[2]
[1]*Department of Psychiatry, Neurobiology, Pharmacology and Biotechnology,*
[2]*Department of Internal Medicine, University of Pisa*
Italy

1. Introduction

In 1990 the American College of Rheumatology (ACR) established criteria for the diagnosis of fibromyalgia (Wolfe et al., 1990), and more recently new criteria have been proposed (Wolfe et al., 2010, 2011). However, the absence of anatomic pathological lesions and of biohumoral abnormalities, demonstrated with classical instrumental methods, has led to considerable difficulties in diagnosis.

Until now, many attempts have been made to search for biomarkers in fibromyalgia, but at present no specific markers have been found (Bazzichi et al., 2010). The problem lies in the presence of too many data, often controversial, rather than in a lack of data

We will discuss methods that are state of the art in searching for biomarkers in fibromyalgia. Furthermore, we will focus on the contribution that proteomics can give in the diagnosis of the disease on the basis of the study we carried out on human whole saliva of patients affected by fibromyalgia.

2. Genetic markers

The presence of fibromyalgia in family clusters and many studies support the idea that genetic factors may predispose to fibromyalgia in combination with environmental triggers such as trauma, infections or emotional stress. The principal gene polymorphisms supposed to be a risk factor for fibromyalgia are those implicated in mood disorders but results are often controversial.

These candidate genes include the serotonin transporter (5-HTT), the serotonin 2A (5-HT2A) receptor, catechol-O-methyltransferase (COMT) and the dopamine receptor.

Offenbaecher and colleagues (Offenbaecher et al., 1999) were the first to analyze the genotypes of the serotonin transporter promoter locus (5HTTLPR) in patients with fibromyalgia. They found that the S/S genotype of 5-HTT occurred more frequently in fibromyalgia patients than in healthy controls. This association is interesting considering that 5-HTT is involved in many conditions that are either risk factors for —or frequent concomitants to— fibromyalgia, such as anxiety, Bipolar Disorder, Psychosis, attention deficit hyperactivity disorder, and Major Depressive Disorder (Maletic & Raison, 2009).

Cohen and collaborators subsequently confirmed this association between fibromyalgia and 5-HTTLPR polymorphism (Cohen et al., 2002).

However, these data were not found in all studies. Recently, Lee and collaborators (Lee et al., 2010) conducted a meta-analysis, and from their results no association was found between 5-HTTLPR and fibromyalgia susceptibility. Potvin et al. found no evidence of a relationship between 5-HTTLPR and fibromyalgia-related psychiatric symptoms (Potvin et al., 2010). Thus, their results are inconsistent with previous reports from Offenbaecher and Cohen but in agreement with those of Gürsoy who concluded that neither 5-HTT nor its polymorphism is associated with fibromyalgia (Gürsoy, 2002).

5-HT2A is also a candidate for involvement in fibromyalgia. Bondy et al., investigated the T102C polymorphism of the 5-HT2A receptor gene and found a significantly different genotype distribution in fibromyalgia patients who had a decrease in T/T and an increase in both T/C and C/C genotypes as compared with the control population (Bondy et al., 1999). In contrast, the pain score was significantly higher for patients with the T/T genotype. It was suggested that the T102 allele might be involved in the complex circuits of nociception. Lee and collaborators using meta-analysis, provided evidence of an association of the 5-HT2A receptor with fibromyalgia susceptibility (Lee et al., 2010). On the contrary, Gürsoy, after studying T102C polymorphism in Turkish patients, concluded that it is not associated with the etiology of fibromyalgia (Gürsoy et al., 2001). Similar results were described by Matsuda et al. in the Brazilian population (Matsuda et al., 2010), and by Tander in a population of eighty patients (Tander et al., 2008).

Another area of investigation is the COMT gene. This gene has long been implicated to be involved in the pathogenesis of neuropsychiatric disorders, schizophrenia, bipolar affective disorder, migraine, and Parkinson's disease. In 2003, Zubieta et al. showed a positive relationship between the COMT gene polymorphism and the experience of pain (Zubieta et al., 2003). Their study focused on the val158met polymorphism, a single nucleotide polymorphism in codon 158 of the COMT gene, which substitutes valine for methionine and resulting in reduced enzyme activity. Individual homozygous for the met158 allele of this polymorphism showed a diminished mu-opioid system response to pain, compared with heterozygotes. Opposite effects on pain and negative effects have been found in val158 homozygotes. The conclusion was that COMT val158met polymorphism influences the human experience of pain and may underlie interindividual differences in the adaptation and responses to pain and other stressful stimuli (Zubieta et al., 2003). On the other hand, an extensive study on the association between this polymorphism and chronic widespread pain concluded that COMT pain sensitivity haplotypes do not play a role in disease susceptibility. The authors proposed that the genetics of chronic widespread pain may differ from that of other pain syndromes (Nicholl et al., 2010).

Cohen et al. observed an association between fibromyalgia and this polymorphism. Moreover, they found that unaffected relatives of fibromyalgia patients had a reduced percentage of the COMT met allele (Cohen et al., 2009). They suggested that this reduced frequency acts as a 'protective' allele in this group and prevents the development of clinical fibromyalgia (Cohen et al., 2009).

Gürsoy and collaborators assessed the significance of COMT polymorphism in fibromyalgia. There were three polymorphisms of the COMT gene: LL, LH, and HH. The analysis of COMT polymorphism was performed using a polymerase chain reaction. The LL and LH genotypes together were more highly represented in patients than in controls. In addition, HH genotypes in patients were significantly lower than in the control group (Gürsoy et al., 2003). It was concluded that COMT polymorphism is of potential pharmacological importance concerning individual differences in the metabolism of catechol drugs. In addition, it may also be involved in the pathogenesis and treatment of fibromyalgia through adrenergic mechanisms as well as in the genetic predisposition to fibromyalgia (Gürsoy et al., 2003). Matsuda et al. confirmed these results in the Brazilian population (Matsuda et al., 2010) showing that the LL genotype was more frequently found among fibromyalgia patients.

On the other hand, Tander and subsequently Lee didn't find any associations between the COMT gene and fibromyalgia (Lee et al., 2010; Tander et al., 2008).

Finally, some studies have established a connection between fibromyalgia and the dopaminergic system.

Dopaminergic neurotransmission was found to be altered in fibromyalgia patients by Malt and colleagues. They suggested an increased sensitivity or density of dopamine D_2 receptors in fibromyalgia patients (Malt et al., 2003). Buskila and collaborators demonstrated an association between fibromyalgia and the 7-repeat allele in exon III of the D_4 receptor gene, specifically that the frequency of this polymorphism was significantly lower in people with fibromyalgia (Buskila et al., 2004). More recently, Treister et al. found an association between the dopamine transporter gene (DAT-1) polymorphism and cold pain tolerance. Their results, together with the known function of the investigated candidate gene polymorphisms, suggest that low dopaminergic activity can be associated with high pain sensitivity (Treister et al., 2009). However, Ablin and co-workers, investigating an association between fibromyalgia and DAT1, found no significant association between fibromyalgia and the genetic marker (Ablin et al., 2009).

All these findings are summarised in Table 1

Reference	Year	Studied gene and findings in fibromyalgia	
Offenbaecher et al.	1999	**5-HTT**	The S/S genotype of 5-HTT occurred more frequently in fibromyalgia patients than in healthy controls
Cohen et al.	2002		Association between fibromyalgia and serotonin transporter promoter region (5-HTTLPR) polymorphism.
Gürsoy	2002		no association
Lee et al.	2010		no association
Potvin et al.	2010		no association
Bondy et al.	1999	**5-HT2A**	They investigated the T102C polymorphism finding a significantly different genotype distribution in fibromyalgia patients with a decrease in T/T and an increase in both T/C and C/C genotypes as compared with the control population.

Lee et al.	2010		They found an association of the 5-HT2A receptor with fibromyalgia susceptibility.
Gürsoy et al.	2001		no association
Tander et al.	2008		no association
Matsuda et al.	2010		no association
Gürsoy et al.	2003	**COMT**	Authors assessed the significance of COMT polymorphism in fibromyalgia. The LL and LH genotypes together were more highly represented in patients than controls. In addition, HH genotypes in patients were significantly lower than in the control group.
Zubieta et al.	2003		COMT val158met polymorphism influences the experience of pain.
Cohen et al.	2009		They found an association between fibromyalgia and the val158met polymorphism of COMT gene.
Matsuda et al.	2010		The L/L genotype was more frequent among fibromyalgia patients.
Tander et al.	2008		no association
Lee et al.	2010		no association
Nicholl et al.	2010		No evidence of association between the COMT pain sensitivity haplotypes and chronic widespread pain.
Malt et al.	2003	**D_2 receptor**	The authors suggested an increased sensitivity or density of dopamine D2 receptors in fibromyalgia patients
Buskila et al.	2004	**D_4 receptor**	They have demonstrated an association between fibromyalgia and the 7 repeat allele in exon III of the D4 receptor gene. Specifically, the frequency of this polymorphism was significantly lower in patients with fibromyalgia.
Treister et al.	2009	**DAT1**	A significant association was found between the quantitative measure of cold tolerance and DAT-1 polymorphism.
Ablin et al.	2009		no association

Table 1. Genetics in fibromyalgia.

In conclusion, all these genetic studies support a role for the polymorphism of genes of the serotoninergic, catecholaminergic and dopaminergic systems in the etiopathogenesis of fibromyalgia.
These polymorphism are often associated with psychiatric disorders, thereby they could be related to psychiatric comorbidities rather than to sole fibromyalgia. Furthermore, genetic results are often controversial, and no specific candidate gene has been closely connected with fibromyalgia.

3. Serologic markers

There is a great interest in using a simple blood test to diagnose fibromyalgia. Thus, many attempts have been carried out in order to identify specific serologic markers. These results, as well as those of genetic studies, are often conflicting and, at present, no clinical test has been validated yet.

3.1 Autoantibodies

Autoimmunity has a central role in pathogenesis of many rheumatic diseases while fibromyalgia is generally considered to be a non-autoimmune disease. Nevertheless, a broad spectrum of autoantibodies has been widely investigated in the sera from patients. The studies have often generated controversial data, and the proposed antibodies don't have diagnostic relevance yet.
The association between the antipolymer antibody (APA) and fibromyalgia was evaluated because APAs were found in the sera of women with silicone breast implants presenting fibromyalgia-like symptoms (Wolfe, 1999). In addition, Wilson et al. found a higher prevalence of APAs (67%) in fibromyalgia patients in the USA population (Wilson et al., 1999). Furthermore, a Danish study showed that fibromyalgia patients tended to have slightly higher APA levels than controls when adjusted for symptom severity (Jensen et al., 2004). However, these results were not confirmed in the Italian population by Bazzichi and colleagues (Bazzichi et al., 2007). Instead, they found a lower percentage of APA seropositivity (23%) in fibromyalgia patients. A correlation between the APA and pain and fatigue severity in fibromyalgia was also proposed by Sarzi-Puttini et al. because the APA test did not distinguish this group from the controls, but instead a positive APA test prevalence increased with less severe pain or fatigue (Sarzi-Puttini et al., 2008). A more recent study evaluated the APA concentration in the serum of patients with fibromyalgia, patients with tension-type headache, and healthy controls. APAs were detected in only 17.6% of fibromyalgia patients confirming the lack of diagnostic relevance of the APA (Iannuccelli et al., 2010).
Two other autoantibodies, the anti-68/48 kDa and the anti-45 kDa, have been described as possible markers for certain clinical subsets of primary fibromyalgia and chronic fatigue syndrome, and for secondary fibromyalgia/psychiatric disorders, respectively (Nishikai et al., 2001).
Antiserotonin, antiganglioside and antiphospholipid antibodies have been shown to be higher in fibromyalgia patients as well as in those with chronic fatigue syndrome, supporting the concept that fibromyalgia and chronic fatigue syndrome may belong to the same clinical entity and may manifest themselves as 'psycho-neuro-endocrinological autoimmune diseases' (Klein & Berg, 1995). A different group, evaluating the prevalence

and potential diagnostic relevance of antiserotonin, antithromboplastin and antiganglioside in patients with fibromyalgia, confirmed an elevated prevalence of antibodies against serotonin and thromboplastin, but they concluded that the measurement of these antibodies had no diagnostic relevance (Werle et al., 2001).

Another important research field is thyroid autoimmunity. Bazzichi and collaborators found that autoimmune thyroiditis is present in an elevated percentage of fibromyalgia patients. In particular, 41% of these patients had at least one thyroid antibody, a percentage that was twice the value of the control group's (Bazzichi et al., 2007). More recently, they confirmed this significant association between thyroid autoimmunity and fibromyalgia (Bazzichi et al., 2010). Pamuk & Cakir also found a higher frequency of thyroid autoimmunity in fibromyalgia patients than in the control group, while the frequency was similar between fibromyalgia patients and the rheumatoid arthritis group (Pamuk & Cakir, 2007). Essentially, the association between rheumatic and thyroid disorders have long been known, the most common being the associations with rheumatoid arthritis, Sjögren's syndrome, and systemic lupus erythematosus.

Therefore, at present, the proposed antibodies have not been validated yet as being strong, useful diagnostic biomarkers for fibromyalgia and further studies are necessary to reach a research end point.

3.2 Neuropeptides

Neuropeptide Y is a neurotransmitter released mainly by sympathetic neurons in the autonomic nervous system, and it has a complex role in mediating analgesia and hyperalgesia. Thus, Neuropeptide Y has been shown, to both reduce and cause pain. Crofford and colleagues were the first to measure plasma Neuropeptide Y levels in fibromyalgia patients and they found that it was significantly lower in fibromyalgia patients than in normal subjects (Crofford et al., 1994). These results are inconsistent with subsequent findings from Anderberg who found significantly elevated plasma Neuropeptide Y levels in fibromyalgia patients compared with the healthy subjects (Anderberg et al, 1999). Furthermore, serum Neuropeptide Y levels have been shown to be significantly higher in fibromyalgia subjects, compared with healthy controls, by two different groups (Di Franco et al., 2009; Iannuccelli et al., 2010).

Substance P is another neuropeptide extensively studied in fibromyalgia patients. Substance P is released in the cerebral spinal fluid when axons are stimulated. An increase in substance P levels in the cerebral spinal fluid can be related to an increase of pain neurotransmitters in the spinal cord. Two studies from two different groups have demonstrated significantly elevated substance P levels in cerebral spinal fluid of fibromyalgia patients (Russell et al., 1994; Vaerøy et al., 1988) but levels of substance P determined in cerebral spinal fluid of patients with chronic fatigue syndrome were within normal range (Evengard et al., 1998). It was proposed that substance P in mice produces disturbances in sleep (Andersen et al., 2006). Subsequently, it is likely that the high levels of substance P found in fibromyalgia patients may be one of the causes of sleep dysfunction.

Brain-derived neurotrophic factor (BDNF) is a member of the neurotrophin family, which also includes the nerve growth factor. BDNF has a wide range of biological activities and is produced by different immune and structural cells (Iughetti et al., 2011). BDNF seems to have a role in Major Depressive Disorder, schizophrenia, and eating disorders such as bulimia and anorexia nervosa, Parkinson's and Alzheimer's diseases, and epileptic and psychogenic nonepileptic seizures (Iughetti et al., 2011). In addition, BDNF is a mediator of pain in the peripheral nerve system, and some findings have also suggested an involvement of BDNF in pain syndrome (Laske et al., 2006). On the basis of this evidence, Laske and collaborators were the first to evaluate BDNF serum concentrations in fibromyalgia patients, and they found that levels significantly increased compared with healthy controls (Laske et al., 2006). Two different groups found significantly higher levels of BDNF in plasma and in the cerebral spinal fluid of fibromyalgia patients than in controls, indicating an involvement of BDNF in the pathophysiology of fibromyalgia (Haas et al., 2010; Sarchielli et al., 2007).

The excitatory neurotransmitter glutamate is known to function in pain neuropathways, and it is suspected to play a role in the pathophysiology of fibromyalgia. This hypothesis is supported by different studies. Harris and collaborators found a correlation between changing levels of insular glutamate and changes in pain in patients with fibromyalgia using proton magnetic resonance spectroscopy and functional magnetic resonance imaging (Harris et al., 2008). Subsequently, they extended these findings by investigating the relationship between insular glutamate and combined glutamate/glutamine in individuals with fibromyalgia and pain free controls (Harris et al., 2009). They concluded that enhanced glutamatergic neurotransmission within the posterior insula is a potential pathologic factor in fibromyalgia. In a pilot study, Fayed and colleagues compared three different magnetic resonance imaging examination methods for the diagnosis of fibromyalgia: magnetic resonance spectroscopy, diffusion-weighted imaging, and diffusion-tensor imaging. One of the principal findings was the increase in the combined glutamate+glutamine levels, and glutamate+glutamine/creatine ratio, in the posterior gyrus in patients with fibromyalgia compared with controls (Fayed et al., 2010). Another group used magnetic resonance spectroscopy techniques to study brain metabolites in the amygdala, thalami, and prefrontal cortex of these women. Patients with fibromyalgia showed higher levels of glutamate compounds in the right amygdala than did healthy controls, and pain was related to increased glutamate levels in the left thalamus (Valdés et al., 2010). These findings have implications for future therapies directed against glutamate receptors, but further studies are desirable to confirm whether these findings are observed in other functional pain syndromes.

3.3 Inflammation

Since fibromyalgia is characterised by widespread pain, and the origin of pain is inflammation, many studies have focused on the inflammatory hypothesis for fibromyalgia, although fibromyalgia is generally regarded as a non-inflammatory disease. Special attention has been paid on circulating pro-inflammatory cytokines as possible markers in fibromyalgia patients.

A connection between the pathogenesis of fibromyalgia and cytokines was suspected when in 1988, after being treated with IL-2 or IFN-α, patients with malignant melanoma, renal cell carcinoma, and hairy cell leukaemia showed fibromyalgia-like symptoms such as myalgias, arthralgias, cognitive impairment, and painful tender points (Wallace et al., 1988, as cited in Di Franco et al., 2010).

Therefore, cytokines have been suggested to be involved in the fibromyalgia syndrome, but results often appear to be controversial, especially because increases, reductions, and no significant changes have been reported. Levels of IL-10, an anti-inflammatory cytokine, were studied by four different groups with different findings. Wallace and Amel Kashipaz didn't find alterations in IL-10 levels, while Bazzichi and co-workers found an increase in plasma levels, and Uçeyler found a reduction (Amel Kashipaz et al., 2003; Bazzichi et al., 2007; Uçeyler et al., 2006; Wallace et al., 2001).

IL-8, a pro-inflammatory cytokine, has been consistently demonstrated to be increased in patients with fibromyalgia. Bearing in mind that IL-8 promotes sympathetic pain, it may play an important role in the occurrence of pain in fibromyalgia. If validated in further studies, the IL-8 expression pattern might help in the diagnosis of fibromyalgia and in the appropriate treatment approach.
Table 2 summarises the main findings for pro- and anti- inflammatory cytokines in fibromyalgia patients.

Reference	**n° of subjects**	**Samples**	**Cytokines**	**Findings in patients**
Wallace et al., 2001	56 fibromyalgia 36 controls	serum	IL-1β, IL-2, IL-10, serum IL-2 receptor (sIL-2R), IFN-γ, TNF-α, IL-8, IL-6, IL-1Ra	↑ IL-8, IL-1Ra
Gür et al., 2002	81 fibromyalgia 32 controls	serum	IL-I, sIL-2R, IL-6, IL-8	↑ IL-8, sIL-2R
Amel Kashipaz et al., 2003	22 chronic fatigue syndrome / fibromyalgia 19 controls	cells	TNF-α, IL-1α, IL-6, IL-10	no differences
Uçeyler et al., 2006	40 chronic fatigue syndrome (26 fibromyalgia) 40 controls	mRNA from blood	IL-2, IL-4, IL-8, IL-10, TNF-α, TGF-β1	↓ IL-4, IL-10
Bazzichi et al., 2007	80 fibromyalgia 45 controls	plasma	IL-1, IL-6, IL-8, IL-10, TNF-α,	↑ IL-8, IL-10, TNF-α
Wang et al., 2008	20 fibromyalgia 80 controls	serum	IL-4, IL-6, IL-8, IL-10, TNF-α,	↑ IL-8, TNF-α
Wang et al., 2009	20 fibromyalgia 80 controls	serum	IL-8	six months after therapy, IL-8 was still significantly higher
Kim et al., 2010	27 fibromyalgia 29 controls	serum	IL-1β, IL-6, IL-8	↑ IL-8

Table 2. Cytokines in Fibromyalgia.

4. Proteomic markers

In the last few years, it has become widely recognized that the genome represents only the first layer of complexity. Biological functions rely on a dynamic population of proteins, and the characterisation of the proteins can reveal posttranslational modifications (e.g., phosphorylation, glycosylation, and methylation), and give insight into protein-protein interactions and functions. For these reasons, there is an increasing interest in the field of proteomics wich is the identification of proteins contained in biological samples such as body fluids and tissue extracts.

In 2005 Baraniuk and collaborators analyzed the cerebral spinal fluid of patients with chronic fatigue syndrome by quadrupole-time-of flight mass spectrometry, compared chronic fatigue syndrome with Persian Gulf War Illness and fibromyalgia, and concluded that the three proteomes overlapped (Baraniuk et al., 2005). Recently, we carried out a study on human whole saliva of patients affected by fibromyalgia. The aim was to identify the protein content of whole saliva and to determine the quantitative or qualitative differences between fibromyalgia patients and healthy subjects. We expected to find multiple biomarkers rather than a single one because a panel of biomarkers may correlate more reliably with fibromyalgia than a single protein (Bazzichi et al., 2009).

In this work we used two-dimensional electrophoresis (2-DE) to obtain the whole saliva protein map of fibromyalgia patients. Figure 1 shows the typical protein pattern obtained from a healthy control subject (figure 1 A) and a fibromyalgia patient (figure 1 B).

Matrix-assisted laser desorption ionization time-of-flight/ time-of-flight mass spectrometry (MALDI-TOF/TOF MS) was used to identify proteins that we found differentially expressed from two analysed groups. In table 3, we report the most interesting proteins which emerged from the data analysis.

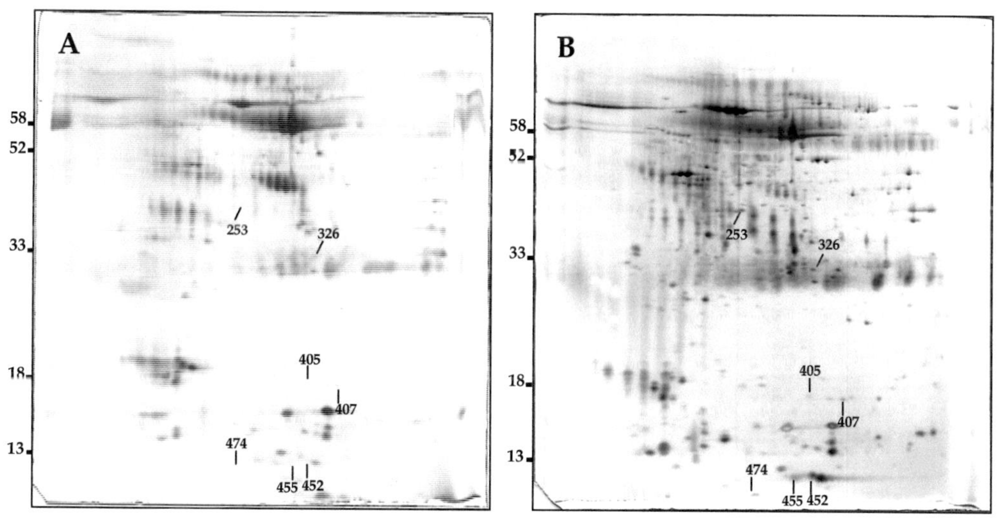

Fig. 1. Representative 2-DE gel map of salivary proteins. Control (A) and a fibromyalgia patient (B).

Spot n°	Accession number	Protein name	MW KDa	pI	score	coverage (%)	p-value
253	P37837	transaldolase	38	6.3	181	42	<0.0001
326	P18669	phosphoglycerate mutase 1	29	6.7	501	44	0.0011
405	P62937	cyclophilin A	18	7.7	242	40	0.05
407	P62937	cyclophilin A	18	7.7	450	58	0.0157
452	P05109	calgranulin A	11	6.5	313	47	0.0036
455	P05109	calgranulin A	11	6.5	187	37	0.0002
474	P80511	calgranulin C	11	5.8	68	7	0.001

Table 3. Protein identification of differentially expressed proteins in whole saliva of fibromyalgia patients by MS/MS

Figure 2 shows the enlarged images of these proteins, histograms show the percentage volumes of the proteins and each bar represents the mean±SD of the mean of each spot. Significant differences from control whole saliva are based on Mann-Whitney test; (*p<0.05, **p<0.01, ***p<0.001).

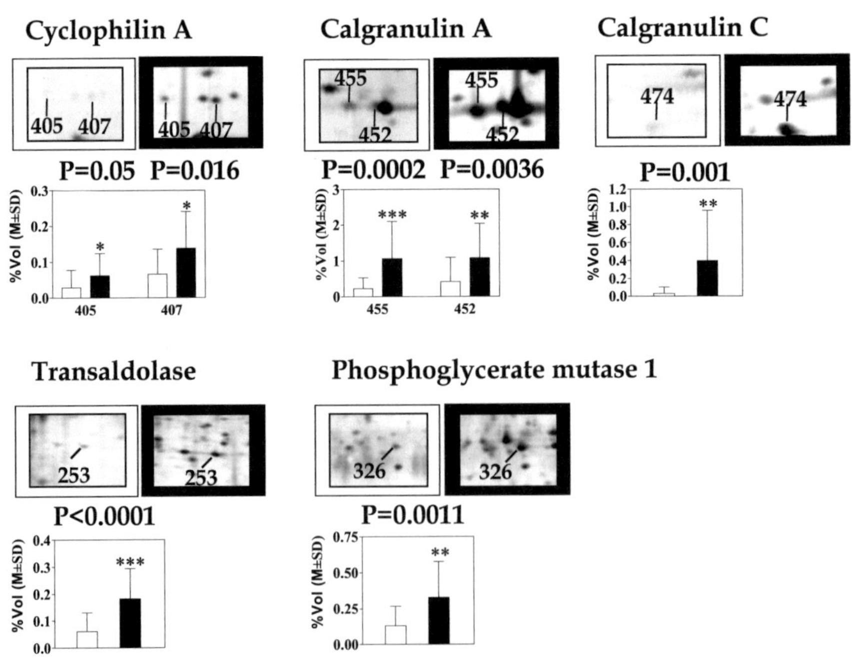

Fig. 2. Enlarged images of the 2-DE gels highlighting some differentially expressed proteins for two representative gels. White: controls, Black: fibromyalgia.

Cyclophilin A resulted over-expressed in fibromyalgia patients in comparison with healthy subjects. Cyclophilin A is a ubiquitously distributed protein belonging to the immunophilin family that can be secreted by cells in response to inflammatory stimuli (Arora et al., 2005). Secreted Cyclophilin A is a potent chemo attractant for monocytes, neutrophils, eosinophils and T cells in vitro. Satoh K and co-workers showed that Cyclophilin A is secreted from smooth muscle cells and macrophages also in response to oxidative stress (Satoh et al., 2008).

Other proteins found to be over expressed in fibromyalgia were calgranulins, belonging to the S100 multigene family implicated in a variety of intracellular activities such as cell proliferation and differentiation, cytoskeletal interactions, rearrangement and structural organization of membranes, intracellular Ca2+ homeostasis, cell migration, inflammation, and protection from oxidative cell damage (Yang et al., 2001). These proteins have already been found in autoinflammatory rheumatic diseases such as Sjögren Syndrome, Scleroderma, and Rheumatoid Arthritis (De Seny et al., 2008; Giusti et al., 2007a, 2007b, 2010). Considering that fibromyalgia is defined as a non-inflammatory disease, we suggested that the over-expression of calgranulins might be related to a sub-inflammatory process, as suggested by the cytokine alterations.

The main finding of our work was the significant over-expression of transaldolase and phosphoglycerate mutase 1 in fibromyalgia patients with respect to controls. The reliability of these potential biomarkers was assessed by receiver operating characteristic curves (figure 3). The sensitivity and the specificity of the transaldolase and phosphoglycerate mutase 1 were 77.3 and 84.6% and 95.5 and 50%, respectively.

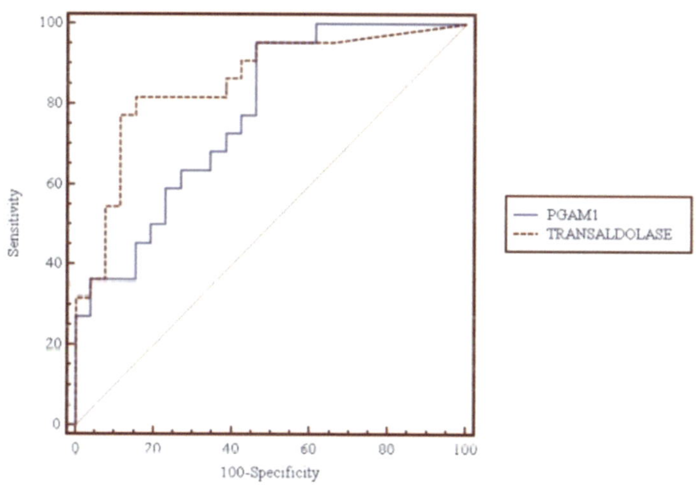

Fig. 3. ROC curve of transaldolase and phosphoglycerate mutase 1.

Furthermore, we validated the results by western blot analysis (figure 4). The statistical analysis of the optical density of specific detected bands confirmed the significant up-regulation of two enzymes in fibromyalgia patients with respect to controls.

Transaldolase is an enzyme of the non-oxidative phase of the pentose phosphate pathway, which is involved in the generation of reduced nicotinamide adenine dinucleotide phosphate (NADPH). There is a lot of evidence have shown that oxidative stress and nitric oxide may play an important role in fibromyalgia pathophysiology (Ozgocmen et al., 2006).

The overexpression of transaldolase might be justified by the NADPH production, which could be involved in limiting oxidative damage to tissues. Moreover, transaldolase links the pentose phosphate pathway to glycolysis. From this point of view, it is interesting to note that phosphoglycerate mutase 1, an enzyme involved in glycolysis, was differently expressed in fibromyalgia patients.

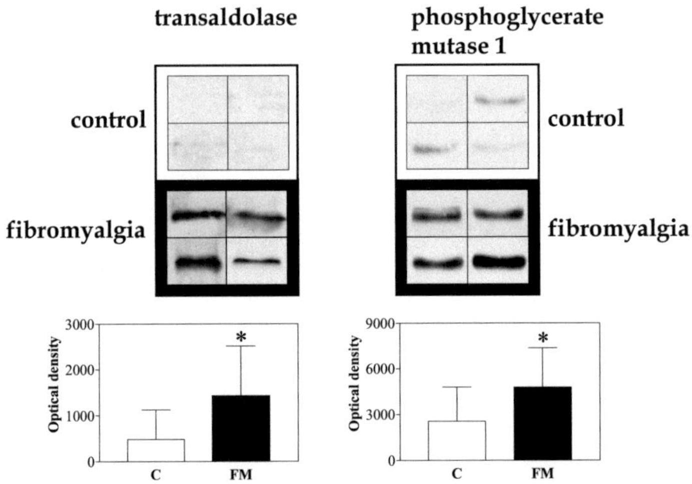

Fig. 4. Western Blot analysis of transaldolase and phosphoglycerate mutase 1 in whole saliva samples from healthy subjects and fibromyalgia patients. Densitometry of the blots is shown. Values that are significantly different ($p<0.05$) as determined by the Mann-Whitney test, are indicated.

In conclusion, our study has attested the potential usefulness of the proteomic characterization of human whole saliva in distinguishing fibromyalgia from healthy subjects. Moreover, the use of saliva may enable the easy characterization of non-invasively collected biological fluids, giving rise to a different approach in the diagnosis of fibromyalgia. The future focus of interest will be to validate the panel of biomarkers in different cohorts of pathological controls in order to identify biomarkers specific to fibromyalgia, and to exclude any interference of concomitant disorder (e.g. psychiatric comorbidities).

However, this study allows us to focus on some of the peculiar pathogenic aspects of fibromyalgia, especially on the oxidative stress which contradistinguishes this condition.

5. Conclusion

The pathogenesis of fibromyalgia is not entirely understood and, at present, the diagnosis is only clinical. Up to now, no objective measures have been determined to be reliable biomarkers, and these measures can only reflect a predisposition to fibromyalgia. However, the different studies offer us insights into the pathophysiology of fibromyalgia. The proteomic analysis, rather than the gene expression profile, might have potential applications as a new tool for the diagnosis of fibromyalgia because proteins are the final effectors that mediate disease pathogenesis. Therefore, proteomic seems to represent a necessary element in the advancement of disease diagnosis and therapeutic targets.

6. Acknowledgement

The authors wish to thank Dr Laura Fatuzzo for her valuable contribution in reviewing the text.

7. References

Ablin, J.N., Bar-Shira, A., Yaron, M. & Orr-Urtreger, A. (2009). Candidate-gene approach in fibromyalgia syndrome: association analysis of the genes encoding substance P receptor, dopamine transporter and alpha1-antitrypsin. *Clinical and Experimental Rheumatology*, Vol. 27, No. 5 suppl 56, (September 2009), pp. S33-38, ISSN 0392-856X.

Amel Kashipaz, M.R., Swinden, D., Todd, I. & Powell, R.J. (2003). Normal production of inflammatory cytokines in chronic fatigue and fibromyalgia syndromes determined by intracellular cytokine staining in short-term cultured blood mononuclear cells. *Clinical & Experimental Immunology*, Vol. 132, No. 2, (May 2003), pp. 360-365, ISSN 0009-9104.

Anderberg, U.M., Liu, Z., Berglund, L. & Nyberg, F. (1999). Elevated plasma levels of neuropeptide Y in female fibromyalgia patients. *European Journal of Pain*, Vol. 3, No. 1, (March 1999), pp. 19-30, ISSN 1090-3801.

Andersen, M.L., Nascimento, D.C., Machado, R.B., Roizenblatt, S., Moldofsky, H. & Tufik, S. (2006). Sleep disturbance induced by substance P in mice. *Behavioural Brain Research*, Vol. 167, No. 2, (February 2006), pp. 212-218, ISSN 0166-4328.

Arora, K., Gwinn, W.M., Bower, M.A., Watson, A., Okwumabua, I., MacDonald, H.R., Bukrinsky, M.I. & Constant, S.L. Extracellular cyclophilins contribute to the regulation of inflammatory responses. *Journal of Rheumatology*, Vol. 175, No. 1, (July 2005), pp. 517-522, ISSN 0315-162X.

Baraniuk, J.N., Casado, B., Maibach, H., Clauw, D.J., Pannell, L.K. & Hess, S.S. (2005). A Chronic Fatigue Syndrome - related proteome in human cerebrospinal fluid. *BMC Neurology*, Vol. 5, No. 22, ISSN 1471-2377.

Bazzichi, L., Rossi, A., Massimetti, G., Giannaccini, G., Giuliano, T., De Feo, F., Ciapparelli, A., Dell'Osso, L. & Bombardieri, S. (2007). Cytokine patterns in fibromyalgia and their correlation with clinical manifestations. *Clinical and Experimental Rheumatology*, Vol. 25, No. 2, (March 2007), pp. 225-230, ISSN 0392-856X.

Bazzichi, L., Giacomelli, C., De Feo, F., Giuliano, T., Rossi, A., Doveri, M., Tani, C., Wilson, R.B. & Bombardieri, S. (2007). Antipolymer antibody in Italian fibromyalgic patients. *Arthritis Research & Therapy*, Vol. 9, No. 5, (September 2007), pp. R86, ISSN 1478-6354.

Bazzichi, L., Rossi, A., Giuliano, T., De Feo, F., Giacomelli, C., Consensi, A., Ciapparelli, A., Consoli, G., Dell'osso, L. & Bombardieri, S. (2007). Association between thyroid autoimmunity and fibromyalgic disease severity. *Clinical Rheumatology*, Vol. 26, No. 12, (December 2007), pp. 2115-2120, ISSN 0770-3198.

Bazzichi, L., Ciregia, F., Giusti, L., Baldini, C., Giannaccini, G., Giacomelli, C., Sernissi, F., Bombardieri, S. & Lucacchini, A. (2009). Detection of potential markers of primary fibromyalgia syndrome in human saliva. *Proteomics Clinical Application*, Vol. 3, No. 11, (November 2009), pp. 1296-1304, ISSN 1862-8346.

Bazzichi, L., Rossi, A., Zirafa, C., Monzani, F., Tognini, S., Dardano, A., Santini, F., Tonacchera, M., De Servi, M., Giacomelli, C., De Feo, F., Doveri, M., Massimetti, G. & Bombardieri, S. (2010). Thyroid autoimmunity may represent a predisposition for the development of fibromyalgia? *Rheumatology International,* DOI 10.1007/s00296-010-1620-1, ISSN 0172-8172.

Bazzichi, L., Rossi, A., Giacomelli, C. & Bombardieri, S. (2010). Exploring the abyss of fibromyalgia biomarkers. *Clinical and Experimental Rheumatology,* Vol. 28, No. 6 Suppl 63, (November-December 2010), pp. S125-130, ISSN 0392-856X.

Bondy, B., Spaeth, M., Offenbaecher, M., Glatzeder, K., Stratz, T., Schwarz, M., de Jonge, S., Krüger, M., Engel, R.R., Färber, L., Pongratz, D.E. & Ackenheil, M. (1999). The T102C polymorphism of the 5-HT2A-receptor gene in fibromyalgia. *Neurobiology of Disease,* Vol. 6, No. 5, (October 1999), pp. 433-439, ISSN 0969-9961.

Buskila, D., Cohen, H., Neumann, L. & Ebstein, R.P. (2004). An association between fibromyalgia and the dopamine D4 receptor exon III repeat polymorphism and relationship to novelty seeking personality traits. *Molecular Psychiatry,* Vol. 9, No. 8, (August 2004), pp. 730-731, ISSN 1359-4184.

Cohen, H., Buskila, D., Neumann, L. & Ebstein, R.P. (2002). Confirmation of an association between fibromyalgia and serotonin transporter promoter region (5-HTTLPR) polymorphism, and relationship to anxiety-related personality traits. *Arthritis and Rheumatism,* Vol. 46, No. 3, (March 2002), pp. 845-847, ISSN 0004-3591.

Cohen, H., Neumann, L., Glazer, Y., Ebstein, R.P. & Buskila, D. (2009). The relationship between a common catechol-O-methyltransferase (COMT) polymorphism val(158) met and fibromyalgia. *Clinical and Experimental Rheumatology,* Vol. 27, No. 5 suppl. 56, (September-October 2009), pp. S51-56, ISSN 0392-856X.

Crofford, L.J., Pillemer, S.R., Kalogeras, K.T., Cash, J.M., Michelson, D., Kling, M.A., Sternberg, E.M., Gold, P.W., Chrousos, G.P. & Wilder, R.L. (1994). Hypothalamic-pituitary-adrenal axis perturbations in patients with fibromyalgia. *Arthritis and Rheumatism,* Vol. 37, No.11, (November 1994), pp. 1583-1592, ISSN 0004-3591.

De Seny, D., Fillet, M., Ribbens, C., Marée, R., Meuwis, M.A., Lutteri, L., Chapelle, J.P., Wehenkel, L., Louis, E., Merville, M.P. & Malaise, M. (2008). Monomeric calgranulins measured by SELDI-TOF mass spectrometry and calprotectin measured by ELISA as biomarkers in arthritis. *Clinical Chemistry,* Vol. 54, No. 6, (June 2008), pp. 1066-1075, ISSN 0009-9147.

Di Franco, M., Iannuccelli, C., Alessandri, C., Paradiso, M., Riccieri, V., Libri, F. & Valesini, G. (2009). Autonomic dysfunction and neuropeptide Y in fibromyalgia. *Clinical and Experimental Rheumatology,* Vol. 27, No. 5 suppl 56, (September 2009), pp. S75-S78, ISSN 0392-856X.

Di Franco, M., Iannuccelli, C. & Valesini, G. (2010). Neuroendocrine immunology of fibromyalgia. *Annals of the New York Academy of Sciences,* Vol. 1193, (April 2010), pp. 84-90, ISSN 0077-8923.

Evengard, B., Nilsson, C.G., Lindh, G., Lindquist, L., Eneroth, P., Fredrikson, S., Terenius, L. & Henriksson, K.G. (1998). Chronic fatigue syndrome differs from fibromyalgia. No evidence for elevated substance P levels in cerebrospinal fluid of patients with chronic fatigue syndrome. *Pain,* Vol. 78, No. 2, (November 1998), pp. 153-155, ISSN 0304-3959.

Fayed, N., Garcia-Campayo, J., Magallón, R., Andrés-Bergareche, H., Luciano, J.V., Andres, E. & Beltrán, J. (2010). Localized 1H-NMR spectroscopy in patients with fibromyalgia: a controlled study of changes in cerebral glutamate/glutamine, inositol, choline, and N-acetylaspartate. *Arthritis Research & Therapy*, Vol. 12, No. 4, (July 2010), pp. R134, ISSN 1478-6354.

Giusti, L., Baldini, C., Bazzichi, L., Ciregia, F., Tonazzini, I., Mascia, G., Giannaccini, G., Bombardieri, S. & Lucacchini A. (2007). Proteome analysis of whole saliva: a new tool for rheumatic diseases--the example of Sjögren's syndrome. *Proteomics*, Vol. 7, No. 10, (May 2007), pp. 1634-1643, ISSN 1615-9853.

Giusti, L., Bazzichi, L., Baldini, C., Ciregia, F., Mascia, G., Giannaccini, G., Del Rosso, M., Bombardieri, S. & Lucacchini, A. (2007). Specific proteins identified in whole saliva from patients with diffuse systemic sclerosis. *Journal of Rheumatology*, Vol. 34, No. 10, (October 2007), pp. 2063-2069, ISSN 0315-162X.

Giusti, L., Baldini, C., Ciregia, F., Giannaccini, G., Giacomelli, C., De Feo, F., Delle Sedie, A., Riente, L., Lucacchini, A., Bazzichi, L. & Bombardieri, S. (2010). Is GRP78/BiP a potential salivary biomarker in patients with rheumatoid arthritis? *Proteomics Clinical applications*, Vol. 4, No. 3, (March 2010), pp. 315-324, ISSN 1862-8346.

Gür, A., Karakoç, M., Nas, K., Remzi, Cevik, Denli, A. & Saraç J. (2002). Cytokines and depression in cases with fibromyalgia. *Journal of Rheumatology*, Vol. 29, No. 2, (February 2002), pp. 358-361, ISSN 0315-162X.

Gürsoy, S., Erdal, E., Herken, H., Madenci, E. & Alaşehirli, B. (2001). Association of T102C polymorphism of the 5-HT2A receptor gene with psychiatric status in fibromyalgia syndrome. *Rheumatology International*, Vol. 21, No. 2, (October 2001), pp. 58-61, ISSN 0172-8172.

Gursoy, S. (2002). Absence of association of the serotonin transporter gene polymorphism with the mentally healthy subset of fibromyalgia patients. *Clinical Rheumatology*, Vol. 21, No. 3 (June 2002), pp. 194-197, ISSN 0770-3198.

Gürsoy, S., Erdal, E., Herken, H., Madenci, E., Alaşehirli, B. & Erdal, N. (2003). Significance of catechol-O-methyltransferase gene polymorphism in fibromyalgia syndrome. *Rheumatology International*, Vol. 23, No. 3, (May 2003), pp. 104-107, ISSN 0172-8172.

Haas, L., Portela, L.V., Böhmer, A.E., Oses, J.P. & Lara, D.R. (2007). Increased plasma levels of brain derived neurotrophic factor (BDNF) in patients with fibromyalgia. *Neurochemical Research*, Vol. 35, No. 5, (May 2010), pp. 830-834, ISSN 0364-3190.

Harris, R.E., Sundgren, P.C., Pang, Y., Hsu, M., Petrou, M., Kim, S.H., McLean, S.A., Gracely, R.H. & Clauw, D.J. (2008). Dynamic levels of glutamate within the insula are associated with improvements in multiple pain domains in fibromyalgia. *Arthritis and Rheumatism*, Vol. 58, No. 3, (March 2008), pp. 903-907, ISSN 0004-3591.

Harris, R.E., Sundgren, P.C., Craig, A.D., Kirshenbaum, E., Sen, A., Napadow, V. & Clauw, D.J. (2009). Elevated insular glutamate in fibromyalgia is associated with experimental pain. *Arthritis and Rheumatism*, Vol. 60, No. 10, (October 2009), pp. 3146-3152, ISSN 0004-3591.

Iannuccelli, C., Di Franco, M., Alessandri, C., Guzzo, M.P., Croia, C., Di Sabato, F., Foti, M. & Valesini, G. (2010). Pathophysiology of fibromyalgia: a comparison with the tension-type headache, a localized pain syndrome. *Annals of the New York Academy of Sciences*, Vol. 1193, (April 2010), pp. 78-83, ISSN 0077-8923.

Iughetti, L., Casarosa, E., Predieri, B., Patianna, V. & Luisi, S. (2011). Plasma brain-derived neurotrophic factor concentrations in children and adolescents. *Neuropeptides*, Vol. 45, No. 3, (June 2011), pp. 205-211, ISSN 0143-4179.

Jensen, B., Wittrup, I.H., Wiik, A., Bliddal, H., Friis, A.S., McLaughlin, J.K., Danneskiold-Samsøe, B. & Olsen, J.H. (2004). Antipolymer antibodies in Danish fibromyalgia patients. *Clinical and Experimental Rheumatology,* Vol. 22, No. 2, (March 2004), pp. 227-229, ISSN 0392-856X.

Kim, S.K., Kim, K.S., Lee, Y.S., Park, S.H. & Choe, J.Y. (2010). Arterial stiffness and proinflammatory cytokines in fibromyalgia syndrome. *Clinical and Experimental Rheumatology,* Vol. 28, No. 6 suppl 63, (November-December 2010), pp. S71-77, ISSN 0392-856X.

Klein, R. & Berg, P.A. (1995). High incidence of antibodies to 5-hydroxytryptamine, gangliosides and phospholipids in patients with chronic fatigue and fibromyalgia syndrome and their relatives: evidence for a clinical entity of both disorders. *European Journal of Medical Research,* Vol. 16, No. 1, (October 1995), pp. 21-26, ISSN 0949-2321.

Laske, C., Stransky, E., Eschweiler, G.W., Klein, R., Wittorf, A., Leyhe, T., Richartz, E., Köhler, N., Bartels, M., Buchkremer, G. & Schott, K. (2006). Increased BDNF serum concentration in fibromyalgia with or without depression or antidepressants. *Journal of Psychiatric Research,* Vol. 41, No. 7, (October 2006), pp. 600-605, ISSN 0022-3956.

Lee, Y.H., Choi, S.J., Ji, J.D. & Song, G.G. (2010). Candidate gene studies of fibromyalgia: a systematic review and meta-analysis. *Rheumatology International,* DOI 10.1007/s00296-010-1678-9, ISSN 0172-8172.

Maletic, V. & Raison, C.L. (2009). Neurobiology of depression, fibromyalgia and neuropathic pain. *Frontiers in Bioscience,* Vol. 1, No. 14, (June 2009), pp. 5291-5338, ISSN 1093-9946.

Malt, E.A., Olafsson, S., Aakvaag, A., Lund, A. & Ursin, H. (2003). Altered dopamine D2 receptor function in fibromyalgia patients: a neuroendocrine study with buspirone in women with fibromyalgia compared to female population based controls. *Journal of Affective Disorders,* Vol. 75, No. 1, (June 2003), pp. 77-82, ISSN 0165-0327.

Matsuda, J.B., Barbosa, F.R., Morel, L.J., França, S.deC., Zingaretti, S.M., da Silva, L.M., Pereira, A.M., Marins, M. & Fachin, AL. (2010). Serotonin receptor (5-HT 2A) and catechol-O-methyltransferase (COMT) gene polymorphisms: triggers of fibromyalgia? *Revista Brasileira de Reumatologia,* Vol. 50, No. 2, (April 2010), pp. 141-149, ISSN 0482-5004.

Nicholl, B.I., Holliday, K.L., Macfarlane, G.J., Thomson, W., Davies, K.A., O'Neill, T.W., Bartfai, G., Boonen, S., Casanueva, F., Finn, J.D., Forti, G., Giwercman, A., Huhtaniemi, I.T., Kula, K., Punab, M., Silman, A,J., Vanderschueren, D., Wu, F.C., McBeth, J. & European Male Ageing Study Group. (2010). No evidence for a role of the catechol-O-methyltransferase pain sensitivity haplotypes in chronic widespread pain. *Annals of the Rheumatic Diseases,* Vol. 69, No. 11, (November 2010), pp. 2009-2012, ISSN 0003-4967.

Nishikai, M., Tomomatsu, S., Hankins, R.W., Takagi, S., Miyachi, K., Kosaka, S. & Akiya, K. (2001). Autoantibodies to a 68/48 kDa protein in chronic fatigue syndrome and primary fibromyalgia: a possible marker for hypersomnia and cognitive disorders. *Rheumatology,* Vol. 40, No. 7, (July 2001), pp. 806-810, ISSN 1462-0324.

Offenbaecher, M., Bondy, B., de Jonge, S., Glatzeder, K., Krüger, M., Schoeps, P. & Ackenheil, M. (1999). Possible association of fibromyalgia with a polymorphism in the serotonin transporter gene regulatory region. *Arthritis and Rheumatism,* Vol. 42, No. 11, (November 1999), pp. 2482-2488, ISSN 0004-3591.

Ozgocmen, S., Ozyurt, H., Sogut, S. & Akyol, O. (2006). Current concepts in the pathophysiology of fibromyalgia: the potential role of oxidative stress and nitric oxide. *Rheumatology International*, Vol. 26, No. 7, (May 2006), pp. 585-597, ISSN 0172-8172.

Pamuk, O.N. & Cakir, N. (2007). The frequency of thyroid antibodies in fibromyalgia patients and their relationship with symptoms. *Clinical Rheumatology*, Vol. 26, No. 1, (January 2007), pp. 55-59, ISSN 0770-3198.

Potvin, S., Larouche, A., Normand, E., de Souza, J.B., Gaumond, I., Marchand, S. & Grignon, S. (2010). No relationship between the ins del polymorphism of the serotonin transporter promoter and pain perception in fibromyalgia patients and healthy controls. *European Journal of Pain*, Vol. 14, No. 7, (August 2010), pp. 742-746, ISSN 1090-3801.

Russell, I.J., Orr, M.D., Littman, B., Vipraio, G.A,, Alboukrek, D., Michalek, J.E., Lopez, Y. & MacKillip, F. (1994). Elevated cerebrospinal fluid levels of substance P in patients with the fibromyalgia syndrome. *Arthritis and Rheumatism*, Vol. 37, No. 11, (November 1994), pp. 1593-1601, ISSN 0004-3591.

Sarchielli, P., Mancini, M.L., Floridi, A., Coppola, F., Rossi, C., Nardi, K., Acciarresi, M., Pini, L.A. & Calabresi, P. (2007). Increased levels of neurotrophins are not specific for chronic migraine: evidence from primary fibromyalgia syndrome. *Journal of Pain*, Vol. 8, No. 9, (September 2007), pp. 737-745, ISSN 1526-5900.

Sarzi-Puttini, P., Atzeni, F., Di Franco, M., Lama, N., Batticciotto, A., Iannuccelli, C., Dell'Acqua, D., de Portu, S., Riccieri, V., Carrabba, M., Buskila, D., Doria, A. & Valesini, G. (2008). Anti-polymer antibodies are correlated with pain and fatigue severity in patients with fibromyalgia syndrome. *Autoimmunity*, Vol. 41, No. 1, (February 2008), pp. 74-9, ISSN 0891-6934.

Satoh, K., Matoba, T., Suzuki, J., O'Dell, M.R., Nigro, P., Cui, Z., Mohan, A., Pan, S., Li, L., Jin, Z.G., Yan, C., Abe, J. & Berk B.C. (2008). Cyclophilin A mediates vascular remodeling by promoting inflammation and vascular smooth muscle cell proliferation. *Circulation*, Vol. 117, No. 24, (June 2008), pp. 3088-3098, ISSN 0009-7322.

Tander, B., Gunes, S., Boke, O., Alayli, G., Kara, N., Bagci, H. & Canturk, F. (2008). Polymorphisms of the serotonin-2A receptor and catechol-O-methyltransferase genes: a study on fibromyalgia susceptibility. *Rheumatology International*, Vol. 28, No. 7, (May 2008), pp. 685-691, ISSN 0172-8172.

Treister, R., Pud, D., Ebstein, R.P., Laiba, E., Gershon, E., Haddad, M. & Eisenberg, E. (2009). Associations between polymorphisms in dopamine neurotransmitter pathway genes and pain response in healthy humans. *Pain*, Vol. 147, No. 1-3, (December 2009), pp. 187-193, ISSN 0304-3959.

Uçeyler, N., Valenza, R., Stock, M., Schedel, R., Sprotte, G. & Sommer, C. (2006). Reduced levels of antiinflammatory cytokines in patients with chronic widespread pain. *Arthritis and Rheumatism*, Vol. 54, No. 8, (August 2006), pp. 2656-2664, ISSN 0004-3591.

Vaerøy, H., Helle, R., Førre, O., Kåss, E. & Terenius, L. (1988). Elevated CSF levels of substance P and high incidence of Raynaud phenomenon in patients with fibromyalgia: new features for diagnosis. *Pain*, Vol. 32, No. 1, (January 1988), pp. 21-26, ISSN 0304-3959.

Valdés, M., Collado, A., Bargalló, N., Vázquez, M., Rami, L., Gómez, E. & Salamero, M. (2010). Increased glutamate/glutamine compounds in the brains of patients with fibromyalgia: a magnetic resonance spectroscopy study. *Arthritis and Rheumatism*, Vol. 62, No 6, (June 2010), pp. 1829-1836, ISSN 0004-3591.

Wallace, D.J., Linker-Israeli, M., Hallegua, D., Silverman, S., Silver, D. & Weisman, M.H. (2001). Cytokines play an aetiopathogenetic role in fibromyalgia: a hypothesis and pilot study. *Rheumatology*, Vol. 40, No. 7, (July 2001), pp. 743-749, ISSN ISSN 1462-0324.

Wang, H., Moser, M., Schiltenwolf, M. & Buchner, M. (2008). Circulating cytokine levels compared to pain in patients with fibromyalgia -- a prospective longitudinal study over 6 months. *Journal of Rheumatology*, Vol. 35, No. 7, (July 2008), pp. 1366-1370, ISSN 0315-162X.

Wang, H., Buchner, M., Moser, M.T., Daniel, V. & Schiltenwolf, M. (2009). The role of IL-8 in patients with fibromyalgia: a prospective longitudinal study of 6 months. *Clinical Journal of Pain*, Vol. 25, No. 1, (January 2009), pp. 1-4, ISSN 0749-8047.

Werle, E., Fischer, H.P., Müller, A., Fiehn, W. & Eich, W. (2001). Antibodies against serotonin have no diagnostic relevance in patients with fibromyalgia syndrome. *Journal of Rheumatology*, Vol. 28, No. 3, (March 2001), pp. 595-600, ISSN 0315-162X.

Wilson, R.B., Gluck, O.S., Tesser, J.R., Rice, J.C., Meyer, A. & Bridges, AJ. (1999). Antipolymer antibody reactivity in a subset of patients with fibromyalgia correlates with severity. *Journal of Rheumatology*, Vol. 26, No. 2, (February 1999), pp. 402-407, ISSN 0315-162X.

Wolfe, F., Smythe, H.A., Yunus, M.B., Bennett, R.M., Bombardier, C., Goldenberg, D.L., Tugwell, P., Campbell, S.M., Abeles, M., Clark, P., Fam, A.G., Farber, S.J., Fiechtner, J.J., Franklin, M., Gatter, R.A., Hamaty, D., Lessard, J., Lichtbroun, A.S., Masi, A.T., Mccain, G.A., Reynolds, W.J., Romano, T.J., Russell, J. & Sheon, R.P. (1990). The American College of Rheumatology 1990 Criteria for the Classification of Fibromyalgia. Report of the Multicenter Criteria Committee. *Arthritis and Rheumatism*, Vol. 33, No. 2, (February 1990), pp. 160-172, ISSN 0004-3591.

Wolfe, F. (1999). "Silicone related symptoms" are common in patients with fibromyalgia: no evidence for a new disease. *Journal of Rheumatology*, Vol. 26, No. 5, (May 1999), pp. 1172-1175, ISSN 0315-162X.

Wolfe, F., Clauw, D.J., Fitzcharles, M.A., Goldenberg, D.L., Katz, R.S., Mease, P., Russell, A.S., Russell, I.J., Winfield, J.B. & Yunus, M.B. (2010). The American College of Rheumatology preliminary diagnostic criteria for fibromyalgia and measurement of symptom severity. *Arthritis Care & Research*, Vol.62, No 5, (May 2010), pp. 600-610, ISSN 2151-464X.

Wolfe, F., Clauw, D.J., Fitzcharles, M.A., Goldenberg, D.L., Häuser, W., Katz, R.S., Mease, P., Russell, A,S., Russell, I.J. & Winfield, J.B. (2011). Fibromyalgia Criteria and Severity Scales for Clinical and Epidemiological Studies: A Modification of the ACR Preliminary Diagnostic Criteria for Fibromyalgia. *Journal of Rheumatology*, Vol. 38, No. 6, (June 2011), pp. 1113-1122, ISSN 0315-162X.

Yang, Z., Tao, T., Raftery, M.J., Youssef, P., Di Girolamo, N. & Geczy, C.L. (2001) Proinflammatory properties of the human S100 protein S100A12. *Journal of Leukocyte Biology*, Vol. 69, No. 6, (June 2001), pp. 986-994, ISSN 0741-5400.

Zubieta, J.K., Heitzeg, M.M., Smith, Y.R., Bueller, J.A., Xu, K., Xu, Y., Koeppe, R.A., Stohler, C.S. & Goldman, D. (2003). COMT val158met genotype affects mu-opioid neurotransmitter responses to a pain stressor. *Science*, Vol. 299, No. 5610, (February 2003), pp. 1240-1243, ISSN 0036-8075.

Part 3

Treatment of Fibromyalgia

10

Mind Body Therapies in the Rehabilitation Program of Fibromyalgia Syndrome

Susanna Maddali Bongi and Angela Del Rosso
Division of Rheumatology, Department of Biomedicine, University of Florence, Italy

1. Introduction

Fibromyalgia syndrome (FMS) is characterised by chronic widespread pain for more than 3 months and bilateral sites of focal tenderness (tender points) (Wolfe et al, 1990) associated with fatigue, stiffness, non restorative sleep, depression, anxiety, difficulty concentrating, forgetfulness, and psychological distress that alter, sometimes severely, quality of life.

Recently, new American College of Rheumatology (ACR) criteria for the diagnosis of FMS were developed, basing on widespread pain and symptoms severity (since, at least, 3 months) and not considering tenderness at tender points (Wolfe et al, 2010). However, in all the studies considered and discussed in this chapter, the patients were enrolled if fulfilling ACR classification criteria (Wolfe et al, 1990).

The optimal care for FMS patients requires a specific and individually tailored multidisciplinary approach combining pharmacological and non pharmacological treatment (Goldenberg et al, 2004; Häuser et al, 2010).

Non pharmacological interventions aim to deal with the long term consequences of FMS, such as disability, psychological distress, muscular deconditioning and fatigue. Overall, reviews of literature showed these approaches to be more effective than pharmacological treatments (Rossy et al, 1999).

Many studies were published about rehabilitation, especially concerning physical training (comprising hydrotherapy, cardiovascular exercises, muscular strengthening or stretching), useful in re-conditioning the patients, in breaking the pain–tension cycle, and in improving disability. Strong evidence of efficacy was shown for cardiovascular exercises (Busch et al, 2008), although their effects were often not maintained beyond the end of the intervention.

Although well defined rehabilitation guidelines for FMS are not validated yet, non pharmacological treatments, including education, self-management programs, exercises and mind body therapies (MBT) are taken into account in sets of guidelines and recommendations for FMS treatment drowned by different Scientific Societies (Burckhardt et al, 2005; Carville et al, 2008; Häuser et al 2005; Häuser et al, 2010).

MBT are useful in modulating central pain processing and pain perception and in dealing both with musculoskeletal pain and tenderness and with FMS central derived symptoms, such as depression, anxiety and fatigue, as they are able to interfere on putative pathogenic mechanisms of FMS.

Although the causes and the mechanisms underlying FMS pathogenesis are not completely known yet, most investigators interpret FMS as led by disordered central pain processing (Cook et al, 2004; Dadabhoy et al, 2008).

According to a recent proposal (Costigan et al, 2009), FMS can be included, as well as irritable bowel syndrome and interstitial cystitis (often associated to FMS), among dysfunctional pain syndromes. Dysfunctional pain reflects malfunctioning sensory processing within the central nervous system and occurs in the absence of identifiable noxious stimuli, inflammation, or damage to the nervous system. Pain induces, and is partially maintained by, a state of central sensitization in which an increased transmission of nociceptive information allows normally non noxious stimuli to be amplified and perceived as noxious. Also peripheral nociceptive inputs may initiate and maintain central sensitizations, ultimately leading to disordered central pain processing and dysfunctional pain (Cook et al, 2004; Costigan et al, 2009).

As the pathogenesis of FMS is not, however, fully understood, the management of the disease represents a huge burden that traditional western medicine is currently failing to approach efficaciously.

MBT, defined as "interventions that use a variety of techniques designed to facilitate the mind's capacity to affect bodily function and symptoms" (Health Information: Mind-Body Medicine. National Center for Complementary and Alternative Medicine, 2006) may be useful in the rehabilitation programs of FMS patients. In fact, despite their conceptual and technical differences, they all yield a global approach and involve both physical and mental dimensions of the subjects by focusing on the relationships among the brain, mind, body, and behaviour, and their effect on health and disease.

Both concentration based and movement based MBT have no or low physical impact and allow the patients themselves to play a more active role in their treatment (Wahbeh et al, 2008).

In FMS patients, MBT, acting by different modalities that depend on the peculiarities of the methods, help to cope with pain and to disconnect the affective response to pain, thus decreasing pain catastrophizing and the potentially associated emotional distress and sympathetic activation.

Thus, MBT are efficacious in approaching chronic pain, the cardinal clinical feature of FMS, as well as its central derived symptoms such as fatigue, difficulty sleeping and relaxing, depression, anxiety, and psychological distress, and, thereby, to improve activities of day living, mood, self efficacy and quality of life, often severely affected, and to reduce disability.

However, all MBT relies on a good compliance and an active participation of the patients to the treatment. Furthermore, especially in concentration based MBT, such as cognitive behavioural therapies (CBTs), that are administered by a psychologist, the patients are needed to be aware of and to cope with the psychological determinants of FMS (Carville et al 2008).

In FMS, MBT should be chosen and tailored according to pain intensity, function and other associated features, such as depression, fatigue and sleep disturbance, according to the preferences and the expectations of the patients (van Koulil et al, 2007). Patients with relatively high levels of psychological or emotional distress seem to benefit most by MBT, whose efficacy may be improved by offering tailored treatment, administered at an early stage by operators skilled in their technique and expert in dealing with FMS (van Koulil et al, 2007).

Different concentration based MBT, such as Cognitive Behavioural Therapies, Hypnosis, Guided Imagery and Mindfulness Meditation have been used in FMS, as well as movement based techniques, such as body awareness technique, Mensedieck system, Yoga, Tai Chi and Qi Gong. Also Pilates method was used in FMS patients (Table 1.).
In our experience, we used with success Rességuier Method, a rehabilitative intervention somewhat resembling MBT, as a novel non pharmacological tool to treat FMS patients.

Concentration based mind body therapies

- Cognitive Behavioural Therapies (CBTs)
  - *Single method CBTs*
    - education programmes
    - relaxation techniques (progressive relaxation, biofeedback, autogenic training)
  - *Multi method CBT*
- Hypnosis
- Guided Imagery
- Mindfulness Meditation

Movement based mind body therapies

- Body Awareness Technique
- Mensedieck system
- Yoga
- Tai Chi
- Qi Gong
- Rességuier Method and Body Movement and Perception method
- Pilates Method

Table 1. Mind body therapies in Fibromyalgia Sindrome.

2. Concentration based mind body therapies

2.1 Cognitive Behavioural Therapies (CBTs)

Cognitive Behavioural Therapies (CBTs) are among the most used treatments for FMS, whose cardinal symptom is chronic and widespread pain.
The transition of acute to chronic pain, independent of a biomedical cause, as in FMS, is described by biopsychosocial models. In acute pain, three response systems are involved: behavioural reactions (e.g avoidance behaviour), cognitive reactions (e.g increased attention to bodily sensations and catastrophizing) and physiological reactions (e.g a high autonomous arousal and muscle tension). These behaviours are appropriate adaptive short-term reactions to acute pain, but they become less functional and also detrimental when applied in a long term period, and in response to chronic pain (Evers et al, 2001; Turk & Flor, 1999; van Koulil et al, 2007). The main objective of CBTs is to change these behaviours into positive attitudes.
A distinction can be made between single method interventions, such as education and relaxation programmes, and multimethod CBTs, incorporating various methods from cognitive behavioural approaches (van Koulil et al, 2007).

2.1.1 Single method Cognitive Behavioural Therapies

Educational programmes include information about pain self-management, coping and relaxation techniques, the importance of physical activity and social support and strategies for behavioural changes.

Educational programmes as single method interventions yield some benefits in FMS patients in pain-coping skills (Burckhardt et al, 1994) and self efficacy (Vlaeyen et al, 1996), but not in reducing pain and disability nor in improving mood (van Koulil et al, 2007). Moreover, the only trial including a follow-up period failed to find any effect of education on pain, disability and mood (King et al, 2002).

Fibromyalgia Self-Help Course (FSHC), an educational programme teaching FMS patients about the condition and self-management skills (to accomplish daily activities, manage symptoms, suggest ways to incorporate wellness activities and exercise into daily life) had minor effects on disability, quality of life, fatigue, depression and self-efficacy than FSHC added to exercises (Rooks et al, 2007).

Other single method CBTs such as **relaxation techniques** (progressive relaxation, biofeedback and autogenic training) were used in FMS to reduce muscular tension and interrupt the pain–tension cycle. Despite the studies could be underpowered due to a low number of enrolled subjects, some improvements were reported on pain, although the effect was not maintained at follow-up. However, no efficacy on disability or mood was shown (Buckelew et al, 1998; Ferraccioli et al, 1987; van Santen et al, 2002).

2.1.2 Multi method Cognitive Behavioural Therapies

Multi method CBTs emphasize the role of cognitive processes in shaping affective experience and hypothesize that problematic emotions, such as anger, depression, and anxiety, result from irrational or faulty thinking.

Differently from classical psychoanalysis, providing 'deep insight' and needing, sometimes, a long time to reach some effect, CBTs offer short-term, goal-oriented psychotherapy, that emphasizes changes in thought patterns and behaviours, with beneficial effects potentially achieved in 10-20 sessions.

They use a combination of various elements, such as cognitive restructuring, pain-coping skills, problem-solving techniques, goal setting, increasing activity levels, activity pacing, stress management adjustment of pain-related medication, and frequently also comprises educational and relaxation components (van Koulil et al, 2007).

Taken together, the results obtained from the several studies conducted on FMS show that multi method CBTs are effective in diminishing pain and depression, in improving disability and mood, with the results, sometimes, maintained also at a long follow-up (van Koulil et al, 2007).

A study assessed, in 40 subjects with FMS admitted in a inpatient clinic, the effects of a 5 week protocol of operant pain treatment comprising reduction of medication; increase of bodily activity; reduction of interference of the pain in the usual activities; reduction of pain behaviours in dealing with the medical system and training in assertive pain incompatible behaviour. The patients reported a significant and stable reduction in pain intensity, interference, solicitous behaviour of the spouse, medication, pain behaviours, and an increase in sleeping time. Also the number of doctor visits, and days at a hospital were reduced in the year after the intervention (Thieme et al, 2003).

In studies combining multi method CBTs and exercises in FMS patients, improvements on pain (confirmed at follow up), disability and mood were shown (Keel et al, 1998; Lemstra et al, 2005).

Basing on the literature and on the opinion of the experts, CBTs (together with exercise) are recommended in the global management of the patients with FMS by different sets of recommendations and guidelines (Burckhardt et al, 2005; Carville et al, 2008; Häuser et al, 2005).

The guidelines for FMS management of the American Pain Society (APS) (Burckhardt et al, 2005) and of the Association of the Scientific Medical Societies in Germany (AWMF) (Häuser et al, 2005) gave the highest level of recommendation to CBTs. In both guidelines, it is suggested to include CBTs into a multimodality treatment approach to reduce pain, enhance self-efficacy, and improve function in FMS patients.

Differently, according to evidence based recommendations of European League Against Rheumatism (EULAR) for the management of FMS, especially focused on pharmacological treatments, CBTs are advised in the management of patients with FMS only funding on expert's opinion (level of evidence "IV", strength of recommendation "D", the lowest grades in both scales, ranging from I to IV and from A to D, respectively) (Carville et al, 2008).

In a systematic review with metaanalysis assessing the efficacy of CBTs in FMS, 14 out of 27 randomised controlled trials were included. CBT reduced depressed mood, improved self-efficacy (with the results maintained at at follow up) but no significant effect was found on pain, fatigue, sleep, and quality of life at post treatment and follow up. Operant behavioral therapy significantly reduced the number of physician visits at follow up. (Bernardy et al, 2010).

However, in a recent metanalysis evaluating the effects of psychological interventions for FMS, CBTs were shown to act significantly better than other psychological treatments in reducing short-term pain and to have the greatest effect sizes on the assessed outcome measures (reduction of long-term pain, sleep problems, depression, functional status and catastrophizing). (Glombiewski et al, 2010).

2.2 Hypnosis and guided imagery

In the field of the concentration based MBT, the use of hypnosis/guided imagery as a complement to pharmacological and non pharmacological treatments was recommended by the German interdisciplinary guideline on FMS, basing on expert consensus (Thieme et al, 2008). The efficacy of the two techniques on FMS symptoms was recently reviewed. (Bernardy et al, 2011).

Hypnosis is used to encourage and to evaluate responses to suggestions. In this procedure, one person (the subject) is guided by another (the hypnotist) to respond to suggestions.

Guided Imagery is a dynamic, psychophysiologic process in which a person, guided by a psychologist, imagines and experiences an internal reality in the absence of external stimuli, with the aim to promote changes in subjective experience, alterations in perception, sensation, emotion, thought or behaviour by suggestion and/or imagination.

All the studies assessing the effects of hypnosis and guided imagery on FMS were performed on small cohorts of patients. Taken together, the results of the reviewed trials indicate that the patients treated with the two techniques, compared to controls, improved in pain perception at the end of the treatment but not in quality of life. Large effects sizes on pain and medium effects sizes on sleep at at the end of treatment and at follow-up were shown.

Basing on their findings and on methodological limitations of the studies, the authors of the review conclude that further studies with adequate sample sizes are necessary to prove treatment efficacy (Bernardy et al, 2011).

2.3 Mindfulness-based stress reduction (MBSR)

Mindfulness-based stress reduction (MBSR) program, centered on the principles and practice of mindfulness meditation and using stress-reduction skills including sitting meditation, hatha yoga, and a somatically focused technique called the "body scan", was developed to relieve suffering in patients with chronic pain (Kabat-Zinn, 1982). MBSR encourages nonjudgmental awareness of one's cognitive and somatic experience on a moment-by moment basis. This decentered stance is thought to disconnect cognitive and affective mental events in an adaptive manner and may reduce the negative impact of thoughts and sensations associated with chronic pain.

In FMS, a 8 week intervention of MBSR, administered by a clinical psychologist, ameliorated (versus a wait-list control group) cognitive and somatic depressive symptoms, with the improvement maintained also at a 2 months follow-up (Sephton et al, 2007) and reduced activation of sympathetic nervous system in patients with FMS (Lush et al, 2009). However, this technique was not different in ameliorating quality of life when compared to an active control procedure (utilised as control treatment in order to check for nonspecific effects of MBSR), although patients in the MBSR group appeared to benefit most. In fact, in a pre-post-analysis, they were improved in quality of life, disability, depressive and anxious symptoms, sleep, pain perception, and physical symptoms (Schmidt et al, 2011).

From these data, MBSR appears to be a promising intervention for FMS, especially to treat depressive symptoms that, as in other chronic pain syndromes, can interact reciprocally with physical symptoms, impairing quality of life.

3. Movement based Mind Body Therapies

3.1 Mensendieck system and body awareness technique

The **Mensendieck System** focuses on teaching patients to understand the concepts of functioning of their bodies by pedagogically designed exercises and aims to enable them to change suboptimal patterns of movement.

Body awareness technique (BAT) combines a series of exercises related to posture, co-ordination, free breathing and awareness. Turning the attention both to the patient's own performance and to what is experienced during the exercises is a central element of BAT , that stimulates mental presence and awareness aiming to provide an increased body consciousness.

In a study comparing the effects of two programs based on Mensendieck system and BAT in 20 female patients with FMS, the subjects treated with Mensendieck system report better scores in fibromyalgia related disability, self-efficacy and coping strategies, with some results maintained at 18 month follow-up, than patients who executed BAT (Kendall et al, 2000).

However patients with irritable colon, often associated to FMS, when treated with a 24 weeks program of BAT, improved significantly in gastrointestinal and psychological symptoms. Concomitantly, body pain as well as coping ability improved (Eriksson et al, 2007).

3.2 Yoga

Yoga is a MBT potentially fulfilling the need for both exercise and coping skills in FMS patients. Yoga varies greatly in the style and comprises, beyond the physical poses identified with it, meditation and breathing exercises.

In FMS, **Relaxing Yoga** (administered once a week for 2 months), consisting in stretching, breathing, and relaxing yogic techniques, improved FMS related pain and disability, with 30% improvement in overall symptoms (da Silva et al, 2007).

Yoga of Awareness is a comprehensive technique that, in a 8-week intervention program, complements yoga poses with mindfulness meditation, breathing exercises, yoga-based coping instructions and group discussion. At post-treatment, subjects assigned to the yoga of awareness program, that was adapted and tailored for FMS patients, showed significantly greater improvements, versus a waiting list control group, on standardized measures of FMS symptoms and functioning, including disability, pain, fatigue, and mood, and in pain catastrophizing, acceptance, and other coping strategies. No effect was shown on tender point counts, muscle strength and balancing (Carson et al, 2010).

From these preliminary results, Yoga seems to be suitable and efficacious in FMS patients, acting both on somatic and on central-derived symptoms.

3.3 Qi Gong

Qi Gong (QG) is an ancient Chinese method, integrating body, energetic, respiratory, and mental training with the aim to achieve optimal status of both mind and body by increasing and restoring the flow of "qi" (vital energy). QG helps physical, psychic and emotional rebalancing, thus improving posture, respiration, and concentration by low impact movements. Although Chinese practitioners have applied various forms of QG for thousands of years to treat diseases and improve health, the term Qi Gong currently used by health care professionals to include all exercises dates since the 1950s.

Internal and external QG can be distinguished: the first is self-directed and involves the use of movements and meditation. External QG is performed by a trained practitioner using the hands and any part of body to direct "qi" energy into the patient.

For its characteristics, QG has potential therapeutic benefits in patients with FMS, but the result shown by different studies are discordant.

In a pilot study, a protocol of external QG lasting 3 weeks improved in FMS patients tender points, pain, disability and depression, with the results maintained at a 3-month follow-up (Chen et al, 2006).

In a subsequent study, FMS patients treated with a 7 weeks protocol of QG significantly improved in pain, psychological health and distress in respect to a control group, with the data confirmed after a 4 months long period (Haak et al, 2008).

However, when tested in children affected with FMS, QG was as effective as aerobic exercise in improving anaerobic function, tender point count, pain, and symptom severity, but did not ameliorate physical function, functional capacity, quality of life, and fatigue, that were improved only in the group performing aerobic exercises (Stephens et al, 2008).

The contradictory results derived from these studies are probably due to the differences on QG techniques used, period of application and selection of patients. However, in our opinion, QG, better if tailored and focused on characteristics of the patients, may be a promising tool in the non pharmacological treatment of FMS.

3.4 Tai Chi

Tai Chi is a MBT that, originated in China as a martial art, is practised since many centuries. It combines meditation with many fundamental postures flowing imperceptibly and smoothly from one to the other through slow, gentle, graceful movements as well as deep diaphragmatic breathing and relaxation, with the aim to move "qi" throughout the body. It may be regarded as a multicomponent intervention integrating physical, psychosocial, emotional, spiritual, and behavioural elements and promoting the mind body interaction (Wang, 2011).

Over the past 2 decades, Tai Chi was shown to provide great benefits for patients with a variety of chronic diseases and, among the others, in rheumatic conditions such as rheumatoid arthritis (Wang, 2008) and osteoarthritis (Wang et al, 2009). In longstanding diseases, Tai Chi improves physical function and psychological well being by reducing stress, anxiety, depression, and mood disturbance and increasing self esteem (Wang, 2011).

The good results on FMS yielded by a Tai Chi program in a previous study (Taggart et al, 2003) were confirmed by a recent work, showing a notable efficacy of a 12 weeks program of the classical Yang-style Tai Chi (with sessions lasting 60 minutes, executed twice a week) in FMS patients. Patients following Tai Chi program ameliorated in all the assessed outcome measures, both at physical and at psychological level. Tai Chi improved FMS related disability, sleep quality, patient and physician global assessment of pain, self efficacy, depression, physical and mental quality of life and, notably, physical performance (assessed by 6-minute walk test), versus a control intervention (consisting of wellness education and stretching). The improvements were maintained also at a 24 weeks follow-up (Wang et al, 2010).

Thus, Tai Chi, integrating physical elements with relaxation, and cognitive behavioural features, may be regarded as an useful potential tool in patients with FMS.

3.5 Rességuier and body, movement and perception methods

Recently, we applied the rehabilitation method conceived by Jean Paul Rességuier (**Rességuier Method** -RM-), inspired by Merleau-Ponty phenomenology (Vitali Rosati, 2007) somewhat resembling MBT, in patients with FMS. The method, executed in individual sessions, aims to obtain patient nonjudgmental awareness and control of bodily perceptions and, in particular, nociception.

Its mainstay is the relationship between therapist and patient based on the continuous attention to the patient during all the session (regarded as "accompanying posture").

During the individual session, the therapist continuously controls patient attention and perception by verbal and manual contacts and leads to perform bodily active and conscious movements and respiratory exercises in different positions (supine, sitting and standing). The exercises are respectful of the pain threshold and tailored for the patients.

In FMS patients, we showed that the application of RM (once a week for 8 weeks) improved significantly quality of life, disability, relaxation and sleep. The most notable result was obtained on perceived pain, significantly reduced in respect to initial values. This datum was confirmed by the reduction of analgesics assumed weekly by the patients. Interestingly, the improvement of all the items evaluated, apart from disability, was confirmed after a follow-up of 6 months.

RM induces self-observation and thoughtful responses to pain, thus, it potentially disconnects the affective response to pain and break the vicious circle of chronic pain-stress typical of the FMS.

For its characteristics, RM is a promising rehabilitative tool in fibromyalgic patients, potentially helpful also as a "first step intervention" when other techniques are difficult to be used (e.g. in subjects that are poorly conditioned and/or with a low pain threshold that avert them from whichever technique implying exercises) (Maddali Bongi et al, 2010a).
In a pilot study on FMS, we also assessed the effects of the "**Body Movement and Perception**" (BMP) method, a group gymnastic applied for 8 weeks (2 sessions-50 minutes each- a week) allowing to treat little groups of patients with RM. This method, based on aware body perception, low impact physical movements, and relaxation integrates the principles of RM with low impact bodily movements. The exercises, always respectful of the pain threshold, partly derives from soft gymnastics and partly are specifically conceived for the BMP. They consist in active movements of the body, more specifically of head, trunk, upper and lower limbs and in exercises of conscious respiration. BMP induces reduction of intensity and duration of pain and awareness to pain, confirmed, as we already showed also for RM, by a significant reduction in analgesic assumption and by improvement of tender point scores. Moreover, patients executing BMP improved in postural control, body alignment, muscle contractures, fatigue, irritability, wellbeing, quality of movement, postural self control, ability to relax, movement perception (Maddali Bongi et al, 2011a).
Resseguièr and Body Movement and Perception methods relies on the principles of patient nonjudgmental awareness and control of nociception, guided by therapist. In our experience, both methods are feasible and useful in the management of FMS patients. Subjects with a low pain threshold and a high grade of hyperalgesia may be approached consequentially with Resseguier method and then with Body Movement and Perception method, that presents low impact movements and may help to gently reconditioning the patients.

3.6 Somatic Practises

The term "Somatics" or "Somatic Practises" was introduced by the philosopher Thomas Hanna in the 1970s to include distinct contemporary mind body techniques sharing a global body centered approach that aim to reach mind body rebalancing, psycho-physical awareness and well-being through movement practices (Eddy, 2009).
The common concepts shared by these techniques are that body and mind are both part of a living process. Thus, body-mind integration and freedom from restrictions in body and mind are common goals. A mainstay of these practises is that growth, change, and transformation are always possible at any age.
The classic somatic techniques began to be developed at the turn of the twentieth century and include, among the others, Mensendieck system, Feldenkrais Method, Pilates Method, Rolfing, Gerda Alexander Euthonie, Alexander Technique, Rosen Method Bodywork (Eddy, 2009).
Although only few studies were undertaken to assess the feasibility and the utility of these techniques in rheumatic diseases and in FMS, they could be suitable for fibromyalgic subjects in improving body awareness and mind body relationship, in reducing the vicious circle of chronic pain-stress and the muscle contractures, by allowing optimal movement pathways and harmony through physical exercises.

3.7 Feldenkrais and Core Integration Methods

Feldenkrais Method is a somatic educational system, designed by Moshe Feldenkrais since 1949, to improve movement repertoire, aiming to expand and refine the use of the self through awareness, in order to reduce pain or limitations in movement, and promote general well-being. Feldenkrais Method tends towards being a form of self-education as opposed to a manipulative therapy (Landi, 2007).

Feldenkrais Method, in patients with chronic musculoskeletal chronic disorders was as efficacious as body awareness therapy and more efficacious than conventional physiotherapy in improving psychological distress, pain and self image (Malmgren-Olsson et al, 2001).

Core Integration Method maintains the principles of Feldenkrais method, but by developing specific pathways of movement, is mainly focused on functional recovery of movement (Landi, 2007)

In our opinion, although not specifically evaluated in clinical trials, but supported by results obtained in patients with chronic pain and on pilot experiences conducted by our group, both methods could be feasible and potentially useful in the management of FMS (Maddali Bongi & Landi, unpublished data).

3.8 Pilates Method

The Pilates or Physicalmind Method, developed in the 1920s, is a low impact, non aerobic fitness routine that combines stretching and strengthening exercises, concerned with economical movement. It relies on kinesthetic monitoring in developing balanced muscle use for ease of motion. The Pilates method promotes balance, strength, flexibility and muscle development, while also helping to relieve muscle tension, focusing to the quality of the movement as opposed to quantity. As an approach that combines sensory awareness with physical training, Pilates can lead to balancing of mental and body relationship and may be potentially included in MBT.

In a group of FMS patients, a 12 week program of Pilates exercises, compared to a home exercises protocol, improved pain and disability, but only the improvement in disability was maintained over time. However, the comparison of the 2 groups showed significantly high improvement in pain and disability for patients who executed Pilates at week 12 but no difference between the groups at a 24 weeks follow-up (Altan et al, 2009).

From these preliminary data, Pilates may be regarded as an effective and safe exercising method for FMS patients. However, given the wide diffusion and visibility of the method, available in a quantity of gymnasium, FMS patients should practise Pilates only under the guide of a certified teacher of the method, that should be also a physiotherapist skilled in rheumatic diseases and, in particular, in FMS management. In fact, fibromyalgic patients need tailored and personalised exercises that should be specifically standardized and validated for the disease.

4. Combination of different Mind Body Therapies in FMS patients

The different studies evaluating the effects of the combination of MBT in FMS, chosen to synergize and integrate their effects potentially acting on different FMS symptoms, yielded interesting but somewhat conflicting results.

A 8 week multimodal mind-body intervention combining the movement based QG with Mindfulness Meditation in a complex mind body approach, was able as well as an educational and informative program on FMS in improving pain, disability, depression and myalgic score. The changes occurring at the end of the protocols were largely maintained by both groups throughout the 6 month follow-up period (Astin et al, 2003).

Concordantly, in a 3 month study, BAT combined with QG, in patients with FMS leads to a significant improvement in movement harmony, while no differences were found in the FMS-related disability and in the functional tests in respect to a control group (Mannerkorpi & Arndorw, 2004).

On the contrary, an open study evaluating a 2 month protocol including education/cognitive-behavioural component, formal relaxation/meditation training, and QG practise in FMS patients improved significantly disability, self-reported physical activity, sleep, depression, quality of life, coping strategies, tender points counts and pain threshold, with the improvements sustained till 4 months after the end of the intervention (Creamer et al, 2000).

MBT somewhat similar in their approach could also be used consequentially, in order to synergize and improve their effect. We used a protocol integrating RM and QG. RM aims to develop body awareness and perception and QG also comprises exercises and adequate postures of the body in the space. From our preliminary experiences, 2 protocol lasting 15 weeks integrating consequentially RM and QG (group 1) and QG and RM (group 2) reduced significantly, with the same effect size, disability, pain, tender point count, and improved sleep and anxious and depressive symptoms, with most of the results confirmed after a 12 week follow-up (Maddali Bongi et al, 2011b).

Despite the non completely positive results reported in these studies, programs including the combination of different MBT should be taken into account in the treatment of FMS.

5. Conclusions

Despite many MBT have been used since many years and, as for Yoga, Qi Gong and Tai Chi, many centuries, they have been standardized and evaluated by scientific methods since few decades, especially in the last twenty years. Thus, although spontaneously sought and utilised by FMS patients, (Baranowsky et al, 2009; Lind et al, 2007) evidence based data on MBT efficacy are not widely available.

The results of the studies are not unequivocal because the clinical trials on MBT suffer from different biases such as: little number of enrolled patients, sometimes also heterogeneous in their clinical characteristics (van Koulil et al, 2007); different outcome measures; the lacking of clinical indicators addressing the patients that may benefit more from a specific treatment; the lacking, sometimes, in movement based MBT of exercises and postures tailored and differentiated for FMS patients (presenting with pain, tenderness, low pain threshold, fatigue); the non definition of an ideal period of treatment, the lacking of congruous follow-up periods.

However, taken together, the effects of MBT in FMS are promising and the clinical trials present in literature address the efficacy and the safety of MBT in FMS.

These data are supported by opinion of the experts, that, in different panels of guidelines and recommendations, advise the use of MBT in the management of FMS patients (Burckhardt et al, 2005; Carville et al, 2008; Häuser et al 2005).

5.1 Future development

Basing on the results of the clinical trials, reviews, and on our experience, different aspects should be taken into account in order to improve the efficacy of MBT interventions in patients affected with FMS.

5.1.1 Early treatment

The non pharmacological treatments, including MBT, in patients early diagnosed with FMS, may prevent the vicious cycle of disability and psychological distress associated with chronic widespread pain, that leads to long-term dysfunction and chronicity. Moreover, an early intervention, especially with concentration-based MBT, may afford and prevent the maladaptive patterns of pain-coping and illness behaviours, arising in patients with a long term disease, that render difficult to modify their comportment (van Koulil et al, 2007).
Some evidence indicates that early intervention is indeed an important factor in improving the outcomes of non pharmacological treatment in FMS. In a study assessing the efficacy of a program integrating CBTs, relaxation, physical exercises and education versus autogenic training, a subgroup of patients with a shorter disease duration responded best to treatment (Keel et al, 1998).

5.1.2 Selection of the patients and tailoring of the treatment

In FMS, the limited effects of non pharmacological interventions may also be attributed to the variability within the patients enrolled in the studies (van Koulil et al, 2007).
It should be better clarified which patients are most likely to respond positively to particular MBT and which psychosocial, contextual, and personal and clinical variables (i.e, emotional distress, readiness to change, desire for control) might be taken into account, so that the treatments could be tailored on the clinical subgroups of FMS (de Souza et al, 2009) and on the patients' principal symptoms and expectations.
In a recent study, by a cluster analysis conducted basing on the Fibromyalgia Impact Questionnaire (FIQ), specifically assessing FMS-related disability, two distinct clusters of FMS patients were identified: FMS-Type I and FMS-Type II. Both clusters reported high levels of pain, fatigue and stiffness, but high levels of anxiety and depressive symptoms were peculiar of FMS-Type II (de Souza et al, 2009).
These findings may help to individualise and to tailor the treatments in FMS. Given that a common set of physical symptoms is reported by all the patients, some aspects of the treatment should be homogeneous and deal with hyperalgesia, stiffness and fatigue.
In fact, it may be necessary to treat depressive and anxious symptoms in FMS-Type II patients with CBTs, mindfulness-based stress reduction program and other meditation based MBT, while movement based MBT could be feasible and potentially useful in all FMS patients (Calandre et al, 2010, de Souza et al, 2009).
On the basis of psychosocial and behavioural characteristics, previous studies identified, among FMS patients, a dysfunctional group characterised by low levels of activity, high levels of pain interference and psychological distress and showed that this kind of patients are likely to benefit most from non pharmacological intervention (van Koulil et al, 2007).
It should be underlined that, in the clinical setting, MBT should be chosen and addressed by the rheumatologist, basing on the psychophysical characteristics and preferences of the patients and on their disease type (Calandre et al, 2010; de Souza et al, 2009). With regard to this, we should consider that not all FMS patients may have an adequate compliance and are prone to participate actively to the treatment.
Thus, the rheumatologist should coordinate the global management of the FMS patient, laying out and following up not only the pharmacological therapy, but also all the rehabilitation program, that include MBT.

The demographic, medical or psychosocial factors of the patients should address also the tailoring and the individualization of the chosen MBT, concordantly with the therapist (Keefe et al, 2004).

Of pivotal importance is that patients affected by FMS and other rheumatic conditions characterized by chronic pain, disability and psychological distress should be approached and treated by operators and therapists skilled in their technique and specialized in the management of rheumatic patients (Maddali Bongi, 2010b), mainly of FMS subjects. This is true especially for movement based MBT. Therapists should firstly assess the patients and then choose the exercises more suitable to their condition and tailor the movements on the individual characteristics, in order to reach optimal compliance and benefits from the treatment.

Predictors of treatment outcome, funded on clinical, clinimetric and psychometric measures, should be also taken into account in order to understand which patients could benefit more from a specific intervention.

Moreover, the outcome measures to be used and the aims to be reached should be differenced according to the different MBT.

In order to work efficaciously in FMS patients, MBT might be combined in protocols in which they could be used together or consequentially and integrated in a global multimodal program of management and care comprising education and movement (Kroese et al, 2009). In this context, also aerobic, stretching and strengthening exercises together with home rehabilitative gymnastique (always respectful of the pain threshold) should be included gradually according to the different preferences of the patients.

5.1.3 Concluding remarks

In our opinion, we'd advise a wide use of MBT in the management of patients affected with FMS. In particular, we recommend regular cycles of MBT, that should be chosen by the rheumatologist, the physiotherapist and the patients, taking into account the clinical and psychological characteristics, the needs and the compliance of the subjects, and that should be potentially integrated and/or alternated with low impact aerobic exercises. The MBT as well as the exercises should be tailored on the characteristics of FMS and integrated with pharmacological treatment and educational measures.

6. References

Altan L, Korkmaz N, Bingol U & Gunay B (2009).Effect of pilates training on people with fibromyalgia syndrome: a pilot study. *Arch Phys Med Rehabil.* Vol 90, No 12, pp.1983-8

Astin JA, Berman BM, Bausell B, & al (2003). The efficacy of mindfulness meditation plus Qigong movement therapy in the treatment of fibromyalgia: a randomized controlled trial. *J Rheumatol*, Vol. 30, No 10, pp. 2257-62.

Baranowsky J, Klose P, Musial F, Häuser W, Dobos G & Langhors I. (2009). Qualitative systemic review of randomized controlled trials on complementary and alternative medicine treatments in fibromyalgia. *J.Rheumatol Int,* Vol. 30, No 1, pp.1-21.

Bernardy K, Füber N, Köllner V, & Häuser W (2010) Efficacy of cognitive beahavioral therapies in fibromyalgia syndrome-a systematic review and metanalysis of randomized controlled trials *J Rheumatol* . Vol 37, No 10, pp.1991-2005.

Bernardy K, Füber N, Klose P, Häuser W (2011). Efficacy of hypnosis/guided imagery in fibromyalgia syndrome - a systematic review and meta-analysis of controlled trials. *BMC Musculoskelet Disord.* Vol 15, No 12 , pp. 133.
Buckelew SP, Conway R, Parker J, Deuser WE, Read J, Witty TE, & al (1998). Biofeedback/relaxation training and exercise interventions for fibromyalgia: a prospective trial. *Arthritis Care Res,* Vol.11, No 3, pp. 196–209.
Burckhardt CS, Mannerkorpi K, Hedenberg L,& Bjelle A (1994). A randomized, controlled clinical trial of education and physical training for women with fibromyalgia. *J Rheumatol,* Vol. 21, No. 4, pp. 714–20.
Burckhardt CS, Goldenberg D, Crofford L, Gerwin R, Gowans S, Kackson & al (2005). Guideline for the management of fibromyalgia syndrome. Pain in adults and children. *APS Clinical Practice Guideline Series* No. 4. Glenview, IL: American Pain Society.
Busch AJ, Schachter CL, Overend TJ, Peloso PM & Barber KA (2008). Exercise for fibromyalgia: a systematic review. *J Rheumatol,* Vol. 35, No. 6, pp.1130-44.
Calandre EP, Garcia-Carrillo J, Garcia-Leiva JM, Rico-Villademoros F, Molina-Barea R & Rodriguez-Lopez CM. (2010). Subgrouping patients with fibromyalgia according to the results of the fibromyalgia impact questionnaire: a replication study. *Rheumatol Int,* May 20. [Epub ahead of print].
Carson JW, Carson KM, Jones KD, & al (2010). A pilot randomized controlled trial of the Yoga of Awareness program in the management of fibromyalgia. *Pain,* Vol. 151, No 2 pp 530-9.
Carville S F, Arendt-Nielsen S, Bliddal H & al (2008). EULAR evidence based recommendations for the management of fibromyalgia syndrome. *Ann Rheum Dis, Vol.*67, No 4: 536-41.
Chen KW, Hassett AL, Hou F, Staller J & Lichtbroun AS (2006). A pilot study of external qigong therapy for patients with fibromyalgia. *J Altern Complement Med,* Vol. 12, No 9, pp. 851-6.
Cook DB, Lange G, Ciccone DS, & al (2004). Functional imaging of pain in patients with primary fibromyalgia. *J Rheumatol,* Vol. 31, No. 2, pp. 364-78.
Costigan M, Scholz J, Woolf CJ (2009). Neuropathic pain: a maladaptive response of the nervous system to damage. *Ann Rev Neurosci,* Vol. 32, pp. 1-32.
Creamer P, Singh BB, Hochberg MC & Berman BM (2000). Sustained improvement produced by non pharmacologic intervention in fibromyalgia: results of a pilot study. *Arthritis Care Res,* Vol. 13, No 4, pp. 198-204.
da Silva GD, Lorenzi-Filho G & Lage LV (2007). Effects of yoga and the addition of Tui Na in patients with fibromyalgia. *J Altern Complement Med,* Vol. 13, No 10 pp. 1107-13.
Dadabhoy D, Crofford LJ & Spaeth M (2008). Evidence-based biomarkers for fibromyalgia syndrome. *Arthritis Res Ther,* Vol.10, No 4, pp. 211.
de Souza JB, Goffaux P, Julien N, et al. (2009). Fibromyalgia subgroups: profiling distinct subgroups using the Fibromyalgia Impact Questionnaire. A preliminary study. *Rheumatol Int.* Vol.29, No 5, pp. 509-15.

Eddy M (2009). A brief history of somatic practices and dance: historical development of the field of somatic education and its relationship to dance. *J of Dance and Somatic Practices*; Vol 1, No 1, pp. 5–27

Eriksson EM, Möller IE, Söderberg RH, Eriksson HT & Kurlberg GK (2007). Body awareness therapy: a new strategy for relief of symptoms in irritable bowel syndrome patients. World J Gastroenterol, Vol. 21, No 13, pp. 3206-14.

Evers AW, Kraaimaat FW, van Riel PL & Bijlsma JW (2001). Cognitive, behavioural and physiological reactivity to pain as a predictor of long-term pain in rheumatoid arthritis patients. *Pain*, Vol. 93, No. 2, pp. 139–46.

Ferraccioli G, Ghirelli L, Scita F, Nolli M, Mozzani M, Fontana S, & al (1987). EMG-biofeedback training in fibromyalgia syndrome, *J Rheumatol*, Vol. 14, No 4, pp. 820–5.

Glombiewski JA, Sawyer AT, Gutermann J, Koenig K, Rief W & Hofmann SG (2010). Psychological treatments for fibromyalgia: a meta-analysis. *Pain*, Vol. 151, No 2, pp. 280-95.

Goldenberg DL, Burckhardt C & Crofford L (2004). Management of fibromyalgia syndrome. *JAMA*, Vol 292, No. 19, pp. 2388–95.

Haak T, & Scott B. The effect of Qigong on fibromyalgia: a controlled randomized study (2008). *Disabil Rehabil*, Vol. 30, No 8, pp. 625-33.

Häuser W, Arnold B, Eich W, Felde E, Flügge C, Henningsen P, Herrmann M, Köllner V, Kühn E, Nutzinger D, Offenbächer M, Schiltenwolf M, Sommer C, Thieme K & Kopp I (2008). Management of fibromyalgia syndrome--an interdisciplinary evidence-based guideline. *Ger Med Sci*. Dec 9;6:Doc14.

Häuser W, Thieme K & Turk DC (2010). Guidelines on the management of fibromyalgia syndrome - a systematic review. *Eur J Pain*, Vol 14, No.1 , pp. 5-10.

Health Information: Mind-Body Medicine. National Center for Complementary and Alternative Medicine (2006) Available from http://nccam.nih.gov/health/backgrounds/mindbody.htm

Kabat-Zinn J (1982). An outpatient program in behavioural medicine for chronic pain patients based on the practice of mindfulness meditation: theoretical considerations and preliminary results. *Gen Hosp Psychiatry*, Vol. 4, No 1, pp. 33–47.

Keefe FJ, Rumble ME, Scipio CD, Giordano LA & Perri LM (2004). Psychological aspects of persistent pain: current state of the science. J Pain, Vol. 5, No 3, pp. 195–211.

Keel PJ, Bodoky C, Gerhard U, &Muller W (1998). Comparison of integrated group therapy and group relaxation training for fibromyalgia. *Clin J Pain*, Vol. 14, No 3, pp. 232–8

Kendall SA, Brolin-Magnusson K, Sören B, & al (2000). A pilot study of body awareness programs in the treatment of fibromyalgia syndrome. *Arthritis Care Res*, Vol. 13, No 5, pp. 304-11.

King SJ, Wessel J, Bhambhani Y, Sholter D & Maksymowych W (2002). The effects of exercise and education, individually or combined, in women with fibromyalgia. *J Rheumatol*, Vol, 29, No. 12, pp. 2620–7.

Kroese M, Schulpen G, Bessems M, Nijhuis F, Severens J & Landewé R (2009). The feasibility and efficacy of a multidisciplinary intervention with aftercare meetings for fibromyalgia. *Clin Rheumatol*, Vol. 28, No 8, pp 923-9.

Landi M (2007). Il metodo Feldenkrais. In *Riabilitazione Reumatologica, approccio multidisciplinare,* S. Maddali Bongi eds, pp 339-364, EDRA SpA, Milano .

Lemstra M, Olszynski WP (2005). The effectiveness of multidisciplinary rehabilitation in the treatment of fibromyalgia: a randomized controlled trial. *Clin J Pain,* Vol. 21, No 2, pp. 166–74.

Lind BK, Lafferty WE, Tyree PT, Diehr PK & Grembowski DE (2007) Use of complementary and alternative medicine providers by fibromyalgia patients under insurance coverage. *Arthritis Rheum,* Vol. 57, No 1, pp. 71–76.

Lush E, Salmon P, Floyd A, Studts JL, Weissbecker I & Sephton SE (2009). Mindfulness meditation for symptom reduction in fibromyalgia: psychophysiological correlates. *J Clin Psychol Med Settings,* Vol.16, No 2, pp. 200-7.

Maddali Bongi S, Di Felice C, Del Rosso A, & al (2010). The efficacy of the Rességuier method in the treatment of fibromyalgia syndrome: a randomized controlled trial. *Clin Exp Rheumatol*, Vol 28, (No 6 Suppl 63), pp. S46-50.

Maddali Bongi S, & Del Rosso A (2010). How to prescribe physical exercise in rheumatology. Reumatismo, Vol. 62, No 1,pp. 4-11.

Maddali Bongi S, Di Felice C, Del Rosso A, & al (2011) Efficacy of the "Body Movement and Perception" method in the treatment of Fibromyalgia Syndrome: an open pilot study. *Clin Exp Rheumatol.* Jul 14. [Epub ahead of print].

Maddali-Bongi S, Del Rosso A, Calà M, Di Felice C, Giambalvo Dal Ben G, Matucci Cerinic M (2011). Resseguièr Method and Qi Gong are efficacious and complementary in the rehabilitation of patients with Fibromyalgia Syndrome. EULAR Congress, London, 25-28 may 2011, Abstract book, Ann Rheum Dis. July 2011; 70 (suppl 3): 698, abstract AB234.

Malmgren-Olsson EB, Armelius BA, Armelius K (2001). A comparative outcome study of Body Awareness Therapy, Feldenkrais, and Conventional Physiotherapy for patients with nonspecic musculoskeletal disorders: changes in psychological symptoms, pain, and self -image. *Physiotherapy Theory and Practice,* Vol.17 , pp. 77 – 95.

Mannerkorpi K & Arndorw M (2004). Efficacy and feasibility of a combination of body awareness therapy and qigong in patients with fibromyalgia: a pilot study. *J Rehabil Med,* Vol. 36, No 6, pp. 279-81 or combined, in women with fibromyalgia. J Rheumatol, Vol, 29, No. 12, pp. 2620–7.

Rooks DS, Gautam S, Romeling M, Cross ML, Stratigakis D, Evans B, Goldenberg DL, Iversen MD, Katz JN (2007). Group exercise, education, and combination self-management in women with fibromyalgia: a randomized trial. *Arch Intern Med,* Vol, 167, No. 20, pp. 2192-200.

Rossy LA, Buckelew SP, Dorr N, Hagglund KJ, Thayer JF, McIntosh MJ, & al (1999). A meta-analysis of fibromyalgia treatment interventions. *Ann Behav Med,* Vol 21, No. 2, pp. 180–91.

Schmidt S, Grossman P, Schwarzer B, Jena S, Naumann J & Walach H (2011). Treating fibromyalgia with mindfulness based stress reduction: results from a 3-armed randomized controlled trial. *Pain,* Vol 152, No 2 pp. 361-9.

Septhon SE, Salmon P, Weissebecker I & al (2007). Mindfulness meditation alleviates depressive symptoms in women with fibromyalgia: results of a randomized clinical trial. *Arthritis Rheum*, Vol.15, No 1, 57, pp. 77-85.

Stephens S, Feldman BM & Bradley N. (2008). Feasibility and effectiveness of an aerobic exercise program in children with fibromyalgia: results of a randomized controlled pilot trial. *Arthritis Rheum,* Vol. 59, No. 10, pp. 1399-406.

Taggart HM, Arslanian CL, Bae S & Singh K (2003). Effects of Tai Chi exercise on fibromyalgia symptoms and health-related quality of life. *Orthop Nurs*, Vol.22, No 5, pp. 353- 60.

Thieme K, Gromnica-Ihle E & Flor H (2003). Operant behavioural treatment of fibromyalgia: a controlled study. *Arthritis Rheum*, Vol. 49, No 3, pp. 314–20.

Thieme K, Häuser W, Batra A, Bernardy K, Felde E, Gesmann M, Illhardt A Settan M, Wörz R, & Köllner V. (2008). [Psychotherapy in patients with fibromyalgia syndrome]. *Schmerz* (German). Vol. 22., No 3 pp. 295-302.

Turk DC, Flor H (1999). Chronic pain: a biobehavioural perspective. In: *Psychosocial factors in pain: critical perspectives.* Gatchel RJ, Turk DC, eds. pp. 18–34, Guilford Press, New York.

van Koulil S, Effting M, Kraaimaat FW & al (2007). Cognitive-behavioural therapies and exercise programmes for patients with fibromyalgia: state of the art and future directions. *Ann Rheum Dis*, Vol.66, No 5, pp. 571-81.

van Santen M, Bolwijn P, Verstappen F, Bakker C, Hidding A, Houben H, & al (2002). A randomized clinical trial comparing fitness and biofeedback training versus basic treatment in patients with fibromyalgia. *J Rheumatol*, Vol. 29, No. 3, pp. 575–81.

Vitali Rosati M (2007). Riflessioni Filosofiche sul Metodo Rességuier. In *Riabilitazione Reumatologica, approccio multidisciplinare*, S. Maddali Bongi eds, pp 391-400, EDRA SpA, Milano

Vlaeyen JW, Teeken-Gruben NJ, Goossens ME, Rutten-van Molken MP, Pelt RA, van Eek H & al (1996). Cognitiveeducational treatment of fibromyalgia: a randomized clinical trial. Clinical effects. *J Rheumatol*, Vol. 23, No. 7, pp. 1237–45.

Wahbeh H, Elsas SM, & Oken BS (2008). Mind-body interventions: applications in neurology. *Neurology*, Vol.70, No 24, pp.2321-2328.

Wang C (2008). Tai Chi improves pain and functional status in adults with rheumatoid arthritis: results of a pilot single blinded randomized controlled trial. *Med Sport Sci*, Vol. 52, 218-29.

Wang C, Schmid CH, Hibberd PL, & al (2009). Tai Chi is effective in treating knee osteoarthritis: a randomized controlled trial. *Arthritis Rheum*, Vol. 61, No 11, pp. 1545-53.

Wang C, Schmid CH, Rones R, & al (2010). A randomized trial of tai chi for fibromyalgia. *N Engl J Med*. Vol. 363 No. 19, pp. 743-54.

Wang C. (2011). Tai Chi and Rheumatic Diseases. *Rheum Dis Clin N Am*. Vol 37, pp 19–32.

Wolfe F, Smythe HA, Yunus MB & al. (1990). The American College of Rheumatology 1990 criteria for classification of fibromyalgia: report of the multicenter criteria committee. *Arthritis Rheum,* Vol. 33, No 2, pp.160-72.

Wolfe F, Clauw DJ, Fitzcharles MA, Goldenberg DL, Katz RS, Mease P, Russell AS, Russell IJ, Winfield JB & Yunus MB (2010). The American College of Rheumatology preliminary diagnostic criteria for fibromyalgia and measurement of symptom severity. *Arthritis Care Res (Hoboken)*, Vol. 62, No 5, pp. 600-10.

11

Influence of Cognitive and Affective Variables in Stress, Functional Limitation and Symptoms in Fibromyalgia

Lilian Velasco, Cecilia Peñacoba, Margarita Cigarán, Carmen Écija and Rafael Guerrero
Universidad Rey Juan Carlos, Madrid, Spain

1. Introduction

The clinical manifestations of Fibromyalgia (FM) have invited extensive inquiry into the potential role of psychological factors associated with this condition (Alegre et al., 2010; Wolfe, et al., 1990). Widespread pain and limitations in physical functioning characterize this condition, and a number of studies have shown that such symptoms are accompanied by considerable psychological distress (Affleck et al., 1996; Esteve-Vives et al., 2010). Some researchers have proposed, however, that deficits in positive affect and cognitions characterize the unique adaptation difficulties of FM patients better than measures of vulnerability (Davis, Zautra, & Smith, 2004; Zautra et al., 2005).
Chronic pain disorders are a source of stress for the patient and demand constant efforts of adaptation. In this context, numerous studies have identified patients´ perceptions of control as significant predictor of health outcomes in chronic pain samples in general and in FMS in particular (Arnstein et al., 1999; Besteiro et al., 2008; Buckelew et al., 1994; Oliver and Cronan, 2005; Wallston, 1989). This perception of control beliefs includes two different constructs: *self-efficacy beliefs* and *locus of control*. Both, previously defined as stress modulator variables, have in common to make up mechanisms that involve a sense of control.
Self-efficacy levels in people living with FM have been shown to predict psychological and physical well-being (e.g. Culos-Reed & Brawley, 2003), and individuals with lower self-efficacy show limitations in their ability to perform everyday tasks required or adaptation to FM. On the other hand, higher levels of pain self-efficacy has been related with better improvement on treatment (Buckelew et al., 1996; Martin et al., 1996; Schachter et al., 2003; Wells-Federman et al., 2003), higher satisfaction levels (Serber et al., 2003), lower pain intensity, less associated symptoms, less functional limitation and psychological distress (Buckelew et al., 1995; Martín-Aragón et al., 2001; Menzies et al., 2006; Miró, 1994; Oliver & Cronan, 2002). Also, self-efficacy is a predictor of physical activity in FM patients (Culos-Reed & Brawley, 2003; Culos-Reed, 2001) and has also been related to a better stress/recovery balance, increasing satisfactory experiences and the involvement in social activities (González-Gutiérrez et al., 2009).
Another perception of control variable that has been studied in FM is locus of control. It seems that FM patients present a more external locus of control (belief that the course of their illness

depends on external factors like other professionals, chance or fate) in comparisson with other chronic pain population or healthy controls (Gustafsson & Gaston-Johansson, 1996). This externality is linked with more pain intensity, worst coping and more psychological distress (Crisson & Keefe, 1988; Härkäpää et al., 1996; Martín-Aragón et al., 2001; Toomey et al., 1991). Oppositely, internal locus of control has been related with better functioning in these patients. This internality may benefit adaptation in FM by providing a cognitive structure that connects actions with expectations of controllable outcomes (Lledó et al., 2010).

It seems that self-efficacy and internal locus of control are two cognitive resources that has been associated to a better control of the symptoms in FM as well as with lower levels of disability, better mood (González-Gutiérrez et al., 2009; Jensen et al., 2001), more treatment adherence (Härkäpää et al., 1991) and use of active coping strategies (Härkäpää et al., 1996, Haythornthwaite et al., 1998; Martín-Aragón et al., 2001; Rudnicki, 2001).

Many researchers (e.g. Cohen et al, 2002; Davis, Zautra & Smith, 2004; Epstein et al, 1999; Malt et al., 2000; McBeth & Silman, 2001; Staud et al., 2003; Walker et al, 1997; Van Houdenhove & Luyten, 2006; Walter et al., 1998; Zautra et al., 2001) suggest that stress and negative affective states such as anxiety and depression may trigger FM symptoms. Some have gone further, proposing that FM is an affective disorder (e.g. Netter & Hennig, 1998). High levels of FM pain itself and the inability to control that pain are sources of stress in themselves that can reduce a person's capacity to cope effectively (Lundberg et al., 2009; Zautra et al., 1999) and lead to depression (Kurtze & Svebak, 2001). It has been found that FM patients show higher levels of negative affect in comparison with osteoarthritis patients (Zautra et al., 2005; Zautra et al., 2007) and this negative affect was related with more pain, fatigue and associated symptoms in FM (Zautra et al., 2001; Davis et al., 2004).

Patients with FM, in comparison with other chronic and painful illnesses, may not have the ability to mobilize sufficient positive cognitive-affective resources to neutralize the experience of pain and the associated negative affect. Recent studies of positive emotion in this population (e.g. Stuifbergen et al., 2010; Zautra et al., 2008, 2010) indicate that patients with FM experience a relative absence of positive emotional resources (Zautra et al., 2005) and researchers have taken an interest in how psychological strengths, such as positive emotion, self-efficacy, may affect the physical and psychological functioning of FM patients (Zautra et al., 2005). Furthermore, cognitive resources may also be in short supply and difficulty in sustaining positive states may present additional challenges to adaptation for FM patients.

In the current study, our first aim is to determine the predictive value that the perception of control variables (pain locus of control and pain self-efficacy) have on stress, affect, functional limitation and symptoms in FM. The second aim is to analyze the mediator effect of positive and negative affect on perception of control variables and functional limitation, anxiety and depression in FM. The third aim is to evaluate a model of cognitive-affective states of FM. In this model, it was defined two correlated but distinct factors: one set of positive features (resources) and another set of negative features (vulnerabilities). Measures of positive affect, self-efficacy over symptoms, self-efficacy over physical activities, self-efficacy over pain and internal pain locus of control were identified a priori as resources. Vulnerabilities included measures of stress responses, external pain locus of control (other professionals, chance and fate) and negative affect. Consistent with prior research, we related the resources factor to better physical functioning and the vulnerabilities factor was related to greater pain intensity, more associated symptoms, anxiety and depression.

2. Methods

2.1 Participants

Participants were recruited from among the patients of a physician in the Pain Clinic at the Foundation Hospital in Alcorcón (Madrid, Spain), and all were members of the Fibromyalgia Association of Madrid (*AFIBROM-Asociación de Fibromialgia de la Comunidad de Madrid*). All participants met the American College of Rheumatology (ACR) criteria for the classification of FM. A high proportion of Pain Clinic patients with Fibromyalgia were female, consistent with prevalence of the condition among both men and women pain patients. We chose to exclude men from the study in order to concentrate limited resources for obtaining a sufficient sample size to test hypotheses with adequate power. Demographic data for the sample are provided in Table 1.

2.2 Procedure

The study followed the ethical principles for research with human participants, and was approved by the University Committee on Ethics. Once participants had signed the forms giving informed consent to take part in the project, they were given a booklet of questionnaires, which took 60–120 minutes to complete. Of the 145 women who signed the consent form, 136 provided complete data: six participants did not complete the questionnaires correctly and three did not agree to participate in the study.

Demographic Characteristics	***N*= 136**
Age (years)	53.18 ± 8.86
Marital status	
Married	72,2 %
Single	9,8 %
Widowed	8,3 %
Divorced	9,8 %
Education	
None	12,2 %
Primary	43,5 %
High School	35,1 %
Higher Education	9,2 %
Employment Status	
Working	27,3 %
Never worked	3,1 %
Inactive for over a year	58,6 %
Inactive within a year	10,9 %
Income / month	
More than 1800 €	6,9 %
Between 1200 € and 1800 €	12,3 %
Between 600 € and 1200 €	21,5 %
Less than 600 €	30 %
None	29,2 %

Table 1. Sample Demographic characteristics.

2.3 Measures

2.3.1 Positive and negative affect

Positive and Negative Affect were measured using the Spanish language version of the 20-item Positive and Negative Affect Schedule (Sandín et al., 1999). Participants were asked to indicate on a five-point scale from 1 (very slightly or not at all) to 5 (extremely) the extent to which they had experienced each type of affect during the past week. The Positive Affect scale included items such as 'interested', 'excited' and 'proud', and the Negative Affect scale included items such as 'distressed', 'nervous' and 'irritable.' Internal consistency reliability for the current sample was α=.83 for the Positive Affect Scale and α=.85 for the Negative scale.

2.3.2 Pain self-efficacy

Pain Self-efficacy was assessed with the Spanish adaptation of the 'Arthritis Self-efficacy Scale' (Martín-Aragón et al., 1999). This scale comprises 19 items measuring three components of self-efficacy: over pain (α=.76 for the current sample), over physical activities (α=.84) and over symptoms (α=.85).

2.3.3 Pain locus of control

It was measured using the Multidimensional Health Locus of Control-Pain Scale (*MHLC-P*; Toomey et al., 1988), adapted from the Multidimensional Health Locus of Control-Form A (Wallston et al., 1976). This scale determines whether respondents believe they are able to achieve control over their pain by their own means ('Internal Locus'), or through other actions ('External Locus'): physicians or health care professionals ('Other Professionals'), or control of pain may also depend on luck, fate or haphazard factors ('Chance' and 'Fate'). These four measures ('Internal', 'Other Professionals', 'Chance' and 'Fate'), are independent factors. Each of one is checked through six different items and rated on a six-point scale (Likert-type format), from 1 ('strongly disagree') to 6 ('strongly agree'), with high scores indicating greater Pain Locus of Control. Internal consistency for this sample was α=.66 for the 'Internal Locus Scale', α=.74 for the 'Other Professionals', α=.59 for the 'Chance Scale' and α=.65 for the 'Fate Scale'.

2.3.4 Stress response

Stress Response (SR) was measured with 68 items developed by Peñacoba (1996) within four domains, each including 17 items: (a) emotional response ('I feel a need to cry, run or hide'), (b) cognitive response ('I am unable to concentrate'), (c) physiological response ('I feel a strong tightening sensation in my chest') and (d) motor response ('I find it impossible to stay still'). The participant read a series of 17 stress responses identified in each domain and was asked to rate them on a five-point scale from 0 (not experienced) to 4 (experienced very frequently). You can obtain specific subscales score and a total Stress Response score was given by the mean score for the four subscales. Cronbach's alpha for this instrument is .97 for all subscales, α=.92 for the emotional subscale, α.=95 for the cognitive subscale, α=.91 for the physiological subscale and α=.87 for the motor subscale.

2.3.5 Anxiety/depression

Self-reports of Anxiety and Depression were assessed from the Fibromyalgia Impact Questionnaire, (*FIQ;* Burckhardt et al., 1991), which has been used in previous work with FM population (e.g. Franks et al., 2004; Gelman et al., 2002). Respondents had to mark a point on a 100-mm Visual Analog Scale, 0-mm (no impact)–100-mm (very severe impact) to indicate how depressed they had felt over the past week, and separately, how anxious they had felt.

2.3.6 Functional limitation

Functional Limitation was assessed using the first 10 items from the Fibromyalgia Impact Questionnaire (*FIQ*, Burckhardt et al., 1991) with which one can obtain a Physical Functioning score from questions assessing the patient's ability to perform 10 different activities. The activities assessed were as follows: shopping, using a washing machine, cooking, washing dishes by hand, vacuuming a rug, making beds, walking 51 km, visiting friends/relatives, gardening and driving. The responses are structured, applying a four-point scale ranging from 0='always able to do' to 3='never able to do'. This first subscale measures how an individual's FM symptoms impact on their daily functions in a typical week. The FIQ is commonly used throughout FM research as an indicator of functional ability (e.g. Franks et al., 2004). Cronbach's alpha for this subscale was .83 for the current sample.

2.3.7 Pain intensity

For measuring Pain Intensity, participants were asked to rate the average amount of pain they experienced in the last week on a scale from 1 to 10, with 1 indicating 'no pain' and 10 indicating 'worst imaginable pain' (Item 5 from the Fibromyalgia Impact Questionnaire, *FIQ*, Burckhardt et al., 1991).

2.3.8 Associated symptoms

FM associated symptoms were assessed with a self-report instrument designed to identify the frequency of FM-related psychosomatic symptoms (González-Alonso, 1999). The patients were asked to indicate whether they presented 26 different symptoms, such as dizziness, apathy, fear, irritability, insomnia, headaches or intestinal disorders. If they did not have the symptom, their score was 0, and if they presented it they scored 1 point. Cronbach's alpha was .83.

3. Results

3.1 Descriptive statistics

Table 2 presents means and standard deviations for all of the continuous variables used in the models.

3.2 Predictive variables of stress, affect and FM variables

To examine the predictive value of the perception of control variables on stress, affect, functional limitation and symptoms in FM a series of hierarchical regressions were conducted. In each regression, sociodemographic variables (age, education, employment status and income) were entered in the first block. The second block consisted of the perception of control variables (pain locus of control and pain self-efficacy). The results of the regression analyses for the Stress Responses are shown in Table 3. It is observed that Self-efficacy Over Symptoms (SOS), Self-efficacy Over Pain (SOP) and Fate Pain Locus of Control (FPLC) were predictors of emotional SR (R^2=.45, F=7.65, P<.01); SOS and SOP were predictors of Cognitive SR (R^2=.42, F=6.66, P<.01) and SOS were predictor of Motor SR (R^2=.31, F=4.24, P<.01). Table 4 shows the results of the regression analyses for the Positive and Negative Affect. SOS were predictor of Negative Affect (R^2=.31, F=4.18, P<.01) and SOS and Self-efficacy Over Physical Activities (SOPA) were predictors of Positive Affect (R^2=.43, F=7.13, P<.01). Results for the regression analysis for functional limitation, associated symptoms, anxiety and depression are shown in Table 5. SOP and FPLC were predictors of associated symptoms (R^2=.29, F=3.86, P<.01); SOPA was predictor of functional limitation (R^2=.24, F=2.87, P<.05); SOP was predictor of Anxiety (R^2=.26, F=3.33, P<.01); and FPLC was predictor of Depression (R^2=.29, F=3.88, P<.01).

Measures	Mean	SD	Skewness	Kurtosis
Positive Affect	28.49	8.43	-0.08	-.48
Negative Affect	27.42	8.09	-.01	-.48
Self-efficacy symptoms	27.25	12.47	-.23	-.007
Self-efficacy activities	30.28	14.40	.03	-.62
Self-efficacy pain	16.87	9.60	.34	.30
Internal Pain Locus of Control	16.79	4.16	-.39	.04
Other Professionals	17.01	5.65	.24	-.30
Chance	7.15	3.43	.78	.59
Fate	10.48	3.63	-.04	-.55
Emotional Stress Response	45.39	15.36	-.62	-.23
Cognitive Stress Response	45.14	16.03	-.80	.05
Physiological Stress Response	41.41	14.94	-.34	-.54
Motor Stress Response	32.97	13.64	-.02	-.47
Anxiety	7.82	2.27	-1.24	1.12
Depression	7.35	2.66	-.91	-.11
Functional limitation	14.08	5.85	.27	-.24
Pain Intensity	7.53	1.69	-.51	-.28
Associated Symptoms	16.30	4.95	-.42	-.53

Table 2. Descriptive statistics of variables measured in the sample.

	F	*R*²	IncR²	Beta	*t*
Emotional Stress Response					
Step 1: Age				-,21	-2,16*
Education				-,08	-,72
Income	3,42*	,13		-,22	-2,13*
Step 2: Other Professionals				-,07	-,85
Internal Pain Locus of Control				,00	,01
Chance				,02	,27
Fate				,17	2,15*
Self-efficacy Over Symptoms				-,37	-3,47**
Self-efficacy Over Physical Activities				,02	,23
Self-efficacy Over Pain	7,65*	,45	,32	-,21	-2,02*
Cognitive Stress Response					
Step 1: Age				-,23	-2,34*
Education				-,01	-,16
Income	2,24	,09		-,14	-1,35
Step 2: Other Professionals				-,06	-,71
Internal Pain Locus of Control				,08	,90
Chance				,10	1,17
Fate				-,03	-,34
Self-efficacy Over Symptoms				-,36	-3,22*

Self-efficacy Over Physical Activities				-,06	-,61
Self-efficacy Over Pain	6,66*	,42	,33	-,25	-2,24*
Motor Stress Response					
Step 1: Age				-,21	-2,11*
Education				-,23	-2,08*
Income	3,06*	,11		-,06	-,58
Step 2: Other Professionals				,03	,41
Internal Pain Locus of Control				-,04	-,46
Chance				,11	1,17
Fate				,02	,28
Self-efficacy Over Symptoms				-,30	-2,51*
Self-efficacy Over Physical Activities				-,08	-,70
Self-efficacy Over Pain	4,24*	,31	,20	-,04	-,36

*p < .05; **p < .01.

Table 3. Regression analysis: Stress Responses and Perception of Control.

	F	R^2	IncR^2	**Beta**	*t*
Positive Affect					
Step 1: Age				,25	2,55*
Education				-,02	-,20
Income	1,89	,07		-,00	-,03
Step 2: Other Professionals				,03	-,42
Internal Pain Locus of Control				,00	,03
Chance				,00	,09
Fate				-,04	-,55
Self-efficacy Over Symptoms				,36	3,29**
Self-efficacy Over Physical Activities				,27	2,47*
Self-efficacy Over Pain	7,13**	,43	,36	,02	,18
Negative Affect					
Step 1: Age				-,21	-2,09*
Education				-,07	-,67
Income	1,16	,05		-,09	-,82
Step 2: Other Professionals				-,16	-1,71
Internal Pain Locus of Control				,00	,02
Chance				,13	1,41
Fate				,13	1,45
Self-efficacy Over Symptoms				-,39	-3,24*
Self-efficacy Over Physical Activities				,12	1,00
Self-efficacy Over Pain	4,18**	,31	,26	-,18	-1,51

*p < .05; **p < .01.

Table 4. Regression analysis: Affect and Perception of Control

	F	R^2	IncR^2	Beta	t
Functional limitation					
Step 1: Age				-,23	-2,30*
Employment Status				-,12	-1,28
Education	1,68	,07		-,10	-,94
Step 2: Other Professionals				-,08	-,83
Internal Pain Locus of Control				,11	1,14
Chance				-,02	-,22
Fate				,05	,60
Self-efficacy Over Symptoms				-,09	-,75
Self-efficacy Over Physical Activities				-,46	-3,66**
Self-efficacy Over Pain	2,87*	,24	,17	,17	1,39
Associated Symptoms					
Step 1: Age				-,17	-1,69
Employment Status				-,06	-,74
Education	2,82*	,11		-,32	-2,88*
Step 2: Other Professionals				-,07	-,80
Internal Pain Locus of Control				,03	,39
Chance				-,08	-,85
Fate				,22	2,40*
Self-efficacy Over Symptoms				-,08	-,67
Self-efficacy Over Physical Activities				-,06	-,54
Self-efficacy Over Pain	3,86**	,29	,18	-,27	-2,23*
Anxiety					
Step 1: Age				-,13	-1,34
Employment Status				-,07	-,78
Education	,86	,03		-,07	-,65
Step 2: Other Professionals				-,11	-1,18
Internal Pain Locus of Control				-,02	-,28
Chance				-,01	-,12
Fate				,16	1,72
Self-efficacy Over Symptoms				-,14	-1,15
Self-efficacy Over Physical Activities				,08	,63
Self-efficacy Over Pain	3,33**	,26	,23	-,37	-3,01*
Depression					
Step 1: Age				-,12	-1,22
Employment Status				,01	,18
Education	3,14*	,12		-,32	-2,94*
Step 2: Other Professionals				-,10	-1,09
Internal Pain Locus of Control				,04	,44
Chance				-,09	-,99
Fate				,23	2,45*
Self-efficacy Over Symptoms				-,10	-,87
Self-efficacy Over Physical Activities				-,06	-,54
Self-efficacy Over Pain	3,88**	,29	,17	-,24	-1,95

*$p < .05$; **$p < .01$.

Table 5. Regression analysis: Functional limitation, associated symptoms, anxiety and depression and Perception of Control.

3.3 Affect as a mediator of perception of control variables and functional limitation, anxiety and depression

To test for the presence of mediating effects, we conducted an ordinary least squares multiple regression analysis and a Sobel test (Baron & Kenny, 1986; Preacher et al., 2007). To assess whether a variable has a "true" mediating effect, the preconditions for a test of mediation are significant relationships between the predictor variable and the mediator, the predictor variable and the outcome variable, and the mediator variable and the outcome variable (Holmbeck, 1997). Partial mediation is demonstrated when the beta weight for the independent variable is reduced (but not to non-significance) as soon as the proposed mediator is added to the equation. Full mediation is demonstrated if the [beta] value for the predictor variable is reduced from significance to non-significance when the proposed mediator is added to the equation (Baron & Kenny, 1986). We conducted hierarchical regression analyses on the perception of control variables (Internal Pain Locus of Control (IPLC), FPLC, SOS, SOPA and SOP). Sociodemographic variables were entered in step 1, perception of control variables were regressed on Step 2 (only those that had shown a significant relationship with the mediator variable), and affect was entered in Step 3.
The analyses showed that positive and negative affect mediated the relationship between IPLC (IPLC) and anxiety and depression (see Table 6). Negative affect mediates the relationship between FPLC and Anxiety and Depression (see Table 7) and between SOS and Functional limitation, anxiety and depression (see Table 8). And negative affect also mediated the relationship between SOPA (see Table 9) and SOP (see Table 10) and Anxiety and Depression.

Outcome Variables	F	R^2	IncR^2	Beta	t
Anxiety					
Step 1: Age	,185	,001		-,038	-,430
Step 2: IPLC	4,602*	,066	,065	,255(,208*)[a]	-3,001**
Step 3: Positive Affect	4,993**	,104	,038	-,208	-2,337*
Depression					
Step 1: Age				-,006	-,204
Education				-,103	-1,052
Income	1.012	,026		-,122	-1,308
Step 2: IPLC	2,493*	,082	,056	-,236 (-,188*)[a]	-2,604*
Step 3: Positive Affect	3,007*	,119	,037	-,204	-2,175*
Anxiety					
Step 1: IPLC	9,042*	,063		-,251(-,116)[a]	-3,007*
Step 2: Negative Affect	35,811**	,350	,287	,552	7,661**
Depression					
Step 1: Education				-,085	-,935
Income	1,510	,026		-,121	-1,303
Step 2: IPLC	3,331*	,081	,055	-,236(-,120)[a]	-2,611*
Step 3: Negative Affect	10,356**	,270	,189	,457**	5,381**

NOTE: Standardized regression coefficients (betas) are reflected when variables are from the step in which they are added to the equation.
a. Beta value after introduction of mediator variable (positive or negative affect). Significant comparisons between standardized regression coefficients after Sobel test are shown in bold.
*$p<0.05$; **$p<0.01$

Table 6. Mediation model: Internal Pain Locus of Control.

Outcome Variables	F	R^2	IncR^2	Beta	t
Anxiety					
Step 1: Education	,458	,004		,062	,677
Step 2: FPLC	3,685*	,060	,056	,243(,148) [a]	2,625*
Step 3: Negative Affect	20,993**	,354	,294	,557	7,235**
Depression					
Step 1: Education				-,097	-1,042
Income	1,510	,026		-,121	-1,303
Step 2: FPLC	2,866*	,071	,045	,216(,136) [a]	2,337*
Step 3: Negative Affect	10,561**	,274	,203	,465	5,598**

NOTE: Standardized regression coefficients (betas) are reflected when variables are from the step in which they are added to the equation.
a. Beta value after introduction of mediator variable (positive or negative affect).
*p<0.05; **p<0.01

Table 7. Mediation model: Fate Pain Locus of Control.

Outcome Variables	F	R^2	IncR^2	Beta	t
Functional limitation					
Step 1: Age	1,68	,07		-,23	-2,30*
Step 2: SOS	10,763**	,074	,067	-,273(-,202*) [a]	-3,281*
Step 3: Negative Affect	7,314**	,099	,025	,172	1,911*
Anxiety					
Step 1: Age	,185	,001		,038	-,430
Step 2: SOS	7,865*	,108	,107	-,333(-,116) [a]	-3,940**
Step 3: Negative Affect	23,339**	,352	,244	,543	6,967**
Depression					
Step 1: Age				-,020	-,204
Education				-,103	-1,052
Income	1,012	,026		-,122	-1,308
Step 2: SOS	3,099*	,100	,074	-,276(-,093) [a]	-3,023*
Step 3: Negative Affect	8,208**	,270	,170	,467	5,088**

NOTE: Standardized regression coefficients (betas) are reflected when variables are from the step in which they are added to the equation.
a. Beta value after introduction of mediator variable (positive or negative affect). Significant comparisons between standardized regression coefficients after Sobel test are shown in bold.
*p<0.05; **p<0.01

Table 8. Mediation model: Self-efficacy Over Symptoms.

Outcome Variables	F	R^2	IncR^2	Beta	t
Anxiety					
Step 1: SOPA	11,443**	,079		-,280(-,125) [a]	-3,383**
Step 2: Negative Affect	36,083**	,352	,273	,545	7,485**
Depression					
Step 1: Education				-,097	-1,042
Income	1,510	,026		-,121	-1,303
Step 2: SOPA	4,314*	,103	,077	-,278(-,147) [a]	-3,113*
Step 3: Negative Affect	10,698**	,276	,173	,444	5,185**

NOTE: Standardized regression coefficients (betas) are reflected when variables are from the step in which they are added to the equation.
a. Beta value after introduction of mediator variable (positive or negative affect).
*p<0.05; **p<0.01

Table 9. Mediation model: Self-efficacy Over Physical Activities.

Outcome Variables	F	R^2	IncR^2	Beta	t
Anxiety					
Step 1: Marital status	,213	,002		-,040	-,461
Step 2: SOP	12,506**	,161	,159	-,402(-,231**) [a]	-4,976**
Step 3: Negative Affect	27,142**	,387	,226	,505	6,890**
Depression					
Step 1: Marital Status				-,132	-1,402
Education				-,099	-1,075
Income	1,671	,042		-,147	-1,557
Step 2: SOP	4,777*	,146	,104	-,325(-,171**) [a]	-3,680**
Step 3: Negative Affect	9,009**	,289	,143	,416	4,723**

NOTE: Standardized regression coefficients (betas) are reflected when variables are from the step in which they are added to the equation.
a. Beta value after introduction of mediator variable (positive or negative affect). Significant comparisons between standardized regression coefficients after Sobel test are shown in bold.
*p<0.05; **p<0.01

Table 10. Mediation model: Self-efficacy Over Pain

3.4 An integrated model

In our sample, multivariate normality was tested using the Mardia test for multivariate kurtosis. Skewness and kurtosis indexes were calculated for each variable (see Table 2) and the Mardia test suggested the presence of non-normality at a multivariate level (critical ratio = 7.83; p< 0.05). Given this, the decision was made to pursue parameter estimation under two scenarios, traditional maximum likelihood analysis and bootstrapping. For the bootstrap analysis we performed 2000 bootstrap replications for purposes of estimating standard errors, p values and confidence intervals. We employed the bias-corrected approach to interval estimation as implemented in AMOS (Arbuckle & Wothke, 1999). Estimation for all 2000 individual bootstrap samples yielded convergence and meaningful solutions. To asses the overall fit of the tested models we used the Bollen Stine bootstrap p-value in addition to the usual maximum likelihood-based p-value (chi square), following Bollen and Stine's recommendations (Bollen & Stine, 1993). In general, conclusions were the same in the two estimation methods. Significance tests and confidence interval reported are from the bootstrap analyses.

3.4.1 Contrasting the two-factor structure: Resources and vulnerabilities

The initial model is evaluated in a structural model that specifies the relationship between two latent variables and a measurement sub-model which relates a series of observable indicators with the latent variables. The two latent variables are as follows: cognitive-affective resources and cognitive-affective vulnerabilities. It can be seen that there is a bidirectional path from the cognitive-affective resources to the cognitive-affective vulnerabilities. In the measurement sub-model, the three observed variables of self-efficacy (over symptoms, over physical activities and over pain) are defined along with the internal pain locus of control and the positive affect as a result of the latent variable "cognitive-affective resources" (plus the random measurement error) that in turn influences the observed endogenous variable "functional limitation". Moreover, the latent variable "cognitive-affective vulnerabilities" is the result of the observed variables relative to the average score of the different stress responses, the external locus of pain control (other professionals, chance and fate) and negative affect, this latent variable in turn influences the associated observed symptom variables; pain intensity, anxiety and depression (See Figure 1).

To test this model a structural equation model (SEM) was carried out which allowed us to make an empirical analysis of the theoretically constructed model and reject it in the case of it being inconsistent with the data. In order to examine the fit of the model the Bollen-Stine p value was taken into account and it turned out to be statistically significant (<.05) which suggested a poor fit for the model. Other commonly used adjustment indices were also taken account of: $\chi 2$, $\chi 2$/g.l., *CFI* (Comparative Fit Index), *IFI* (Incremental Fit Index), *TLI* (Tucker Lewis Index) and *RMSEA* (Root Mean Square Error of Approximation). The $\chi 2$ should have non-significant values (p>.05) but as it is very sensitive to sample size (Jöreskog & Sörbom, 1989), it is recommended to calculate $\chi 2$/g.l. which is considered to be acceptable when it is less than 5 (Bentler, 1989). The incremental models (CFI, IFI and TLI) are based on a comparison between the hypothesized model and the null model and are not affected by the size of the sample with values greater than .90 being considered acceptable (Hu & Bentler, 1995). Values of .08 or less are considered acceptable for RMSEA (Browne and Cudeck, 1993). These indices offer results which tend to show an insufficiently robust fit ($\chi 2$= 161,756; $\chi 2$/g.l.= 1,81; *CFI*=0,87; *IFI*=0,88; *TLI*= 0,85; *RMSEA*=0,08). For this reason it was decided to examine the coefficient paths in order to eliminate those not statistically significant. However, no path was eliminated by following this criterion. The test used next was that of Lagrange multipliers (LM) with the objective of identifying possible paths or covariances among the measurement errors, which could improve the fit of the model (Aitchinson & Silvey, 1958). In this way two covariances were added from among the measurement errors, the first from between the measurement error of the FPLC and Chance

Pain Locus of Control (CHPLC) (modification index=25.,03) and the second between the measurement errors of anxiety and depression (modification index=16.25). All of this gave rise to the model presented in Figure 2, with a Bollen-Stine p value of 0.25 while the rest of the indices of global fit were: $\chi2$= 113,198; $\chi2$/g.l.= 1,30; *CFI*=0,95; *IFI*=0,95; *TLI*= 0,94 and *RMSEA*=0,05 all of which points to an appropriate degree of fit for the model. All of the coefficient paths were statistically significant ($p < 0.05$). Figure 2 provides a graphic representation of the final model.

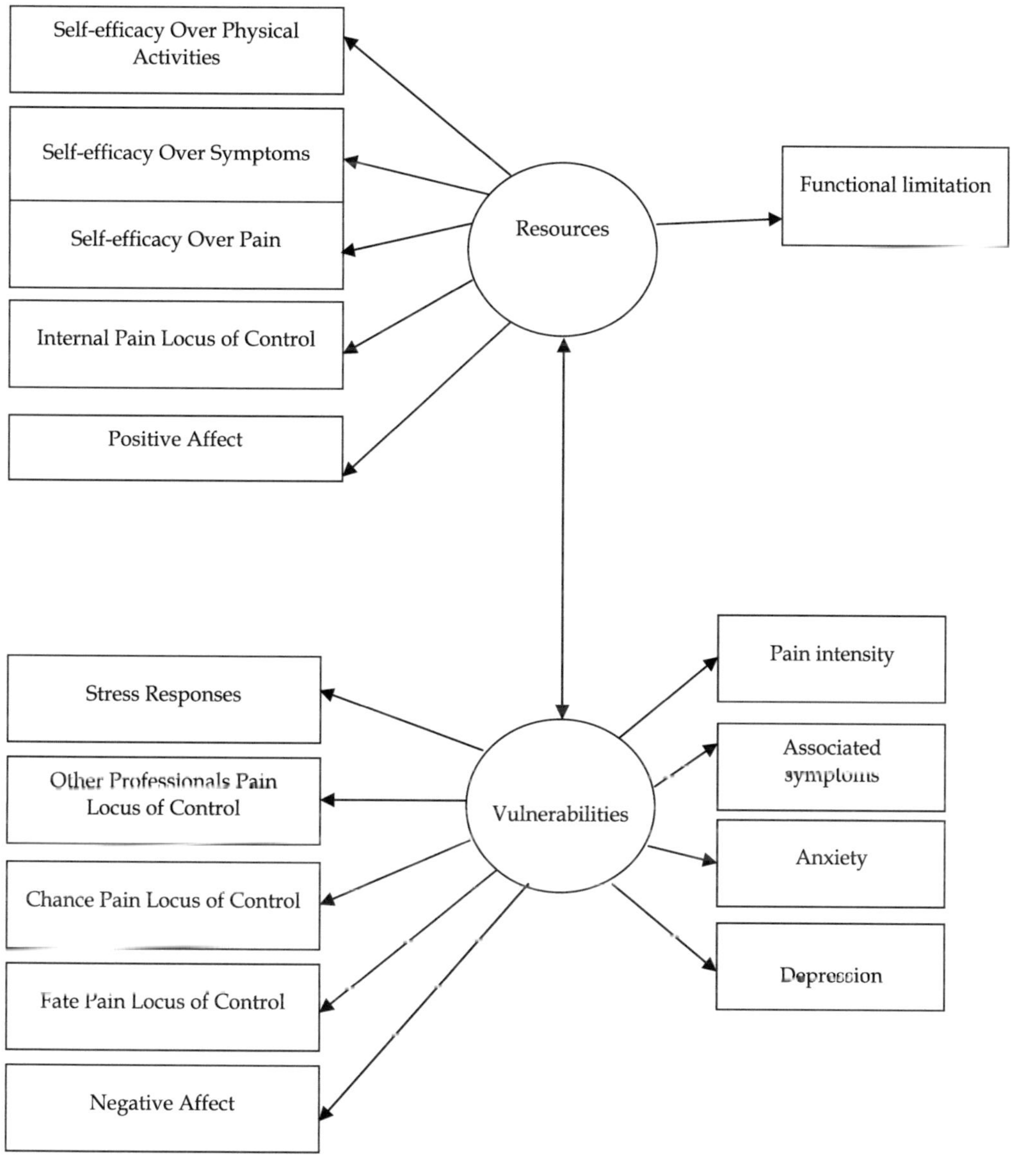

Fig. 1. Initial trajectories model hypothesized. The rectangles represent observed variables (measured), circles represent latent variables (unobserved). For the sake of clarity the error variances of the endogenous variables are not shown.

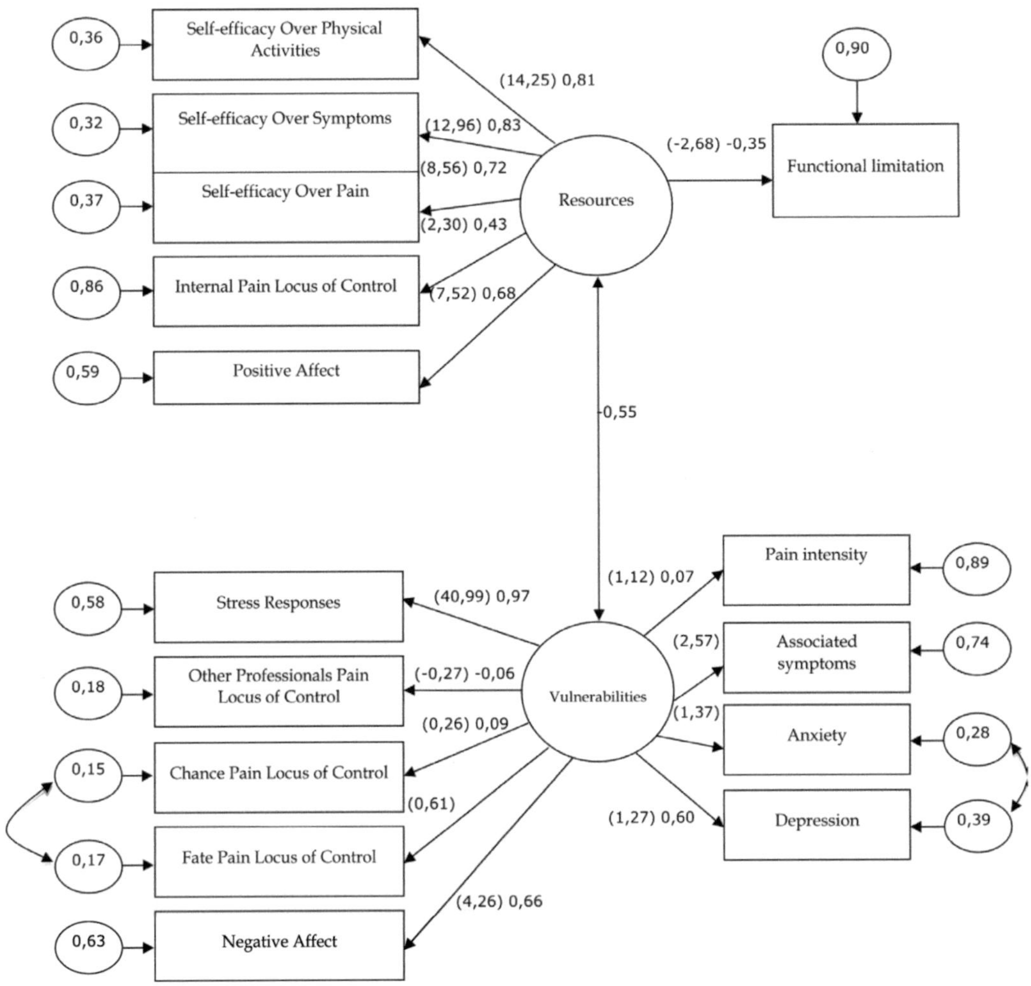

Fig. 2. Final Model. The circles represent latent variables (unobserved), the ellipses are the standardized variances of error and the rectangles are the observed endogenous variables. Values in parentheses are unstandardized path coefficients, while the values outside parentheses are standardized path coefficients. The values attached to the bidirectional lines are the coefficients of correlation between the error variances.

4. Discussion

4.1 Predictive variables of stress, affect and FM variables

The results show that the variables of perceived pain control play an important role in stress responses in women with FM. In particular, self-efficacy over pain and over symptoms play a key role since they are predictors in the cognitive-emotional patterns of response. In the case of the motor response, predictive capability is only seen for SOS. Moreover, in the case of emotional response, FPLC is included as a predictor variable.

Our results show that external locus of control (Fate) are predictors of stress responses consistent with previous studies (Crisson & Keefe, 1988; Härkäpää et al., 1996). Thus, the belief that one's illness or behavior is determined by external agents, in this case by others or by one's own fate or destiny, increases the stress response in their emotional and physiological modalities. Different studies have shown the positive role of self-efficacy for over pain and over symptoms associated with FM (Buckelew et al., 1995, Martin-Aragon et al., 2001, Menzies et al., 2006 ; Miró, 1994), among them stress. It seems that our data support the idea that low pain self-efficacy predicts the evaluated stress responses.

As for the explanatory models of Affect, it appears that perception of control variables, particularly pain self-efficacy play an important role. These results show the important connection between emotional and cognitive aspects as has been demonstrated in previous studies (v.g. Palomino et al., 2007). High self-efficacy is related to positive affect, whereas low self-efficacy goes with negative affect. In addition, the results point to the independence of both types of affectivity, so that the predictors differ in both cases. According to this data, women with FM can regulate their negative affect by increasing their self-efficacy regarding associated symptoms. However, to achieve positive emotions associated with affect they also need to increase their self-efficacy related to the carrying out of activities. Although previous studies have shown the relationship between negative affect, perceived pain, stressful experiences, low self-efficacy and symptoms associated with FM (Davis et al., 2001, Davis et al. 2004; Kersh et al. 2001; Kurtz & Svebak, 2005, Potter et al., 2000, Zautra et al., 1999), there is little literature on the variables with influence on positive affect, as has been pointing out by positive psychology. This differentiation has important practical implications, since according to the objectives of the intervention programs, the activity carried out should be differential on the basis of the variables that are really involved. In this context it is particularly important to work both with the illness and health variables, with the strengths and weaknesses, with positive and negative affect, with all health problems in general and particular with FM. With regard to affect, studies such as Zautra et al. (2005) have shown that FM patients have emotional equilibrium problems; they have lower levels of positive affect though they do not have high levels of negative affect, unlike what has been claimed by the majority of studies (Zautra et al., 2005). The problem of emotional regulation in these patients is not explained by negative affect but by a certain vulnerability in the system of positive affect: these patients show little responsiveness to positive emotions or events, focusing instead on the negatives. These results highlight the importance of targeting interventions toward increasing positive resources in these patients.

In this vein, and in order to differentiate between positive affect and negative effect, the research team (Velasco et al., 2009) evaluated a causal model using structural equations defining the presence of two dimensions: cognitive-affective resources and cognitive-affective vulnerabilities. The resources are made up of positive affect, self-efficacy over symptoms, self-efficacy over physical activities, self-efficacy over pain and internal pain locus of control. The vulnerabilities were stress responses (emotional, physiological, cognitive and motor), anxiety, depression and negative affect. The resources showed a direct influence on physical functioning, which in turn affected the thermal tolerance to pain. This highlights the capacity of the patients to retain wellness in spite of pain; the presence of positive emotions helps them to not feel dysfunctional, on the contrary, it helps them adapt to the pain. Positive resources are the key for better tolerance of pain. Vulnerabilities, on the other hand, predicted pain intensity and associated symptoms. These results allow us to reflect on the close relationship between cognitive and emotional variables, in their classification as positive and negative variables. The participation of the patient in stress processes, as an active and transforming agent in his or her surroundings, is fundamental for the resolution of conflicts and dealing with situations of disruption. This participation functions to cushion the impact of stressful situations and in turn could act as an enhancer of the health and welfare of the individual (Söderfeldt et al., 2000). Thus, within this role of agent, perception of control variables would have an important role in alleviating stressful experience and the negative affect as well as promoting positive emotional resources. From the perspective of a process model of positive personality, this capacity of the active subject is related to emotional coping skills that seek to regulate the emotional response and enhance emotional resistance. Perceived self-efficacy in stressful or threatening situations generates expectations about the results based on the individual's conviction that he or she not only possesses the necessary coping resources but is also able to maintain an adaptive flexibility that allows him or her to control the demands of the situation (Idel et al., 2003). If there are expectations of self-efficacy this will facilitate the choice of healthy behaviors and appropriate coping strategies to deal with a difficult situation like a disabling illness. At the same time the emotional effect of self-efficacy will shape a sense of control which will see health as being an important good for the subject and as something capable of being positively modified by his or her behavior (see Reich & Zautra, 1981). All of this in such a way that by being able to carry out tasks and achieve goals, the subjective state of well being will be reflected in a positive mood as well. By the same token, the internal locus of control, seen as a tendency to attribute results more to oneself that to external factors, has been related to subjective well being (Cooper et al., 1995) which would explain its predictive role with regard to affect.

With regard to the specific variables of FM, namely the associated symptoms and emotional aspects, we found that both self-efficacy over pain in the case of anxiety and the fate pain locus of control in the case of depressive symptoms are relevant variables and both for the associated symptoms. Taking as a reference point the risk personality category defined by Folkman (1984), certain similarities can be found with patients with FM; individuals inclined to the perception of threat and danger and the perception of a lack of resources and lack of personal control, with a greater probability of emotional coping and the feeling of more negative emotions and anxiety (Lundberg et al., 2009). These factors would place the person in a continuous situation of stress, anxiety and depression. This lack of resources would be reflected in low levels of self-efficacy and an external locus of control (Buckelew et al., 1990; Crisson & Keefe, 1988; Skevington, 1990).

Finally, with respect to functional limitation, it has been shown to be a predictor variable for self-Efficacy over physical Activities. Previous studies have shown that self-efficacy is related to greater physical activity in FM (Culos-Reed & Brawley; 2003). Confidence in the carrying out of certain activities may lead to participation in them and also participation in activities which require less effort such as those valued by this present study (going shopping, going for a walk etc.) Furthermore, the specific self-efficacy necessary for the activities predicts the physical activity and functioning of these patients (Culos-Reed, 2001). Different treatment programs of an inter-disciplinary or cognitive-behavioral nature have demonstrated the necessity of working with self-efficacy expectations to diminish the symptoms and pain associated with FM (Aspegren et al., 2000; Bailey et al., 1999; Buckelew et al., 1995; 1996; Lera et al., 2009; Schachter et al., 2003; White & Nielson, 1995).

4.2 Affect as a mediator between perception of control and functional limitation, anxiety and depression

From the results obtained from mediation analysis a major mediating role can be seen for negative affect between perception of control and the outcome measures of FM. Furthermore, in the majority of cases, negative affect has a total mediator role and so its predictive capacity for the perception of control variables considered here can only be explained by its influence. The outcomes explained by models of mediation were primarily anxiety and depression, and to a lesser degree functional limitation. Thus negative affect acts as a mediator among IPLC, FPLC, SOS, SOPS, SOP as predictor variables and Anxiety and Depression as outcome variables. In all cases the mediation is total except that of self-Efficacy over pain where it was partial. At the same time negative affect acted as a mediator, in this case partial between self-Efficacy over symptoms and Depression. These results are consistent with previous literature (González-Gutiérrez et al. 2009; Kurtze & Svebak, 2005, Potter et al., 2000), which shows how, in relation to the explanation of anxiety, depression and functional limitation on the basis of perception of control and affective variables, negative affect would be the best predictor of these outcomes. The mediating role of positive affect is clearly lower and when associated with an internal perception of control helps to reduce levels of anxiety and depression.

FM has often been associated with negative affect (Kurtz & Svebak, 2005, Potter et al., 2000) and even authors such as Davis et al. (2001), Potter et al. (2000), Staud et al. (2003), Zautra et al. (1999; 2005; 2007), consider it as an important variable in the consolidation and development of the disease. Furthermore, as part of a vicious circle, these negative emotions in turn influence the symptoms associated with FM (Ramírez-Maestre et al., 2001).

One of the limitations in interpreting the results found is the assessment of anxiety and depression through individual FIQ items and its possible contamination with positive and negative affect, precisely because of the measurement procedure used. However, while recognizing these limitations, it is necessary to clarify that in this case the measures of anxiety and depression related to emotional symptoms rather than to psychiatric diagnosis. In this context, Pérez-Pareja et al. (2004) differentiate the concept of psychiatric disorder from that of mood and indicate the importance of this difference not only in the context of the diagnosis of the illness but also in relation to the possible treatments for it. In the authors view the depressed mood of these patients is a consequence of the symptoms of the disease itself as well as of pain interferences, although other authors (O'Brien et al., 2010) does not agree. Thus, not being able to carry out previously habitual activities or not being able to control one's own pain or its interference provokes more negative symptoms, in line with the mediating effects of negative affect found in the present study.

The partial mediation role played by negative affect between SOS and functional limitation must also be highlighted. As has previously been mentioned, self-efficacy beliefs play an important role in the physical functioning of patients with FM. In this sense, lower self-efficacy regarding symptoms lead to greater negative affect which in turn affect functional limitation. Previous studies have established a relationship between self-efficacy for pain and functional limitation (Buckelew et al. 1995; Jegede, 2006; Martin-Aragon et al., 2001M Mueller et al. 2003; Oliver & Cronan, 2002; Theadom et al., 2007; Velasco et al., 2008), although they did not include affect as a mediating variable.

Functional capacity is a major source of concern in patients with FM as they face limitations in carrying out different activities, especially in activities of daily life which they regard as fundamental. The inability to carry out, or the difficulties encountered in carrying out, such tasks results in a series of negative emotions, which is reflected in the high levels of negative affect and low positive affect of these patients, as well as poor self-efficacy of pain both for carrying out activities as well as for symptom control and pain control. This functional impairment is also often compounded by a lack of comprehension on the part of those around the sufferer due to the nature of the disease (absence of demonstrable organic changes) (Esteve-Vives et al., 2010).

Various studies (Alonso et al., 2004, Davis et al., 2001, Hassett et al., 2008, Zautra et al., 1999) have assessed the relationship of positive and negative affect with the symptoms of FM, using the same instrument used in our study (PANAS). The results show that negative affect has more to do with the symptoms of the disease and functional limitation than positive affect. Although the results for positive affect are more limited, we should remember also the tendency to use negative indicators of disease and outcome variables while ignoring positive outcomes such as satisfaction, well being or quality of life. In the same vein, authors such as Lyubomirsky et al. (2005) consider that positive emotions are more than the absence of negative emotionality, so it is important to consider the potential utility of positive emotions to achieve high levels of psychological well-being, even in times of illness.

This "lack of prominence" as a mediator of positive affect with regard to the effect of perception of control variables could be understood on the basis, mentioned above, of the inability of FM patients to mobilize positive resources to neutralize the experience of pain and negative affect associated with it (Zautra et al., 2005). That is why their difficulty in maintaining a positive mood is an additional challenge for patients with FM. In addition, there may be a vicious circle connecting lower positive affect and fewer satisfactory social relationships, which could also explain the results in relation to interpersonal stress responses. As already noted, these patients have low responsiveness to positive emotions or events, they focus more on the negative, and have a relative absence of positive emotional and cognitive resources (Potter et al., 2000; Zautra et al., 2001).

Other studies (Arnold et al., 2006; Hassett et al. 2008; Porter-Moffitt et al. 2006; Suhr, 2003) have confirmed this affective pattern (low positive affect and high negative affect) in FM patients and its impact on lower physical functioning and increased clinical symptoms. Some authors (Clauw & Chrousos, 1997, Martinez-Lavin, 2004; Sarzi-Puttini et al., 2006) have brought up the study of the role of affective variables via physiological correlates in this regard.

In any case, it seems clear that interventions aimed at decreasing negative affect may contribute to improving the quality of life of these patients by optimizing their own perception of control variables. Similarly, one of the fundamental aspects of disease treatment would aim to design interventions that enhance positive resources. Positive experiences would be a key to increase the quality of life and physical functioning of these patients. One of the practical implications involved would be the inclusion, in psychotherapy, of programs for emotional regulation, currently almost nonexistent in the approach to FM, as a vehicle for improving these patients and their ability to cope with the disease.

4.3 An integral explanatory model for fibromyalgia

In the final phase of our study it was evaluated a causal model of the relationship between cognitive-affective resources and cognitive-affective vulnerabilities in patients with fibromyalgia, with the researchers interest focused on the relationship between these variables and a number of disease outcomes, such as functional limitation, pain intensity, associated symptoms and anxiety and depression. Techniques derived from structural equation modeling (SEM) were used and indices obtained suggested an adequate fit of the model evaluated.

The model evaluated in this study of the relationship between resources and cognitive-affective vulnerabilities supports the hypothesis of the existence of a number of variables that function as resources including cognitive processes of control perception as well as other emotional resources like positive affect (Thompson, 2006). Thus, self-efficacy over pain and control expectancies are two constructs whose positive effects have been demonstrated in areas such as coping with stress and adherence to therapy. Both variables have shown a positive role in health (Dwyer, 1997; Pastor et al., 1999). Regarding the study of these control beliefs in FM, studies that focus on self-efficacy are more numerous, being fewer focused on locus of control, although the internal locus of control has been associated with fewer FM symptoms and greater physical functioning improving the quality of life of the patients (Zaharoff, 2005). Several studies have shown the positive role of self-efficacy in managing pain and symptoms in people with FM (Menzies et al., 2006). In addition to these cognitive variables, we found that positive affect is also included in this set of "resources" that acting together, have a major influence on the physical functioning of patients with FM. Thus, the use of these resources would reduce functional limitation. In this regard, previous studies have demonstrated the influence of these variables on physical functioning in FM (Buckelew et al. 1995; Culos-Reed & Brawley, 2003; Fontaine et al., 2010), and various treatment programs have mentioned that self-efficacy is fundamental to predict physical functioning (Buckelew et al., 1995; Oliver & Cronan, 2002; Rivera et al. 2004; Schachter et al., 2003). The presence of positive emotions and cognitive resources reflect better physical functioning and reduced clinical symptoms in fibromyalgia (Arnold et al., 2006; Hassett et al., 2000; 2008; Porter-Moffitt et al. 2006; Suhr, 2003).

In this present model we found a series of cognitive-affective vulnerabilities that directly affect the symptoms presented in FM. These vulnerabilities are the different responses to stress, the external pain locus of control and negative affect. Stress has been associated with high levels of psychological distress, including symptoms related to anxiety, depression or negative affect. These types of affective symptoms appear on most occasions together with stress processes in different contexts and populations (Fresco, 2000; Stillerman, 2007). As already mentioned, there are numerous studies that have found relationships between stress (studied as a stimulus, response or consequence) and fibromyalgia as well as between emotional distress and symptoms of the disease. These cognitive-affective vulnerabilities impact on the intensity of pain and symptoms associated with FM. It seems that people with fibromyalgia have a more external pain locus of control than people with other types of chronic pain and healthy people (Gustafsson & Gaston-Johansson, 1996), although this has not been confirmed in our work. This externality of the locus of pain is in turn associated with greater intensity of pain, worse coping and psychological distress in fibromyalgia (Crisson & Keefe, 1988; Härkäpää et al., 1996; Martin-Aragon et al., 2001; Toomey et al., 1991). It seems that negative affect is associated with increased symptoms and pain in FM (Alexander et al., 1998; Potter et al., 2000; Velasco et al., 2006). In subjects with high negative affect an increased perception of threat has been suggested in relation to which the subjects experience an increase in the perception of pain intensity (sensory and affective) when presented with an ambiguous stimulus, if their pain expectation is manipulated (Alexander et al., 1998). Staud et al. (2004) found an association between negative affect, the number of tender points and pain intensity in fibromyalgia. There are numerous studies that relate affective unease and intensity of pain in fibromyalgia (Alexander et al. 1998; Kurtze & Svebak, 2005, Potter et al., 2000) and there are several hypotheses about the temporal relationship between the two elements (the emotional distress as a precedent and as a consequence of chronic pain) and empirical evidence can be found in favor of both approaches (e.g. Gaskin et al., 1992, Sullivan et al., 2001), which seems to suggest the presence of a bidirectional relationship. However, as a piece of isolated data, a relationship between negative affect and pain intensity has not been found in our work. In our model, we took anxiety and depression to be direct consequences of this set of "vulnerabilities" that could be considered within the same construct. This allows us to understand the correlation between the measurement errors of these variables in the model analysis. Future studies should establish control mechanisms in order to check and eliminate this type of response bias. It would also be interesting to check the fit of this model to other samples of patients with chronic pain (with either a known or unknown etiology), in order to test its generalizability to other disorders and to examine the differential characteristics with respect to FM.

In previous studies the research team evaluated two models in samples of women with FM, on similar lines. Firstly, as has already been pointed out when talking about the independence between positive and negative affect, one of the models (Velasco et al., 2009), based itself on the presence of two dimensions: cognitive-affective resources and cognitive-affective vulnerabilities. The resources include positive affect, self-efficacy over symptoms, self-efficacy over physical activities and over pain and internal pain locus of control, while the vulnerabilities include stress responses, anxiety, depression and negative affect. The resources have a direct influence on physical functioning and the vulnerabilities predict pain intensity and associated symptoms. Unlike the model set out in this present study, pain and anxiety were considered as variables within this set of vulnerabilities, given the bidirectional relationship between emotional variables and the results indicated throughout this present study. The second of the proposals (González-Gutiérrez et al., 2010) evaluated a pattern of relationships between a number of cognitive resources (self-efficacy expectancies and internal pain locus of control), the stress-recovery process and emotional distress in 130 women with fibromyalgia. The results showed that the stress-recovery balance mediates the relationship between cognitive resources and emotional distress. The presence of a direct effect of cognitive resources on functional limitation was also noted, while the intensity of pain and other symptoms of illness were directly predictive of emotional distress.

Other recent study (Lledó et al., 2010) has highlighted the relationship between control beliefs, coping focused on the disease (most used by these patients) and emotions to explain the health outcomes of FM patients. It shows cognitive and emotional resources to be important variables in explaining the impact on health of FM. In this regard, perceived competence in health (the perception that one is able to interact effectively in situations of health) and self-efficacy of pain were the most important variables in explaining the health status of FM. The perceived competence predicted the intensity of pain while self-efficacy and depressed mood accounted for physical activity, with the cognitive component being more important than the emotional one, the perception of competence and anxiety and depression predicted the psychosocial impact of disease, self-efficacy and perceived competence in health predicted depression, with the first having a greater effect than the first, and the use of behavioral or passive coping strategies (or strategies centered on the disease), such as the protection of areas or movements, asking for help or and rest, only had an effect on the degree of physical activity.

We would like to point out some limitations of this study which need to be considered. First, its correlational cross-sectional nature made it unable to establish cause-effect relationships. To establish these, longitudinal studies are necessary. Moreover, the biases associated with the application procedure of the questionnaires, especially the self-report questionnaires that were used in this study and their own nature, may have interfered with the results found. Another limitation is the excessive focus on disease variables to the exclusion of positive health variables. The focus of the study on personality as a state and not as a process was also a limitation. We believed this latter approach would provide key findings for understanding how certain variables operate and to design appropriate intervention programs. This limitation is especially relevant for the assessment of stress and emotion.

5. Conclusions and implications

This study shows a predictive capacity of pain perception of control on stress responses and affect in people with FM. As for the associated disease variables (associated symptoms, functional limitation, anxiety and depression), the study shows self-efficacy over physical activities, over pain and external pain locus of control (other professionals and fate) to be predictors for them. Likewise, negative affect (compared to positive) is a powerful mediator between the variables of perceived pain control and the outcomes of fibromyalgia variables (anxiety and depression and to a lesser extent, functional limitation). Positive affect partially mediates between internal pain locus of control and anxiety and depression. The study also demonstrated an integrative model of FM which confirms the existence of a series of variables that function as resources (cognitive processes of perception of control and positive affect) that affect functional limitation, and a series of cognitive-affective vulnerabilities (stress responses, external pain locus of control and negative affect) that directly affect pain and associated symptoms presented by these patients.

The conclusions drawn from this study may yield important practical implications. Although protocols have been designed for more complete care of FM, there are few interventions that address the main areas affected by the disease. Thus, the objectives that should be set out would focus on the integral development of four areas: emotional (anxiety and depression in particular), cognitive (perception of self-efficacy, beliefs and expectancies), behavioral (daily activities that are reduced or eliminated as result of FM) and social (impact of the disease in the patient's social and family sphere).

Thus, in accordance with the above and consistent with the results obtained in this work, we propose that these intervention programs should offer at least three fundamental objectives: 1) promote beliefs related to perceived control, 2) promote strategies of emotional regulation aimed mainly at promoting positive emotions and 3) provide active coping strategies aimed at proper management of stress responses. Changing beliefs about control of the disease are key elements to ensure the positive development of the patients. Lledó et al. (2010) suggest that promoting successful experiences in managing the problem is a therapeutic resource to enhance the perception of control. Therefore, it is important to consider small achievable goals that involve successful experiences in confronting the problem. Perhaps the key would be to carry out treatment programs aimed at changing attitudes regarding pain

management. To the extent that patients own self-control of the disease increases, their beliefs associated with control will be modified and they will have greater involvement in the management of the disease (Biurrun-Unzué et al., 2002).
Another line of action should focus on the relationship between health professional and patient. Bearing in mind the importance of the influence of the health professional and the healthy role of "other professionals pain locus of control" (in combination with "internal locus of control") any strategy to facilitate the patient's confidence in their health professional is specially needed in the case of this disorder. Effective and educational communication could also have a very positive effect in reinforcing control beliefs. In the same way, by working on control, negative emotions can be reduced, another therapeutic objective to consider, not only because they themselves are the health area most affected, but also because they are a resource to increase physical and social activity, and thus prevent further deterioration. It has been shown that positive emotions play an important role in the recovery from painful episodes of FM. In addition, these emotions might modulate negative affect present in the moments of greatest pain. The positive experiences and positive affect on the FM are a key point in preserving the quality of life and physical functioning of these patients. That is why the regulation of the activity should be adjusted to promote activities suitable for abilities of the person, their needs and their particular circumstances. The activity should be managed in such a way as to motivate the person to continue participating, maintaining the improvement and adapting the task to the relapse (Turner, Foster y Johnson, 2003).
Another area to intervene on these patients is their social and ocupational adaptation as their impact is such that it is reflected in the negative affect experienced. Behavioral strategies, organization, time planning, delegation and enjoyment of pleasurable activities, are key to better adaptation. Furthermore, family support is also important so programs of family education about the disease should be carried out, promoting commitment, understanding and family collaboration with the patient. Finally, as has been shown throughout the present study, stress plays a fundamental role in FM and is seen as a possible etiological hypothesis, although it also the case that the very experience of FM, from the beginning of symptoms until treatment is also a clear stress factor. That is why there should be a complete and comprehensive assessment of the stress process and its influence on the symptoms of the disease (Salaffi et al., 2009). It would be highly appropriate to set up standardized programs of stress management which would have among their objectives the provision of information about and psychoeducation regarding the stress process and the teaching of techniques, strategies and tools to promote problem solving and coping strategies as well as the learning of relaxation and self-control techniques. The implementation of these techniques would help in the management of pain, increasing physical and social functioning and reducing the associated symptoms in Fibromyalgia.

6. Acknowledgment

This research was supported by the Ministerio de Educación y Ciencia (Plan Nacional I+D+I, ref SEJ2004-08171/PSIC). The authors are grateful to the staff of the Pain Clinic in Fundación Hospital Alcorcón and to AFIBROM. Special thanks are extended to the participants of the study.

7. References

Affleck, G., Urrows, S., Tennen, H., Higgins, P., Abeles, M. (1996). Sequential daily relations of sleep, pain intensity, and attention to pain among women with fibromyalgia. *Pain*, 68, 363–368.

Alegre de Miquel, C., García Campayo, J., Tomás Flórez, M., Gómez Arguelles, J.M., Blanco Tarrio, E., Gobbo Montoya, M., Pérez Martin, A., Martínez Salio, A., Vidal Fuentes, J., Altarriba Alberch, E. Gómez de la Cámara, A. (2010). Documento de Consenso interdisciplinar para el tratamiento de la fibromialgia. *Actas Españolas de Psiquiatría*, 38(2), 108-121.

Alexander R.W., Bradley, L.A., Alarcon, G.S., Triana-Alexander, M.,Aaron, L.A., Alberts, K.R., Martin, M.Y., Stewart, K.E. (1998). Sexual and physical abuse in women with fibromyalgia: association with outpatient health care utilization and pain medication usage. *Arthritis Care & Research*, 11(2), 102-15.

Alonso, C., Loevinger, B.L., Muller, D., Coe, C.L. (2004). Menstrual cycle influences on pain and emotion in women with fibromyalgia. *Journal of Psychosomatic Research*, 57, 451-458.

Angst, F., Brioschi, R., Main, C.J., Lehmann, S., Aeschlimann, A. (2006). Interdisciplinary Rehabilitation in Fibromyalgia and Chronic Back Pain: A Prospective Outcome Study. *The Journal of Pain*, 7 (11), 807-815.

Arbuckle, A.J., & Wothke, W. (1999). *AMOS 4.0 user's guide*. Chicago: Small Waters Corp.

Arnold, L.M., Hudson, J.I., Keck P.E., Auchenbach, M.B., Javaras, K.N., Hess, E.V. (2006). Comorbidity of Fibromyalgia and psychiatric disorders. *Journal of Clinical Psychiatry*, 67, 1219-1225.

Arnstein, M., Caudill, C., Mandle, A., Norris, R., Beasley. (1999). Self-efficacy as a mediator of the relationship between pain intensity disability and depression in chronic pain patients. *Pain*, 80, 483–491.

Aspegren Kendall, S., Brolin-Magnusson, B., Sören, B., Gerdle, B., Henriksson, K.G. (2000). A pilot study od body awareness programs in the treatment of Fibromyalgia Syndrome. *Arthritis Care and Research*, 13(5), 304-311.

Bailey, A., Starr, L., Alderson, M., Moreland, J. (1999). A comparative evaluation of a Fibromyalgia rehabilitation program. *Arthritis Care and Research*, 12 (5), 336-340.

Baron, R.M., Kenny, D.A. (1986). The moderator-mediator variable distinction in social psychological research: conceptual, strategic and statistical considerations. *Journal of Personality and Social Psychology*, 51, 1173-1182.

Bennett, R.M., Burckhardt, C.S., Clark, S.R., O'Reilly, C.A., Wiens, A.N., Campbell, S.M. (1996). Group treatment of fibromyalgia: A 6-month outpatient program. *Journal of Rheumatology*, 23, 521-528.

Besteiro, J.A., Lemos, S., Muñiz, J., Costas, C., & Weruaga, A. (2008). Dimensiones de personalidad, sentido de coherencia y salud percibida en pacientes con un síndrome fibromiálgico. *International Journal of Clinical and Health Psychology, 8*, 411-427.

Biurrún-Unzué, A., Fernández Cuadrado, M.V., Jusué Erro, G. (2002). Efectos de la aplicación de un programa de tratamiento cognitivo-conductual en el locus de control de pacientes con dolor crónico. *Anales de Psiquiatría*, 18 (9), 407-410.

Bollen, K.A., & Stine, R.A. (1993). Bootstrapping goodness-of-fit measures in structural equation modeling. In K.A. Bollen, & J.S. Long (Eds.), Testing structural equation models. Newbury Park, CA: Sage.

Buckelew, S.P., Huyser, B., Hewett, J.E., Parker, J.C., Johnson, J., Conway, R., Kay, D.R. (1996) Self-efficacy predicting outcome among fibromyalgia subjects. *Arthritis Care and Research.* 9 (2), 97-104.

Buckelew, S.P., Murray, S.E., Hewett, J.E., Johnson, J. y Huyser, B. (1995). Self-efficacy, pain and physical activity among fibromyalgia subjects. *Arthritis Care and Research,* 8, 43-50.

Buckelew, S.P., Parker, J.C., Keefe, F.J., Deuser, W.E., Crews, T.M., Conway, R. (1994). Self-efficacy and pain behavior among subjects with fibromyalgia. *Pain,* 59, 377–84.

Burckhardt, C.S., Clark, S.R., Bennett, R.M. (1991). The fibromyalgia impact questionnaire: development and validation. *Journal of Rheumatology, 18,* 728–733.

Carver, C.S., Lawrence, J.W. y Scheier, M.F. (1996). A control-process perspective on the origins of affect. En L.L. Martin y A. Tesser (Eds.), *Striving and feeling: Interactions among goals, affect and regulation* (pp. 11-52). Mahwah: Erlbaum.

Clauw, D.J., Chorusos, G.P. (1997). Chronic pain and fatigue syndromes: overlapping clinical and neuroendocrine features and potential pathogenic mechanisms. *Neuroinmunology,* 4: 134-53.

Cohen, H., Neumann, L., Haiman, Y., Matar, M. A., Press, J., Buskila, D. (2002). Prevalence of post-traumatic stress disorder in fibromyalgia patients: overlapping syndromes or post-traumatic fibromyalgia syndrome?, *Seminars .Arthritis & Rheumatism,* 32, (1), 38-50.

Collado Cruz, A., Torresi Mata, X., Arias i Gassol, A., Cerdà Garbaroi, D., Vilarrasa, R., Valdés Miyar,M. (2001). Eficacia del tratamiento multidisciplinario del dolor crónico incapacitante del aparato locomotor. *Medicina Clínica* (Barc), 117, 401-5.

Cooper, H., Okamura, L. y McNeil, P. (1996). Situation and personality correlates of psychological well-being social activity and personal control. *Journal of Research in Personality,* 29(4), 395-417.

Crisson, J.E., Keefe, F.J. (1988). The relationship of locus of control to pain coping strategies and psychological distress in chronic pain patients. *Pain,* 35, 147-154.

Culos-Reed, S.N. (2001). Use of social-cognitive theories in the study of physical activity and fibromyalgia: Self-efficacy theory and the theory of planned behaviour. *Dissertation Abstracts International: Section B: The Sciences and Engineering,* 61(10-B), 5267.

Culos-Reed, S.N., Brawley, L. R. (2003). Self-efficacy predicts physical activity in individuals with fibromyalgia. *Journal of Applied Biobehavioral Research,* 8 (1), 27-41.

Davis, M. C., Zautra, A. J., Reich, J. W. (2001). Vulnerability to stress among women in chronic pain from fibromyalgia and osteoarthritis. *Annals of Behavioral Medicine,* 23 (3), 215-226.

Davis, M.C., Zautra, A. J., Smith, B.W. (2004). Chronic pain, stress, and the Dynamics of affective differentiation. *Journal of Personality,* 72, 1133-1159.

Dwyer, K.A. (1997). Psychosocial factors and health status in women with rheumatoid arthritis: predictive models. *American Journal of Preventive Medicine;* 13(1): 66-72.

Epstein, S.A., Kay, G., Clauw, D., Heaton, R., Klein, D., Krupp, L. (1999) Psychiatric disorders in patients with fibromyalgia: A multicenter investigation. *Psychosomatic Medicine,* 40, 57-63.

Esteve-Vives, J., Rivera Redondo, J., Vallejo Pareja, M.A. (2010). Evaluacion de la capacidad funcional en fibromialgia. Análisis comparativo de la validez de constructo de tres escalas. *Reumatología Clinica*, 6(3), 141–144.

Folkman, S. (1984). Personal control and stress and coping processes: a theoretical analysis. *Journal of Personality and Social Psychology*, 46, 839-852.

Fontaine, K.R., Conn, L., Clauw, D.J. (2010). Effects of lifestyle physical activity on perceived symptoms and physical function in adults with fibromyalgia: results of a randomized trial. *Arthritis Research & Therapy*, 12, R55.

Franks, H.M., Cronan, T.A., Oliver, K. (2004). Social support in women with Fibromyalgia: is quality more important than quantity? *Journal of Community Psychology*, 32, 425–438.

Fresco, D.M. (2000). Cognitive styles as moderators of the relationship of life stress to depression and anxiety. *Dissertation Abstracts International: Section B: The Sciences and Engineering*, 60 (12-B), 6361.

Gaskin, M.E., Greene, A.F., Robinson, M.E., Geisser, M.E. (1992). Negative affect and the experience of chronic pain. *Journal of Psychosomatic Research*, 36 (8), 707-713.

Gelman, S.M., Lera, S., Caballero, F., López, M.J. (2002). Tratamiento multidisciplinario de la fibromialgia. Estudio piloto prospectivo controlado. *Revista Española de Reumatología*, 29(7), 323-329.

Glombiewski, J.A., Sawyer, A.T., Gutermann, J., Koenig, K., Rief, W. y Hofmann, S.G. (2010). Psychological trearments for fibromyalgia: a meta-analysis. *Pain*, 151 (2), 280-295.

González-Alonso, A. I. (1999). Evaluación y tratamiento del sindrome fibromiálgico. *Tesis doctoral no publicada*. Madrid: Universidad Complutense de Madrid.

González-Gutiérrez, J.L., Peñacoba, C., Velasco, L., López-López, A., Mercado, F., Barjola, P. (2009). Recursos cognitivos de percepción de control, procesos de estrés-recuperación y malestar afectivo en la fibromialgia. *Psicothema*, 21(3), 359-368.

Gustafsson, M., Gaston-Johansson, F. (1996). Pain intensity and health locus of control: a comparison of patients with fibromiálgia syndrome and rheumatoid arthritis. *Patient Education and Counseling* (29), 179-188.

Härkäpää K. (1991). Relationships of psychological distress and health locus of control beliefs with the use of cognitive and behavioral coping strategies in low back pain patients. *Clinical Journal of Pain*, 7(4): 275-282.

Härkäpää, K., Järvikoski, A. y Estlander, A.M. (1996). Health optimism and control beliefs as predictors for treatment outcome of a multimodal back treatment program. *Psychology and Health*, (12), 123-134.

Hassett, A.L., Simonelli, L.E., Radvanski, D.C., Buyske, S., Savage, S., Sigal, L.H. (2008). The relationship between affect balance style and clinical outcomes in Fibromyalgia, *Arthritis & Rheumatism*, 59 (6), 833-840.

Häuser, W., Bernardy, K., Arnold, B., Offenbaecher, M., Schiltenwolf, M. (2009). Efficacy of Multicomponent Treatment in Fibromyalgia Syndrome: A Meta-Analysis of Randomized Controlled Clinical Trials. *Arthritis & Rheumatism (Arthritis Care & Research)*, 61 (2), 216–224.

Haythornthwaite, J.A., Menefee, L.A., Heinberg, L.J., Clark, M.R. (1998). Pain coping strategies predict perceived control over pain. *Pain*, 77 (1), 33-39.

Idel, M., Melamed, S., Merlob, P., Yahav, J., Hendel, T. y Kaplan, B. (2003). *Journal of Nursing Management*, 11(1), 59-63.

Jegede, A.B. (2006). Muscle characteristics of persons with fibromyalgia syndrome. *Dissertation Abstracts International: Section B: The Sciences and Engineering,* 66 (12-B), 6925.

Jensen, M.P., Turner, J., Romano, J.M. (2001). Changes in beliefs, catastrophizing, and coping are associated with improvement in multidisciplinary pain treatment. *Journal of Consulting and Clinical Psychology*, 69 (4), 655-62.

Keefe, F.J., Lumley, M., Anderson, T., Lynch, T., Carson, K.L. (2001). Pain and emotion: New research directions. *Journal of Clinical Psychology*, 57, 587-607.

Kersh, B. C., Bradley, L. A., Alarcón, G., Alberts, K. R., Sotolongo, A., Martin, M. Y., Aaron, L. A., Dewaal, D. F., Domino, M. I., Chaplin, W. F. (2001). Psychosocial and health status variables independently predict health care seeking in fibromyalgia, *Arthritis care and research*, 45, 362-371.

Kurtze, N., Svebak, S. (2005). A County Population of Males Given the Diagnosis of Fibromyalgia Syndrome: Comparison with Fibromyalgia Syndrome Females Regarding Pain, Fatigue, Anxiety, and Depression: The Nord Trondelag Health Study [The HUNT Study]. *Journal of Musculoskeletal Pain*, 13 (3), 11-18.

Lera, S., Gelman, S.M., López, M.J., Abenoza, M., Zorrilla, J.G., Castro-Fornieles, J., Salamero, M. (2009). Multidisciplinary treatment of fibromyalgia: does cognitive behavior tharapy increase the response to treatment? *Journal of Psychosomatic Research*, 67, 433-441.

Lundberg, G., Anderberg, U.M., Gerdle, B. (2009). Personality features in female Fibromyalgia Syndrome. *Journal of Musculoskeletal Pain*, 17 (2), 117-130.

Lyubomirsky, S., Sheldon, K.M., Schkade, D. (2005). Pursuing hapiness: the architecture of sustainable change. *Review of General Psychology*, 9, 111-131.

Lledó, A., Pastor, M.A., Pons, N., López-Roig, S., Rodríguez-Marín, J., Bruehl, S. (2010). Control beliefs, coping and emotions: exploring relarionships to explain fibromyalgia health outcomes. *International Journal of Clinical and Health Psychology*, 10(3), 459-476.

Malt, E.A., Berle, J., Olafsson, S., Lund, A., Ursin, H. (2000). Fibromyalgia is associated with panic disorder and functional dyspepsia with mood disorders. A study of women with random sample population controls. *Journal of Psychosomatic Research*, 49, 285–9.

Martín-Aragón, M., Pastor, M. A., Lledó, A., López-Roig, S., Perol, M. C., Rodríguez-Marín, J. (2001). Percepción de control en el síndrome fibromiálgico: variables relacionadas. *Psicothema*. 13 (4), 586-591.

Martín-Aragón, M., Pastor, M.A., Rodríguez-Marín, J., March, M.J., Lledó,A., López-Roig, S. y Terol, M.C. (1999). Percepción de autoeficacia en dolor crónico: adaptación y validación de la Chronic Pain Self-Efficacy Scale. *Revista de Psicología de la Salud*, 11 (1/2), 53-76.

Martin, M.Y., Bradley, L.A., Alexander, R.W., Alarcón, G.S., Triana-Alexander, M., Aaron, L.A., Alberts, K.R. (1996). Coping strategies predict disability in patients with primary fibromyalgia. *Pain*, 68 (1), 45-53.

Martinez-Lavin, M. (2004). Fibromyalgia as a sympathetically maintained pain syndrome [review]. *Current Pain Headache Reports*, 8, 385–9.

McBeth, J., Silman, A.J. (2001). The role of psychiatric disorders in Fibromyalgia. *Current Rheumatology Reports*, 3, 157-164.

Menzies, V., Taylor, A. G., Bourguignon, C. (2006). Effects of guided imagery on outcomes of pain, functional status, and self-efficacy in persons diagnosed with fibromyalgia. *Journal of Alternative and Complementary Medicine*, 12 (1), 23-30.

Miró, J. (1994). Papel de las expectativas de autoeficacia y del autocontrol en la experiencia de dolor. *Dolor*, 9, 186-190.

Mueller, A., Hartmann, M., Mueller, K., Eich, W. (2003). Validation of the arthritis Self-efficacy shortform scale in German fibromyalgia patients. *European Journal of Pain*, 7: 163-171.

Netter, P., Hennig, J. (1998). The fibromyalgia syndrome as a manifestation of neuroticism? *Zeitschrift fur Rheumatologie*, 57(Suppl 2), 105–108.

Nielson, W.R., Jensen, M.P. (2004). Relationship between changes in coping and treatment outcome in patients with Fibromyalgia Syndrome. *Pain*, 109, 233-241.

O'Brien, E.M., Waxenberg, L.B., Atchison, J.W., Gremillion, H.A., Staud, R.M., McCrae, C.S., Robinson, M.E. (2010). Intraindividual Variability in Daily Sleep and Pain Ratings Among Chronic Pain Patients: Bidirectional Association and the Role of Negative Mood. *Clinical Journal of Pain*, 27(5), 425-433.

Oliver, K., Cronan, T. (2002). Predictors of exercise behaviours among fibromyalgia patients. *Preventive Medicine*, 35, 383-389.

Oliver, K. and Cronan, T.A. (2005). Correlates of physical activity among women with fibromyalgia syndrome. *Annals of Behavioral Medicine, 29*, 44-53.

Palomino, R.A, Nicassio P.M., Greenberg M.A., Medina E.P. Jr.(2007). Helplessness and loss as mediators between pain and depressive symptoms in fibromyalgia. *Pain*, 129, 185–94.

Pastor, M.A., López-Roig, S., Rodríquez-Marín, J., Martín-Aragón, M., Terol, MC. y Pons, N. (1999). Percepción de control, Impacto de la enfermedad y ajuste emocional en enfermos crónicos. *Ansiedad y Estrés*; 5(2-3), 299-311.

Peñacoba, C. (1996). Estrés, salud y calidad de vida: influencia de la dimensión cognitivo-emocional. *Tesis doctoral no publicada*. Madrid: Universidad Autónoma de Madrid.

Pérez-Pareja, J., Borrás, C., Palmer, A., Sesé, A., Molina, F., Gonzalvo, J. (2004). Fibromialgia y emociones negativas. *Psicothema*, 16(3), 415-420.

Porter-Moffitt, S., Gatchel, R.J., Robinson, R.C., Deschner, M., Posamentier, M., Polatin, P., Lou, L. (2006). Biopsychosocial profiles of different pain diagnosis groups. *The Journal of Pain*, 7(5), 308-318.

Potter, P. T., Zautra, A. J., Reich, J. W. (2000). Stressful events and information processing dispositions moderate the relationship between positive and negative affect: Implications for pain patients. *Annals of Behavioral Medicine*. 22 (3), 191-198.

Preacher, K.J., Rucker, D.D., Hayes, A.F. (2007). Addressing Moderated Mediation Hypotheses: Theory, Methods, and Prescriptions. *Multivariate Behavioral Research*, 42(1), 185–227.

Ramírez-Maestre, C., Zarazaga R., López-Martínez, A.E. (2001). Neuroticismo, afrontamiento y dolor crónico. *Anales de Psicología*, 17, 129-137.

Reich, J.W. & Zautra, A. (1981). Life events and personal causations: Some influences on well-being. *Journal of Personality and Social Psychology*, 41, 1002-1012.

Rivera, J, Moratalla, C., Valdepeñas, F., García, Y., Osés, J.J., Ruiz, J., González, T., Carmona, L., Vallejo, M.A. (2004). Fibromyalgia: A Physical Exercise-Based Program and a Cognitive-Behavioral Approach. *Arthritis & Rheumatism (Arthritis Care & Research)*, 51 (2), 184–192.

Rossy, L.A., Buckelew, S.P., Doit, N., Hagglund, K.J., Thayer, J.F., McIntosh, M.J., Hewett, J.E., Johnson, J.C. (1999). A meta-analysis of fibromyalgia treatment interventions. *Annals of Behavioral Medicine*, 21, 180-191.

Rudnicki, S. R. (2001). A model of psychosocial factors and pain in fibromyalgia: An integrative approach. *Dissertation Abstracts International: Section B: The Sciences and Engineering*, 62 (2-B), 1098.

Salaffi, F., Sarzi-Puttini, P., Ciapetti, A., Atzeni, F. (2009). Assessment instruments for patients with Fibromyalgia: properties, applications and interpretation. *Clinical and Experimental Rheumatology*, 27(5), S92-105.

Sandín, B., Chorot, P., Lostao, L., Joiner, T. E., Santed, M. A., Valiente, R. M. (1999). Escala PANAS de afecto positivo y negativo: validación factorial y convergencia transcultural. *Psicothema*, 11, 37-51.

Sarzi-Puttini, P., Atzeni, F., Diana, A., Doria, A., Furlan, R. (2006). Increased neural sympathetic activation in fibromyalgia syndrome. *Annals New York Academic Science*, 1069, 109–17.

Schachter, C.L., Busch, A.J., Peloso, P.M., Sheppard, M.S. (2003). Effects of short versus long bouts of aerobic exercise in sedentary women with fibromyalgia: a randomized controlled trial. *Physical Therapy*, 83(4), 340-358.

Serber, E.R., Cronan, T. A., Walen, H.R. (2003). Predictors of patient satisfaction and health care costs for patients with fibromyalgia. *Psychology and Health*, 18(6), 771-787.

Skevington, S.M. (1990). A standarised scala to measure beliefs about controlling pain (B.P.C.Q.): a preliminary study. *Psychology and Health*, 2, 221-232.

Söderfeldt, M; Söderfeldt, B.; Ohlson, C.G.; Theorell, T., Jones, I. (2000). The impact of coherence and high-demand / low-control job environment on self-reported health, burnout and psychophysiological stress indicators. *Work & Stress*, Vol. 14, (1), 1-15.

Staud, R., Price, D.D., Robinson, M.E., Vierck, C.J Jr. (2004). Body pain area and pain-related negative affect predict clinical pain intensity in patients with fibromyalgia. *Journal of Pain*, 5, 338–43.

Staud, R., Robinson, M.E., Vierck, C.J., Cannon ,R.C., Mauderli, A.P., Price, D,D. (2003). Ratings of experimental pain and pain-related negative affect predict clinical pain in patients with fibromyalgia syndrome. *Pain*, 105, 215–22.

Stillerman, L. (2007). Humor in relation to anxiety, depression, and stress for college students. *Dissertation Abstracts International: Section B: The Sciences and Engineering*, 67 (9-B), 5424.

Stuifbergen, A.K., Blozis, S.A., Becker, H., Phillips, L., Timmerman, G., Kullberg, V., Taxis, C., Morrison, J. (2010). A randomized controlled trial of a wellness intervention for women with fibromyalgia syndrome. *Clinical Rehabilitation*, 24, 305-318.

Suhr, J.A. (2003). Neuropsychological impairment in fibromyalgia: relation to depression, fatigue, and pain. *Journal of Psychosomatic Research*, 55, 321–9.

Sullivan, M. J. L., Rodgers, W. M., Kirsch, I. (2001). Catastrophizing, depression and expectancies for pain and emotional distress. *Pain*, 91, 147-154.

Theadom, A., Cropley, M., Humphrey, K. L. (2007). Exploring the role of sleep and coping in quality of life in fibromyalgia. *Journal of Psychosomatic Research*, 62 (2), 145-151.

Thieme, K., Gromnica-Ihle, E., Flor, H. (2003). Operant behavioral treatment of fibromyalgia. A controlled study. *Arthritis & Rheumatism*, 49, 314-320.

Thompson, S. (2006). The relationship of self-efficacy, internal/external locus of control, achievement goal orientation, and academic performance. *Dissertation Abstracts International: Section A: Humanities and Social Sciences*, 62 (2-B), 1098.

Toomey, T.C., Lundeen, T.F., Mann, J.D. Abashian, S. (1988). *The pain locus of control scale: a comparison of chronic pain patients and normals.* Comunicación presentada en la Reunión Anual de las Asociaciones de dolor de América y Canadá, Toronto.

Toomey, T.C.; Mann, J.D.; Abashian, S. and Thompson- Pope, S. (1991). Relationship between perceived self- control of pain, pain description and functioning. *Pain, 45*: 129-133.

Turner, A., Foster, M., Johnson, S.E. (2003). *Terapi a ocupacional y disfunción física: principios, técnicas y práctica*. España: Elsevier.

Van Houdenhove, B., Luyten, P. (2006). Stress, depression and fibromyalgia. *Acta neurologica belgica*, 106, 149-156.

Velasco, L., Zautra, A., Peñacoba, C., López-López, A., Barjola, P. (2008). Cognitive-affective assets and vulnerabilities: two factors influencing adaptation to Fibromyalgia. *Psychology & Health*, 25, 2, 197-212.

Walker, E.A., Keegan, D., Gardner, G., Sullivan, M., Bernstein, D., Katon, W.J. (1997). Psychosocial factors in fibromyalgia compared with rheumatoid arthritis: II. Sexual, physical and emotion abuse and neglect. *Psychosomatic Medicine*, 59, 572-577.

Wallston, K.A. (1989). Assessment of control in health-care settings. En: Andrew Steptoe y ad Appels (Eds): *Stress, personal control and health.* John Wiley & Sons.

Wallston, B., Wallston, K., Kaplan, G., Maides, S. (1976). Development and validation of the Health Locus of Control (HLC) scale. *Journal of Consulting and Clinical Psychology*, 44:580-585.

Walter, B., Vaitl, D., Frank, R. (1998). Affective distress in fibromyalgia syndrome is associated with pain severity. *Z Rheumatol*, 57(Suppl. 2), 101 –4.

Wells-Federman, C., Arnstein, P., Caudill-Slosberg, M., (2003). Comparing Patients with Fibromyalgia and Chronic Low Back Pain Participating in an Outpatient Cognitive-Behavioral Treatment Program. *Journal of Musculoskeletal Pain*, 11(3), 5-12.

White, K.P., Nielson, W.R. (1995). Cognitive behavioral treatment of fibromyalgia syndrome: A follow-up assessment. *Journal of Rheumatology*, 22, 717-721.

Williams, D.A., Clauw, D.J. (2009). Understanding Fibromyalgia: Lessons from the Broader Pain Research Community. *The Journal of Pain*, 10(8), 777-791.

Wolfe, F., Smythe, H.A., Yunus, M.B., Bennett, R.M., Bombardier, C., Goldenberg, D.L., Tugwell, P., Campbell, S.M., Abeles, M., Clark, P., Fam, A.G., Farber, S.J., Fiechtner, J.J., Franklin, C.M., Gatter, R.A., Hamaty, D., Lessard, J., Lichtbroun, A.S., Masi, A.T., McCain, G.A., Reynolds, W.J., Romano, T.J., Rusell, I.J., Sheon, R.P. (1990). The American College of Rheumatology. 1990. Criteria for the classification of fibromyalgia: report of the Multicenter Criteria Committee. *Arthritis & Rheumatism*, 33, 160-172.

Worrel, L.M., Krahn, L.E., Sletten, C.D., Pond, G.R. (2001). Treating fibromyalgia with a brief interdisciplinary program: Initial outcomes and predictors of response. *Mayo Clinical Procedures*, 76, 384-390.

Zaharoff, A.D. (2005). The relationship between fibromyalgia and emotional expressivity and its influence on locus of control, ways of coping, and quality of life. *Dissertation Abstracts International: Section B: The Sciences and Engineering,* 65(8-B), 4311.

Zautra, A.J., Davis, M.C., Reich, J.W., Nicassario, P., Tennen, H., Finan, P., Kratz, A, Parrish, B., Irwin, M.R. (2008). Comparison of cognitive behavioral and mindfulness meditation interventions on adaptation to rheumatoid arthritis for patients with and without history of recurrent depression. *Journal of Consulting and Clinical Psychol.ogy,* 76(3), 408-21.

Zautra, A.J., Fasman, R., Davis, M.C., Craig, A.D. (2010). The effects of slow breathing on affective responses to pain stimuli: An experimental study. Pain, 149 (1), 12-18.

Zautra, A.J., Fasman, R., Parish, B.P., Davis, M.C. (2007). Daily fatigue in women with osteoarthritis, rheumatoid arthritis, and fibromyalgia. *Pain,* 128, 128–35.

Zautra, A.J, Fasman, R., Reich, J.W., Harakas, P., Johnson, L.M., Olmsted, M.E., Davis, M.C. (2005). Fibromyalgia: evidence for deficits in positive affective regulation. *Psychosomatic Medicine,* 67, 147-155.

Zautra, A.J., Hamilton, N.A., Burke, H.M. (1999). Comparison of stress responses in women with two types of chronic pain: fibromyalgia and osteoarthritis. *Cognitive Therapy and Research,* 23 (2), 209-30.

Zautra, A.J, Johnson, L., Davis, M. (2005). Positive Affect as a source of resilience for woman in chronic pain. Journal of Consulting and Clinical psychology, 73, 212-220.

Zautra, A.J., Smith, B., Affleck, G., & Tennen, H. (2001). Examination of chronic pain and affect relationship: application of a dynamic model of affect. Journal of Consulting and Clinical Psychology, 69, 786 -795.